Handbook of Optoelectronics

Handbook of Optoelectronics

Edited by Rodney Lappin

Larsen & Keller
www.larsen-keller.com

Handbook of Optoelectronics
Edited by Rodney Lappin
ISBN: 979-8-88836-000-2 (Hardback)

Larsen & Keller
Published by Larsen and Keller Education,
5 Penn Plaza,
19th Floor,
New York, NY 10001, USA

Cataloging-in-Publication Data

Handbook of optoelectronics / edited by Rodney Lappin.
p. cm.
Includes bibliographical references and index.
ISBN 979-8-88836-000-2
1. Optoelectronics. 2. Photonics. 3. Electronics. I. Lappin, Rodney.
TA1750 .068 2023
621.381 045--dc23

Contents

Preface

This book was inspired by the evolution of our times; to answer the curiosity of inquisitive minds. Many developments have occurred across the globe in the recent past which has transformed the progress in the field.

Optoelectronics refers to the study and use of electronic devices and systems for finding, detecting and controlling light. It is commonly regarded as a subfield of photonics. Optoelectronic devices are electrical-to-optical or optical-to-electrical transducers, or instruments that utilize such devices. Optoelectronics is based on the quantum mechanical effects of light on electronic materials, particularly semiconductors in the presence of electric fields. It has significant applications in optical fiber communications and optocouplers. This book will unfold the innovative aspects of optoelectronics which will be crucial for the progress of this field in the future. From theories to research to practical applications, studies related to all contemporary topics of relevance to this field have been included herein. The book will serve as a valuable source of reference for graduate and post graduate students.

This book was developed from a mere concept to drafts to chapters and finally compiled together as a complete text to benefit the readers across all nations. To ensure the quality of the content we instilled two significant steps in our procedure. The first was to appoint an editorial team that would verify the data and statistics provided in the book and also select the most appropriate and valuable contributions from the plentiful contributions we received from authors worldwide. The next step was to appoint an expert of the topic as the Editor-in-Chief, who would head the project and finally make the necessary amendments and modifications to make the text reader-friendly. I was then commissioned to examine all the material to present the topics in the most comprehensible and productive format.

I would like to take this opportunity to thank all the contributing authors who were supportive enough to contribute their time and knowledge to this project. I also wish to convey my regards to my family who have been extremely supportive during the entire project.

Editor

1

Effects of Quantum-Well Base Geometry on Optoelectronic Characteristics of Transistor Laser

Iman Taghavi and Hassan Kaatuzian
Amirkabir University of Technology (Tehran Polytechnic)
Iran

1. Introduction

Since the discovery of the transistor, the base current (minority carrier recombination) has been the key to the device operation. Among all developments through which Bipolar Junction Transistors (BJT) have progressed, the most innovating modification may be the replacement of the homojunction emitter material by a larger-energy-gap material, thus forming a Heterojunction Bipolar Transistor (HBT). In a silicon bipolar junction transistor, both homo and hetero structures, the base current is dissipated as heat, i.e. through nonradiative recombination, but can yield substantial radiative recombination for a direct-bandgap semiconductor like GaAs. Thus we can reinvent the base region and its mechanism (i.e. carrier recombination and transport function) to decrease current gain significantly and achieve stimulated recombination.

Researches on III–V high-current-density high-speed HBT, revealed the ability of the transistor to operate electrically and optically in the same time. The modified transistor, called in literature as light emitting transistor (LET), works as a three port device with an electrical and an optical outputs (Feng et al. 2004a). Further improvements including wavelength tunability obtained by incorporating a place for better carrier confinement, called a Quantum-Well (QW) (Feng et al. 2004b). Carrier recombination in quantum well (QW) can be modified with a reflecting cavity changing the optoelectronic properties of QWLET (Walter et al. 2004). Room temperature continuous wave operation of such a device at GHz develops a novel three-terminal device, called a HBT Transistor Laser (HBTL) or briefly TL, in which laser emission produces by stimulated recombination (Feng et al. 2005).

In the TL, the usual transistor electrical collector is accompanied with an optical collector, i.e. the above mentioned QW, inserted in the base region of the HBT. Stimulated recombination, unique in the TL, causes "compression" in the collector I-V characteristics and decrease in gain. Combined functionality of an HBT, i.e. amplification of a weak electrical signal, and that of a diode laser, i.e. generating laser emission, is observed in the TL. In other words, a modulating base current leads to modulated signals of the laser output power and collector current. It raises the possibility of replacing some metal wiring between components on a circuit board or wafer chip with optical interconnections, thus providing more flexibility and capability in optoelectronic integrated circuits (OEIC) (Feng et al. 2006a). It has been planned that TL is appropriate for telecommunication and other applications because of its capability of achieving a large optical bandwidth (BW) and a

frequency response without sharp resonance. Processing speed can increase when the TL-equipped microprocessors are commercialized in the future.
Utilizing the I-V characteristics of a TL (Then et al., 2007a) together with common transistor charge analysis, dynamic properties and collector current map are extracted where the above mentioned "compression" is observed in the common emitter gain ($\beta \equiv I_C/I_B$)(Chan et al. 2006). A Transistor Laser with increased (external) mirror reflection demonstrates empirically (Walter et al. 2006) lower threshold current and higher collector breakdown voltage while increased breakdown at lower currents is observed on the collector I-V characteristics if the base region cavity Q is spoiled. Electrical properties of TL will therefore become similar to those of normal transistor ($\beta stim \rightarrow \beta spon$) (Chan et al. 2006). Trade-off between collector current gain and the differential optical gain of a heterojunction bipolar transistor laser TL has been demonstrated analytically before (Then et al. 2009). The electrical-optical gain relationship shows that a reduction in the transistor current gain is accompanied by an increase in the differential optical gain of the TL and, as a consequence, results in a larger optical modulation bandwidth. Another trade-off, i.e. electrical gain-optical bandwidth, has also been reported for which one can utilize in order to extract the maximum possible optical bandwidth of TL (Taghavi & Kaatuzain, 2010).
We proceed in section 2 by introducing the TL crystal structure and fabrication issues, while details are left with proper references for an interested reader. In section 3 carrier dynamics and charge control model in TL will be discussed which is necessary for interpretation of device physics. This section determines a simplified method for analytical modelling of TL and their modifications through the years TL has been invented. TL optical bandwidth and current gain are formulized in the section 4. The method will be utilized in section 5 to simulate QW-base geometry effects on optoelectronic characteristics of the TL. The investigated transistor laser has an electrical bandwidth of more than 100GHz. Thus the structure can be modified, utilizing the methods mentioned in the section 5, to equalize optical and electrical cut-off frequencies as much as possible. Indeed, any improvement in optoelectronic characteristics of TL, i.e. optical bandwidth and current gain, is currently necessary in order to incorporate the TL in OEIC`s. The most urgent work needed in the future is an investigation of other quantum-well parameters effect on TL characteristics. Among these parameters are width and number of quantum wells. Section 6 is devoted to discussions about new researches in the field of TL including MQWTL. The chapter will be completed with conclusion in this section.

2. TL structure

The device studied here is based on n-p-n heterojunction bipolar transistor (n-InGaP/ p-GaAs +InGaAs quantum well/n-GaAs). The epitaxial structure of the crystal used for the HBTL, demonstrated in Fig. 1, consist of a 5000 Å n-type heavily doped GaAs buffer layer, followed by three layers of n-type forming the bottom cladding layers. These layers are followed by an n-type subcollector layer, a 600 Å undoped GaAs collector layer, and an 880 Å p-type GaAs base layer (the active layer), which includes (in the base region) a 120 Å InGaAs QW (designed for $\lambda \approx 1000$ nm). The base area is $\approx 4000 \mu m^2$ and we can treat the QW base region as three distinct subregions: Two 10^{19} cm^{-3} doped GaAs regions and one 160Å undoped InGaAs QW within them. The epitaxial HBTL structure is completed with the growth of the n-type emitter and upper cladding layers. Finally, the HBTL structure is capped with a 1000 Å heavily doped n-type GaAs contact layer. The HBTL fabrication process utilizes eight mask layers for three wet chemical etching steps, three dry (plasma) etching steps, and three metallization steps for top-side contacts. The HBTL sample is cleaved front-to-back using conventional etching methods to a length of 150μm between Fabry-Perot facets. Detailed fabrication process of HBTL has been described in (Chan et al., 2006; Feng et al., 2005; Feng et al., 2006a).

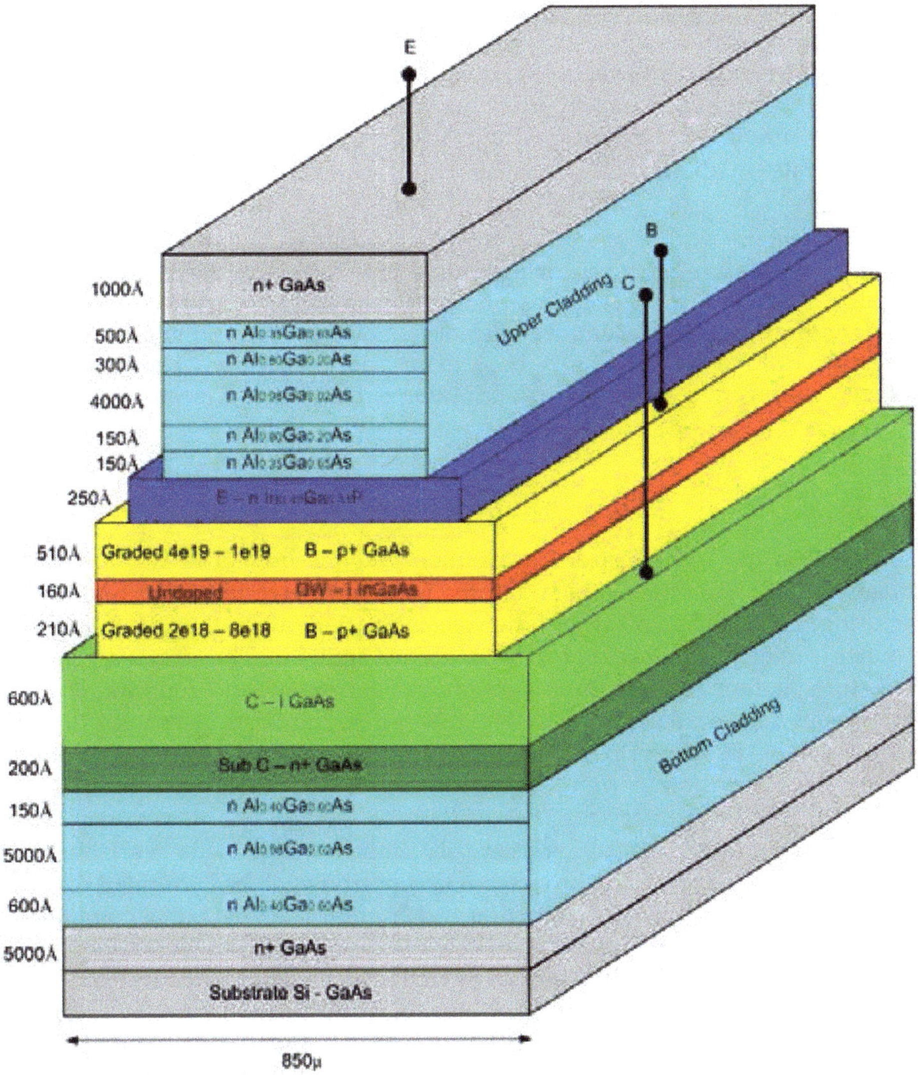

Fig. 1. Epitaxial structure of heterojunction bipolar Transistor Laser (HBTL). A QW is incorporated within the base to improve the recombination efficiency.

Although other device structures like Tunnel Junction Transistor Laser (TJTL)(Feng et al., 2009), VSCEL (Shi et al., 2008), Heterojunction Field effect (HFETL)(Suzuki et al., 1990) and Distributed Feedback Transistor Laser (DFBTL)(Dixon, 2010) have been proposed in the literature, we focus solely on Heterostructure Bipolar Junction transistors.

3. Charge analysis

The three-terminal TL is fundamentally different from a diode laser (single p-n junction), i.e. a two-terminal device, where the injection current is carried totally by recombination. There is no third terminal and collector current "competition". In TL the collector current I_C (I_E=I_B+I_C) is the fraction of emitter current "collected" at the reverse-biased base collector junction, with the remainder I_B supplying base recombination radiation. We show here that the carrier lifetime can be extracted from the HBT I-V characteristics, thus allowing the intrinsic optical frequency response of the TL to be derived by modifying the coupled carrier-photon rate equations.

3.1 Carrier dynamics

Being an essential feature of transistor operation, QW-base recombination of electron-hole supported by the base current I_B causes photon generation while carrier injection into the base region is provided by the emitter current I_E. Solving the continuity equation for the above mentioned three regions of QW-base, expressed in equations 1 and 2 for base bulk and QW region respectively, can shape distribution of base injected electrons from emitter.

$$\partial n/\partial t = D\,\partial^2 n/\partial x^2 - n/\tau_{bulk} \quad (bulk\ base) \tag{1}$$

$$\partial n/\partial t = D\,\partial^2 n/\partial x^2 - n/\tau_{qw} \quad (QW) \tag{2}$$

Where n=n(x,t) is the base electron distribution and the quantities τ_{bulk} and τ_{qw} are the recombination lifetimes in the GaAs and in the InGaAs QW, respectively. Sudden decrease in τ_{qw} at lasing threshold is a result of faster recombination rate in the QW thus base carrier tilt at emitter is intensified compared to spontaneous emission. The term n/τ_{qw} in equation 2 represents for spontaneous and stimulated photon generation and coupling of the optical field into the device operation. The bulk lifetime, τ_{bulk}, in the GaAs layer with 10^{19} doping is approximated as 193 ps. Uniform diffusion constant, $D\approx 26\ cm^2/s$, is assumed throughout the base region and diffusion current, $I=qAD\partial n/\partial x$, is supposed continuous across the GaAs/InGaAs-QW/GaAs interface. The position of QW, W_{EQW}, is assessed here by its central location distance from the emitter-base interface and is the variable parameter during analysis of section 5 where the effect of QW position will be investigated.

Zero charge density at collector-base junction and $\partial n/\partial t=0$ form the initial conditions for the continuity equations 1 and 2. The calculated charge distribution for increasing base current (I_B) corresponding to the I-V characteristics of HBTL are shown plotted in Fig.2. The device currents are then calculated from common formula for diffusion current. Deviation from triangular approximation is observed in Fig. 2 and the profile is inclined more steeply upward near the emitter-base interface due to faster recombination in QW compared with the bulk base and reduced base width (W_{QW}<W_{EC}). At the lasing threshold, τ_{qw} decreases abruptly agreeing with β_{stim}< β_{spon} as stimulated recombination speeds up the overall rate of recombination in the QW. Consequently, the emitter-end "tilt" of the carrier population is more noticeable beyond the threshold.

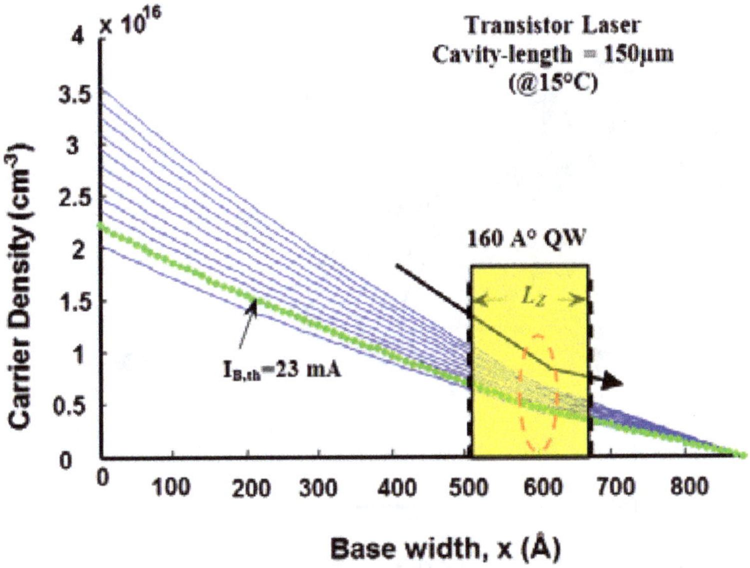

Fig. 2. Calculated minority carrier distribution for increasing I_B corresponding to the I-V characteristics. The shape tilts inside the QW due to faster recombination and reduced base width.

3.2 Charge control model

Charge control model for TL consists of superposition of two different triangular charge populations named as Q_1 and Q_2 (Feng et al., 2007). As TL integrates two distinguish operations, i.e. HBT electrical and semiconductor diode laser optical outputs, it might be supposed as a two-collector device. One is the base-collector junction and the other is QW called correspondingly electrical and optical collector. In the model of charge control Q_1 and Q_2 are responsible for transporting minority carriers to the QW (for optical operation) and keeping transport of carriers to the electrical collector (for electrical operation) respectively. Q_1 and Q_2 are obtained from carrier density profile, Fig. 3 and calculated so that

$$Q_1 = q\Delta n_1 A W_{EQW}/2 \tag{3}$$

$$Q_2 = q\Delta n_2 A W_B/2 \tag{4}$$

Where Δn_1 and Δn_2 are extracted from bias data, i.e. emitter base bias voltage (v_{BE}) and I_C, so that ($\Delta n_1 + \Delta n_2$) exhibits total carrier density at the emitter-base interface. Deriving I_C and I_B from I-V characteristics of TL (W_{EQW}=590 Å), we can estimate Q_1 and Q_2 as

$$I_C = Q_2/\tau_{t,2} \tag{5}$$

$$I_B = (Q_1 + Q_2)/\tau_{bulk} + Q_1/\tau_{t,1} \tag{6}$$

Where $\tau_{t,1}$ is the transient time from emitter to QW and $\tau_{t,2}$ is the transient time across the entire base (W_B=880Å) and calculated as below.

$$\tau_{t,1} = {W_{EQW}}^2/2D \approx 0.67ps \tag{7}$$

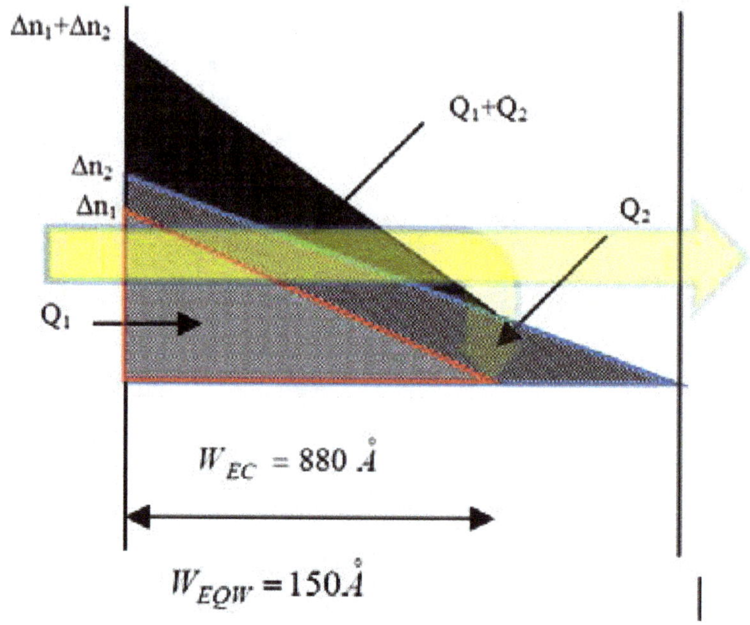

Fig. 3. Charge Control Model illustrates the role of Q_1 and Q_2 triangles in the entire base carrier distribution. Large amount of carriers enter the base while a little of them survive to the collector.

$$\tau_{t,2} = {W_B}^2/2D \approx 1.5ps \tag{8}$$

Two base carrier populations, Q_1 and Q_2, are combined to form the overall effective base recombination lifetime, τ_B, of the entire base population which is expressed as

$$1/\tau_B = 1/\tau_{bulk} + \tau_{t,2}\, Q_1/(Q_1 + Q_2) \tag{9}$$

Since $\tau_{bulk}>>\tau_{t,1}$, $\tau_B\approx(1+Q_2/Q_1).\tau_{t,1}$. Beyond the lasing threshold, τ_B decreases when stimulated emission accelerates the overall rate of recombination in the base of the TL. This is obvious in the steeper slope of the Q_1 population, transporting a larger proportion of the base carriers to a faster ($\tau_{B,stim}$) optical collector (QW), leading to a shorter overall base recombination lifetime, τ_B, and decreased gain β. τ_B values for charge density distributions near the threshold and beyond, plotted in Fig. 4 of (Feng et al., 2007).

4. Device optoelectronic characteristics

Among several characteristics, either optical or electrical, of the TL, small signal optical frequency response and current gain are most interesting. The reason arises from the fact that the TL needs suitable optoelectronic characteristics to be incorporated in an Optoelectronic Integrated Circuit (OEIC). The intrinsic frequency response of the TL can be improved towards 100 GHz owing to carrier lifetime determined by a thin base and the ability of the TL to inject and withdraw stored charge within picoseconds (forcing recombination to compete with E-C transport). Methods are required to improve the optical bandwidth to equalize both bandwidths as much as possible. The following describes a brief analysis of these characteristics.

4.1 Small-signal analysis of optical frequency response

When integrated over the entire base width, continuity equations (1) and (2) can result in an approximation for optical frequency response. The result for the TL is a modification of the coupled carrier-photon equations, formulated by Statz & deMars (Feng et al., 2007) which is expressed as

$$dN/dt = I_E/q - (I_C/q) + N/\tau_{B,spon} + \upsilon g\Gamma N_p \quad (10)$$

$$dN_p/dt = \upsilon g\Gamma N_p - N_p/\tau_p \quad (11)$$

Where $N \approx (Q_1+Q_2)/q$ is the total base minority carrier population, υ is the photon group velocity, Γ is the optical confinement factor of active medium (i.e. QW), τ_P is the photon lifetime, g is the gain per unit length of the active medium, and N_p is the total generated photon number. The small signal optical frequency response is obtained as below

$$\frac{\Delta P_m(\omega)}{\Delta I_B(\omega)} = \frac{\Gamma_b\tau_p/(qAW_B)}{[1-(\omega/\omega_n)^2]+j2(\omega/\omega_n)\xi} \quad (12)$$

Where $\Delta P_m(\omega)$ is the modulated photon density, $\Delta I_B(\omega)$ is the modulated base current and Γ_b is the optical confinement factor of the waveguide. The natural frequency and damping ratio are:

$$\omega_n^2 \approx (1/\tau_p\tau_{B,spon})(I_B/I_{B,th} - 1) \quad (13)$$

$$\xi = (2\omega_n\tau_{B,spon})^{-1} + 0.5(\omega_n\tau_p) \quad (14)$$

For $\zeta \leq 0.7$ the resonance frequency and bandwidth are:

$$\omega_r = \omega_n\sqrt{(1-2\xi^2)} \quad (15)$$

$$\omega_{-3db} = \omega_n\sqrt{(1-2\xi^2) + \sqrt{4\xi^4 - 4\xi^2 + 4}} \quad (16)$$

While these values for $\zeta \geq 0.7$ are expressed as:

$$\omega_r = 0 \quad (17)$$

$$\omega_{-3db} = \omega_n\sqrt{(1-2\xi^2) + \sqrt{4\xi^4 - 4\xi^2 + 2}} \quad (18)$$

We can obtain τ_P from the common equation:

$$1/\tau_p = \left(\frac{c}{n}\right)\{\alpha_i + (1/2L)\ln(1/R_1R_2)\} \quad (19)$$

For cleaved cavity laser of lengths L=150 and 400(μm) with $R_1=R_2=0.32$ and a photon absorption factor (α) of 5cm^{-1}, the approximate photon lifetime are τ_P=1.5 and 3.6(ps) respectively. It should be noticed that we neglect the capture and escape lifetimes in this small signal analysis. Optical gain is also assumed independent on photon and carrier populations. Although the difference in final results is negligible, an interested reader can refer to (Faraji et al., 2009) for more information.

In equations (13) and (14) the parameter $\tau_{B,spon}$ is a specific value of overall base recombination lifetime, τ_B, evaluated in spontaneous emission region at lasing threshold. According to Fig. 4

of (Feng et al.,2007) (not shown here) this value for a TL with W_{EQW}=590Å is about 2.5ps. Under bias condition such that I_B=33mA ($I_B/I_{B,th}$=1.5), small signal optical frequency response for two TL with above mentioned cavity length is sketched in Fig. 4. For comparison, the curve also contains optical response of an ordinary diode laser (DL) with a large base recombination lifetime of $\tau_{B,spon}$=100ps. When compared with traditional diode laser, fast modulation response of TL is due to this difference in their optical frequency response which suffers a resonance peak in low frequencies in the case of diode laser.

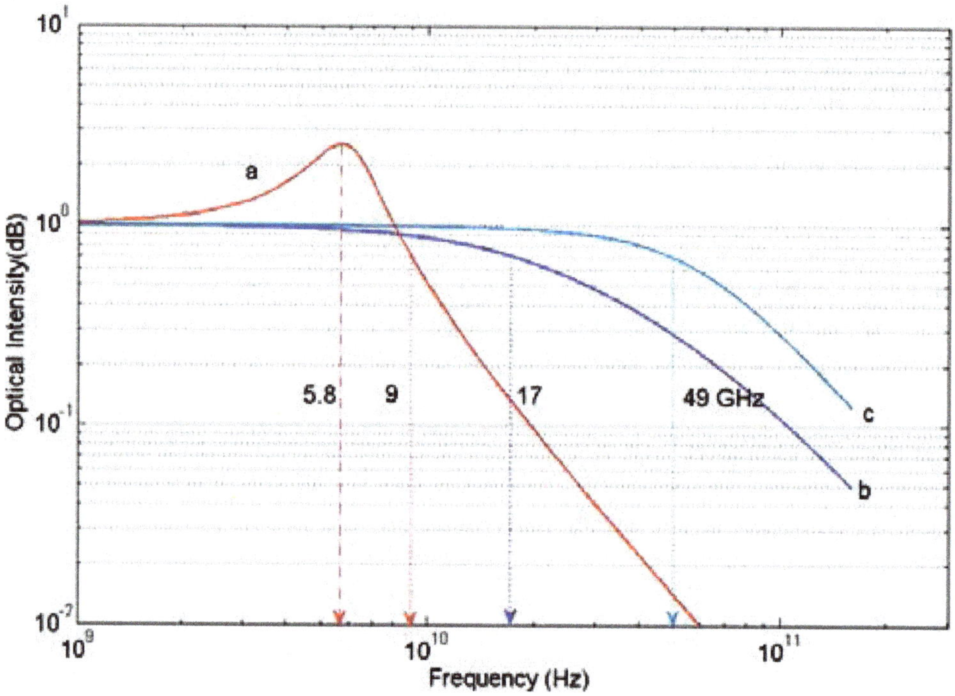

Fig. 4. Small signal optical frequency response of (a) Diode laser with L=400μm ($\tau_{B,spon}$=100ps), (b) TL with L=400μm ($\tau_{B,spon}$=2.5ps), (c) TL with L=150μm ($\tau_{B,spon}$=2.5ps)

The very short effective carrier lifetime of a TL, i.e. the same order of magnitude as the cavity photon lifetime, results in a near unity resonance peak and provides in effect a significantly larger bandwidth , which is typically not available when the bandwidth is limited by carrier-photon density resonance, as in a diode with a large carrier lifetime (0.1–1ns). We see that by charge analysis, if a QW base HBT is capable of laser operation, its "speed" of operation is not affected by common recombination, and the limitation of spatial charge "pileup" as in a diode, but by the base dynamics of charge transport described in this section. Indeed, only fast recombination is able to compete with fast collection of carriers.

4.2 Current gain

Thanks to special features of QW incorporated in its base region, the most recent form of laser, the Transistor Laser, govern carrier recombination and hence electrical current gain , $\beta=\Delta I_C/\Delta I_B$. This gain therefore becomes a crucial parameter in both transistor and laser operations of TL. As previously mentioned, stimulated emission, an extra feature of transistor, makes the collector I-V characteristics "compressed" and decreases β. The gain value for both the transistor laser and an ordinary HBT (i.e. a Q-spoiled TL) was demonstrated in Fig. 6 of (Walter et al., 2006). Charge control analysis can be utilized in conjunction with these values of current gain to determine the TL terminal currents and estimate base transit and recombination lifetimes. Conversely, one can calculate theoretically β for a TL as described below.

The base and collector currents and the static common-emitter current gain are given for an HBT as

$$I_E = I_B + I_C = Q_B/\tau_B + Q_B/\tau_t \tag{20}$$

$$\beta \equiv \Delta I_C/\Delta I_B = I_C/I_B = \tau_B/\tau_t \tag{21}$$

where Q_B is the charge stored in the base, τ_B is the base recombination lifetime, and τ_t is the base transit time. For the same HBT with a quantum well inserted in the base, a QWHBT, the base current and current gain can be modified and expressed as

$$I_B \equiv I_{BQW} = \left(Q_B/\tau_B + Q_B/\tau_{QW}\right) = (Q_B/\tau_{BQW}) \tag{22}$$

$$\beta_{QW} = \tau_{BQW}/\tau_t \tag{23}$$

where τ_{QW} is the recombination lifetime of the quantum well while τ_{BQW} is the effective recombination lifetime of the base and quantum well. For an QWHBT transistor laser with stimulated base recombination, a QWHBTL, the base current and current gain can be further modified and expressed as

$$I_B \equiv I_{BTL} = \left(Q_B/\tau_B + Q_B/\tau_{QW} + Q_B/\tau_{st}\right) = (Q_B/\tau_{TL}) \tag{24}$$

$$\beta_{TL} = \tau_{TL}/\tau_t \tag{25}$$

where τ_{TL} is the effective base recombination lifetime of the transistor laser including τ_{st}, the stimulated recombination lifetime.

5. QW-base geometry effects

Since the first report of successful operation of TL, a number of publications reported different material combinations (Dixon et al., 2006), incorporation of Quantum Well(s) (QW) in the base, modulation characteristics (Then et al., 2009)(Feng et al.,2006)(Then et al., 2007b), use of tunnel junctions (Feng et al., 2009), electrical-optical signal mixing and multiplication (Feng et al., 2006a), etc. Various schemes have also been employed to increase the modulation bandwidth of the TL (Then et al., 2008)(Taghavi & Kaatuzian, 2010). A proposal has also been put forward to use the gain medium as an optical amplifier. Several models have appeared in the literature for terminal currents, gain, modulation bandwidth, etc. of TL (Feng et al., 2007)(Faraji et al., 2008)(Basu et al., 2009)(Then et al., 2010)(Zhang & Leburton, 2009). However, these models based on rate equations, give values for optoelectronic characteristics of TL. In addition, the TL has been simulated both numerically (Shi et al., 2008)(Kaatuzian & Taghavi, 2009) and by CAD, i.e. software packages, (Shi et al., 2008)(Duan et al., 2010).

The electron recombination plays an important role in determining the base current due to spontaneous and stimulated emission in the QW, so that the electronic characteristics of the

TL have a strong dependence on its laser operation. This is proved by the decrease in the electrical gain $\beta=\Delta I_C/\Delta I_B$ due to enhanced carrier recombination in the base region when stimulated emission occurs. According to the model described in section 4, effects of geometrical parameters of QW-base on small signal optical frequency response and current gain should be significant. In other words optoelectronic characteristics of TL are directly affected by parameters like QW position in the base (W_{EQW}), well width (W_{QW}) (and barrier width in the case of multiple quantum-wells TL), number of QWs, base width (W_B), etc. For instance, TL characteristics depend on the relative position of QW because of the quasi-linear density profile of the base minority carriers (Fig. 2). Although, the mentioned first order model of charge control does not distinguish between bulk and QW carriers, TL characteristics dependency on QW and base width are still acceptable. Herein, we utilize analysis described in section before to model the effect of QW position on optical bandwidth and current gain of TL. As a goal, we look for an enhanced performance for the TL by "restructuring" it.

5.1 QW location effect

Being integrated inside the base region of the TL, the quantum-well structural parameters have certainly significant influence on both optical and electrical properties of TL, e.g. gain and bandwidth for optical and electrical outputs. This subsection is dedicated to the mentioned effects while other parameters will be studied later.

5.1.1 Carrier profile

According to the model, constant base-emitter voltage and collector current makes (Δn_1+ Δn_2), Δn_1, Q_2 unchanged. Altering the position of QW within the base can therefore change Q_1, i.e. portion of carriers responsible for lasing. The new values for Q_1 and $\tau_{t,1}$ (base transit time) when QW moves are as below

$$Q_{1,new} = Q_1 * (W_{EQW,new}/W_{EQW,old}) \tag{26}$$

$$\tau_{t,1,new} = \tau_{t,1,old} * (W_{EQW,new}/W_{EQW,old})^2 \tag{27}$$

For the TL described in previous sections we may set $W_{EQW,old}$=590Å. As a result, minority carrier profile of base region solved analytically for W_{EQW}=150 Å is sketched in Fig. 5. Also Fig. 6 demonstrates the modified charge control model. Displacement of QW position, W_{EQW}, towards emitter can cause two noteworthy effects simultaneously. First, τ_B falls due to reduction of Q_1 while Q_2 is constant. Second, in accordance to the diffusion formula for I_E and Fig. 6, I_E would increase significantly as a result of steeper charge density profile at the base emitter interface.

These conditions make base current (I_B) to rise abruptly beyond the threshold current (22 mA). Due to (Kaatuzian, 2005) the threshold current is constant during this movement of QW. In other words, moving the QW in this direction decreases the electrons which are "trapped" and recombined with holes within it. Thus more carriers are allocated to electrical operation and fewer involved in light generation, i.e. fewer photons created within the well. This phenomenon, exhibited by animation in Fig. 7, means that carriers arrive sooner at QW. As a result of "faster rich time" of minority carriers to the well, it is predicted that both τ_B and β decreases drastically (by about 4 or 5 times for τ_B).

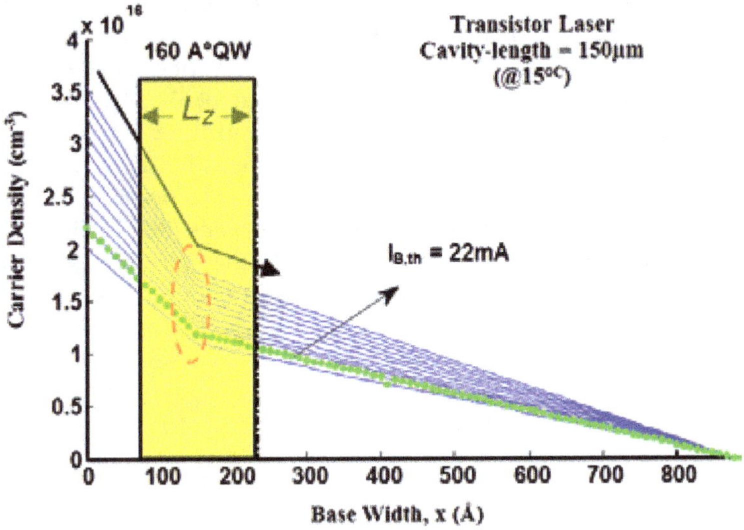

Fig. 5. Calculated minority carrier distribution for QW moved toward emitter; the profile deviates more from triangular form of regular HBT.

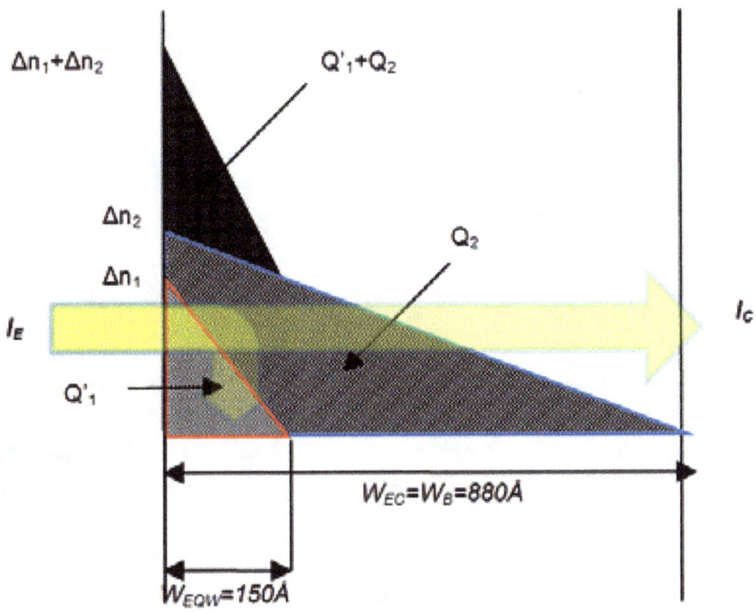

Fig. 6. Modified charge control model for W_{EQW}=150Å. Q_2 is constant while Q_1 decreased to Q_1`.

5.1.2 Base recombination lifetime

Overall effective base recombination lifetime of the entire base population, τ_B, can be calculated using the concept of charge control model and is changed for different QW locations, i.e. W_{EQW}. The special calculation method of τ_B value for different base currents is based on a reverse estimation of them from experimental data for W_{EQW}=590Å and the charge relations (physical model). The method involves a set of calculations which generates new values of τ_B using old values of τ_B (i.e. Fig. 4 in (Feng et al., 2007)) and charge relations according to the charge control model of section 3. The result for new τ_B values at I_B=22mA, i.e. $\tau_{B,spon}$ which equals τ_B evaluated at I_{Bth}, for QW located in different places through the bulk base is sketched in Fig. 8. To show the reliance of base recombination lifetime on I_B and W_{EQW} simultaneously, τ_B has been drawn in three dimensions (3D) in Fig. 9. Assessing these two figures, it is obvious that the closer to the collector the QW, the less variant with I_B the τ_B. Spontaneous base recombination lifetime (τ_{Bspon}) is obtained by cross section of 3-D scheme of τ_B and I_B=I_{th}=22mA plane as well.

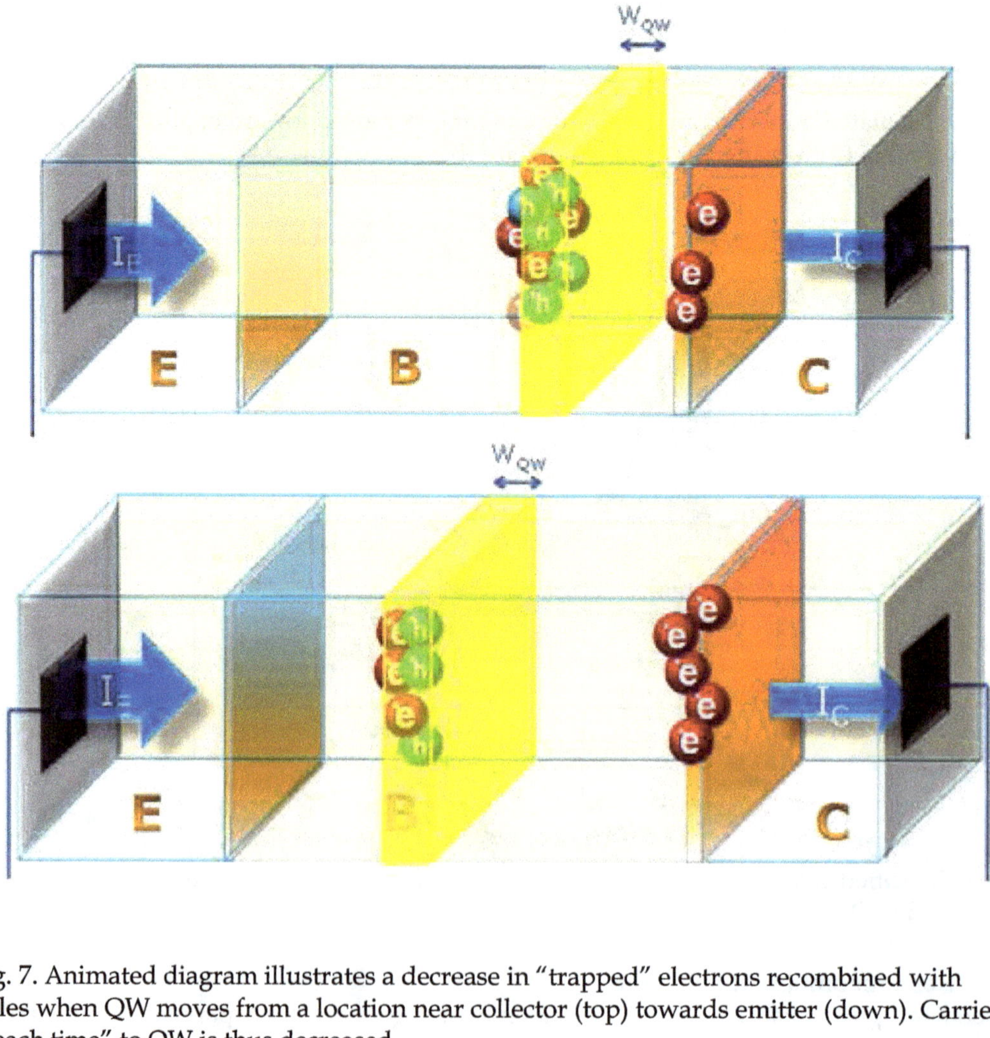

Fig. 7. Animated diagram illustrates a decrease in "trapped" electrons recombined with holes when QW moves from a location near collector (top) towards emitter (down). Carrier "reach time" to QW is thus decreased.

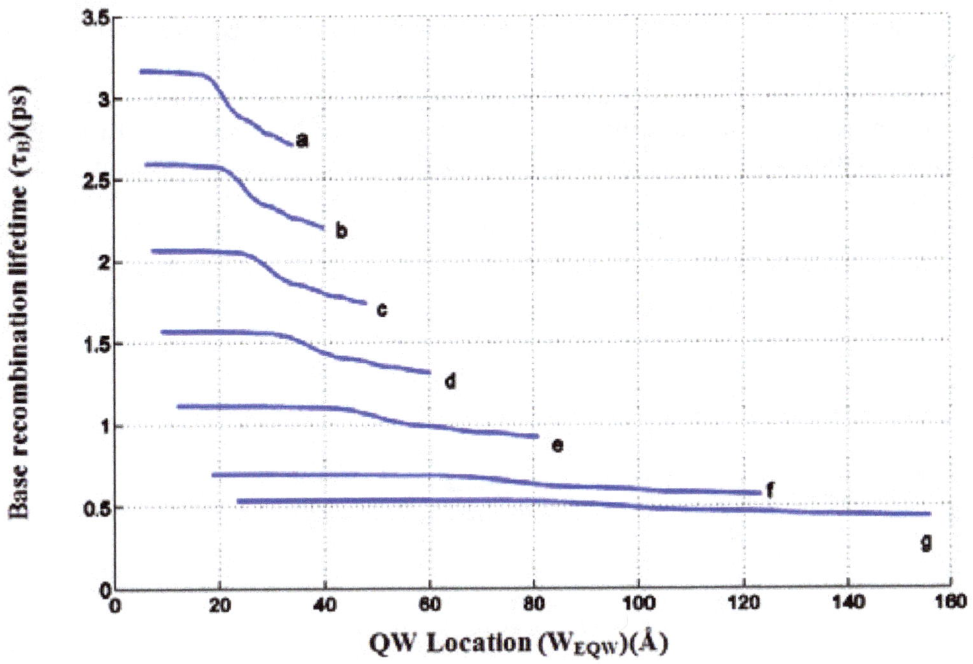

Fig. 8. Calculated base recombination lifetime (τ_B) versus W_{EQW} (a)690 (b)590 (c)490 (d)390 (e)290 (f)190 (g)150(Å).

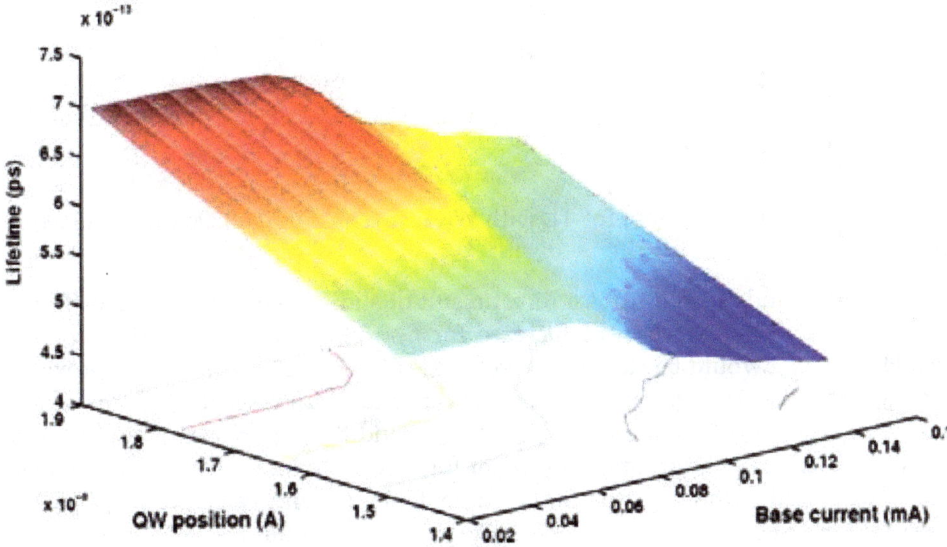

Fig. 9. 3-D scheme of base recombination lifetime (τ_B) versus I_B and W_{EQW} placed in different positions through the base.

5.1.3 Small-signal optical frequency response

Calculating the optical frequency response of transistor laser in 4.1, one can sketch the response for different QW locations. Table 1 shows the simulation results of device parameters for different QW locations through the base region while Fig. 10 and Fig. 11 show optical response and bandwidth dependence on QW location, respectively. As it is obvious from the values for simulated ζ, all of them exceed 0.7 and no resonance peak, the limiting factor for diode lasers, occurs due to QW dislocation through such a thin base layer. Non unity resonance peak in diode lasers with a large carrier lifetime (0.1 - 1 ns), sets limit on f_{-3db} and restricts performance of diode laser.

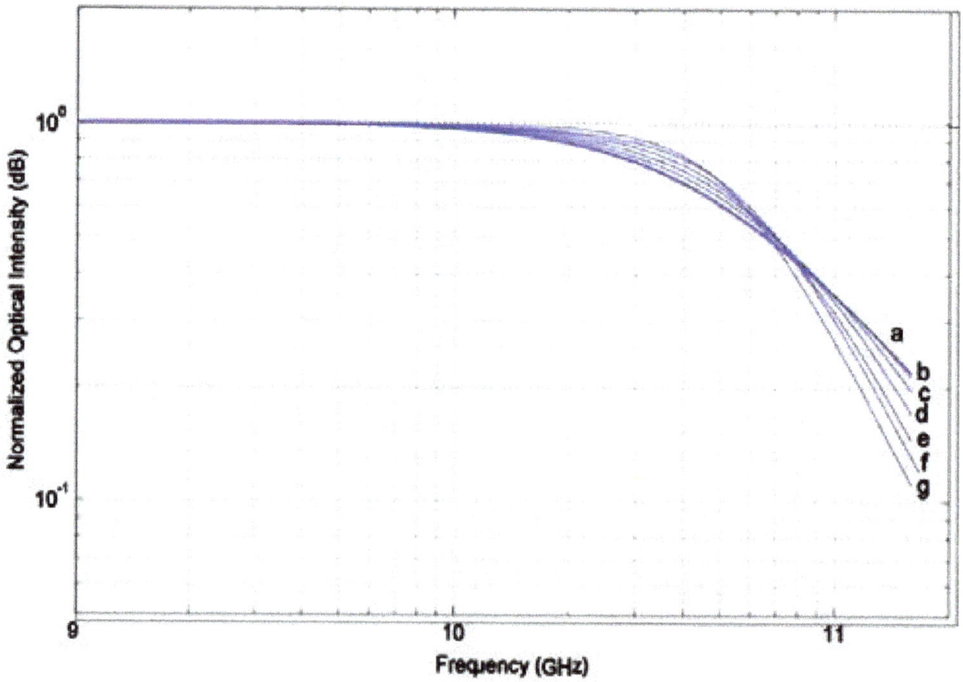

Fig. 10. Optical frequency response for TL with W_{EQW} (a)150 (b)190 (c)290 (d)390 (e)490 (f)590 (g)690(Å).

Increase in optical bandwidth (f_{-3db}) is a direct result of sliding the QW toward collector. At the first glance opposite direction of increase for f_{-3db} and $\tau_{B,spon}$ is confusing and seems mistaken. However, it would be clear if we sketch the f_{-3db} versus $\tau_{B,spon}$ (using approximation of f_{-3db} by equation 18). As Fig. 12 demonstrates, the bandwidth rises abnormally with $\tau_{B,spon}$ increases while reach a peak at about $\tau_{B,spon}$=3.37 ps and then decreases. This fact shines a novel idea that there may be an optimum point for QW to be located in order to maximize the optical bandwidth. Utilizing this figure in conjunction with calculated $\tau_{B,spon}$, one can move the QW toward collector to W_{EQW}=730Å to achieve maximum optical bandwidth (~54 GHz, taking into account the error due to approximate equation). As a result, the QW should be placed closed to collector as much as possible in order to achieve the maximum bandwidth.

Quantum-Well Position (Å)	$\tau_{B,spon}$ (ps)	*fn* (undamed natural ferquency)(GHz)	ξ (damping ratio)	Simulated -3dB Bandwidth(GHz)
150	0.54	125.8	1.7675	38.7
190	0.7	109.7	1.5595	39.3
290	1.11	87.5	1.2288	42.3
390	1.57	74	1.0365	45.2
490	2.1	64	0.9063	47.4
590	2.57	57	0.8103	48.9
690	3.03	53.5	0.7424	50.9

Table 1. Simulated device parameters for different W_{EQW}; No resonance peak due to QW movement when ξ≥0.7.

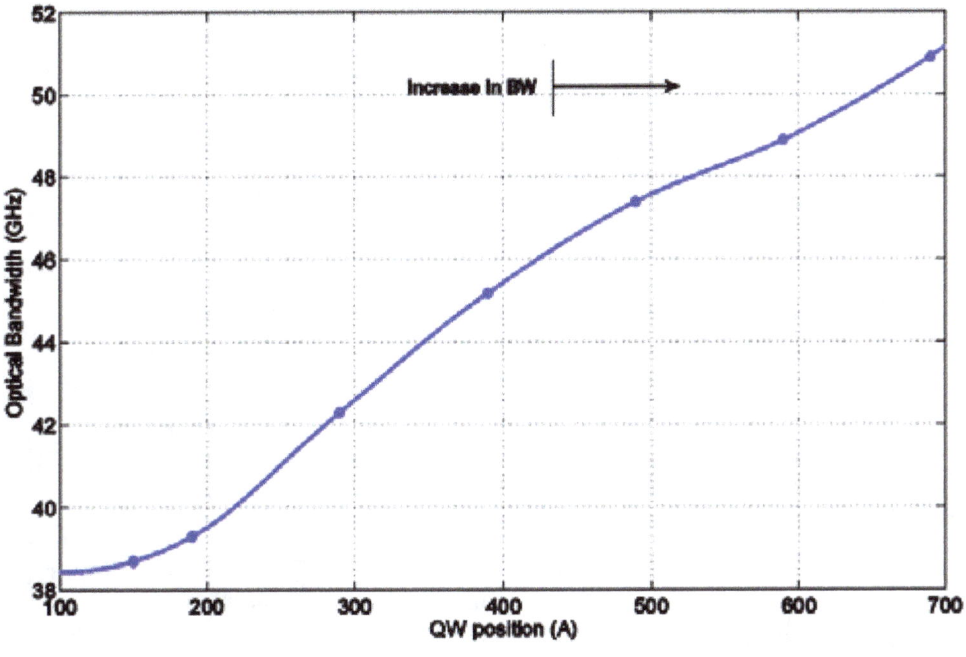

Fig. 11. QW location effect on optical bandwidth; an HBTL with well near its collector has larger optical bandwidth while electrical bandwidth is the same.

5.1.4 Current gain

We predicted previously that optical bandwidth increment would be at cost of sacrificing the current gain (β). Further investigation of QW effect on TL optoelectronic characteristics, this time electrical, leads in an interesting result that matures the previous founding. Collector current gain for an HBTL was calculated in subsection 4.2 in equations (24) and (25). As a result of QW movement towards collector, simulation at constant bias currents shows that τ_{TL} and consequently β declines. Fig. 13 shows this change in optical bandwidth and current gain versus displacement of QW while bias voltage, i.e. v_{be}, forces base current to be constant at I_B=33mA for W_{EQW}=590 Å. This bias enforcement does not disturb generality of the simulation results.

The opposite dependence of BW and β to W_{EQW}, is a trade-off between TL optoelectronic characteristics. Other experimental and theoretical works proved the described "trade-off" between β and f_{-3db} as well (Then et al.,2009),(faraji et al., 2009). They also predict analytically the above mentioned direct dependence of f_{-3db} on $\tau_{B,spon}$. In (Then et al., 2008) the authors utilized an auxiliary base signal to enhance the optical bandwidth. As a merge of their work and the present analysis we can find the optimum place for QW that leads to better results for both β and f_{-3db} of a TL. It means we can use both method, i.e. auxiliary base signal and QW dislocation method, simultaneously. A suggestion for finding an "optimum" QW location consists of two steps . First we focus on β and make it larger by locating the QW close to emitter, e.g. W_{EQW}<300 Å, which results in BW<43 GHz. Then we use auxiliary AC bias signal to trade some gain for BW. It should be noted that β less than unity is not generally accepted if TL is supposed to work as an electrical amplifier.

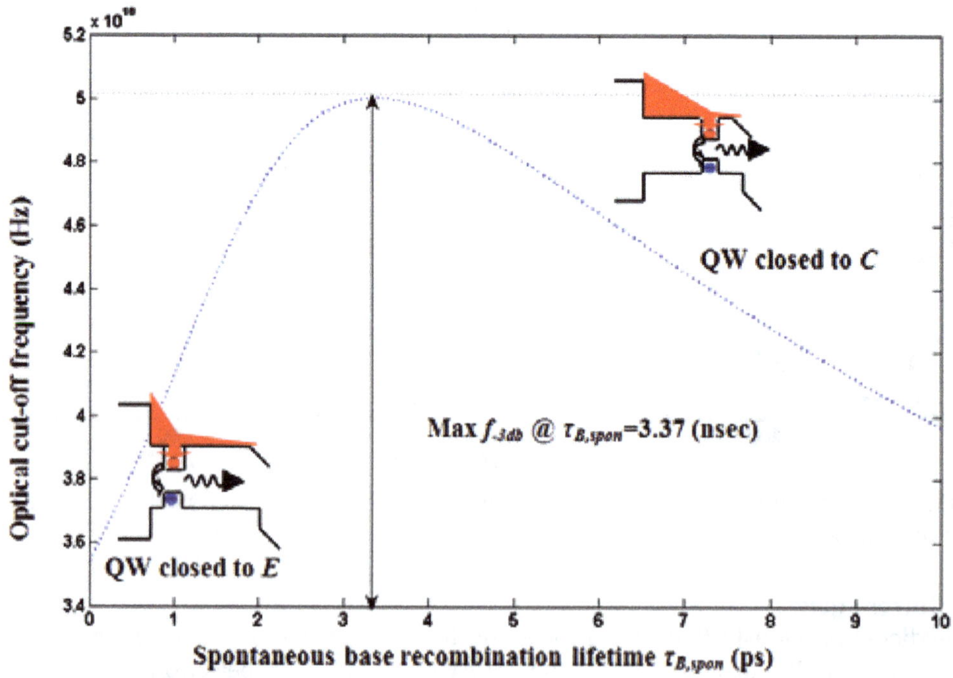

Fig. 12. Calculated optical cut-off frequency (f_{-3db}) versus $\tau_{B,spon}$. Bandwidth is maximum for $\tau_{B,spon}$ corresponding to W_{EQW}≈730Å.

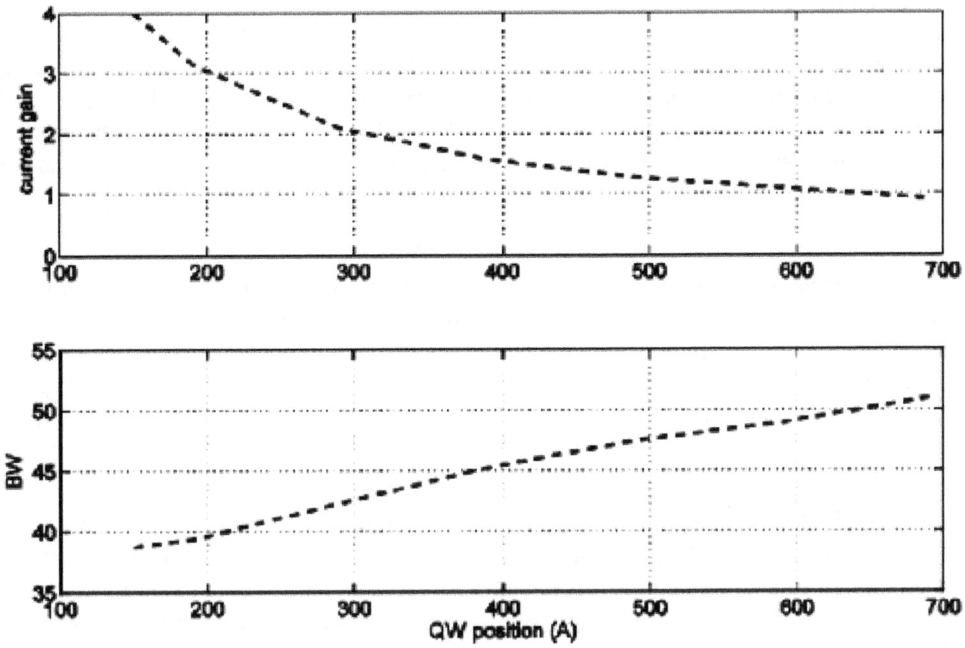

Fig. 13. Electrical gain-optical bandwidth trade-off in an HBTL.

5.2 Quantum-well width

Being an optical collector, the QW plays a significant role as it governs both optical and electrical characteristics of TL. Among QW-base geometry parameters, other than QW location, one can investigate the width of QW incorporated within the base region. For instance, we can change the well width while other geometrical parameters, like QW location, are left unchanged in order to examine how the optical frequency response and current gain alter. Base minority carrier recombination lifetime was calculated before as in equation (9) which was an independent function of well width. Charge control model and charge analysis based on this model, described in previous sections, should be completed in order to do this analysis. The base carrier lifetime, τ_B, can be written as below (Then et al., 2007c)

$$1/\tau_B = \sigma \upsilon_{th} N_r \tag{28}$$

Where υ_{th} is the thermal velocity of carriers, N_r is the density of possible recombination sites and σ is the cross section of carrier capture. σ is a measure of the region that an electron has the possibility to capture and recombine with a hole and is proportional to well width (W_{QW}). In the other hand, N_r depends on the hole concentration, i.e. N_A of the base region. So we can evaluate τ_B as

$$1/\tau_B = G W_{QW} N_A \tag{29}$$

where G is a proportionality factor defined by other geometrical properties of the base. Using this equation one can extract base recombination lifetime of base minority carriers for different base doping densities. Calculations exhibit an indirect relation between τ_B and well

width, agreeing with a larger QW width enhancing the capture cross section for electrons. Moreover, the larger N_A, the greater the recombination and hence the smaller τ_B.
The results for optical frequency response based on Statz-deMars equations of section 4 can be utilized to evaluate the optical properties of a TL for varying well width. Indeed, equations (13) and (14) require $\tau_{B,spon}$ not τ_B, as described above, therefore threshold current should be calculated for different QW widths. An expression for base threshold current of TL is as below

$$J_{th} = qn_0\tau_{cap}/v\tau_{qw}\tau_{rb} = \frac{qn_0}{\tau_{qw}}\left[1 + \left(\frac{1-v}{v}\right)\frac{\tau_{cap}}{\tau_{rbo}}\right] \tag{30}$$

where n_0 is minority carrier density in steady-state (under dc base current density of J_0), τ_{cap} is the electron capture time by QW (not included in charge control model for simplicity), τ_{qw} is the QW recombination lifetime of electron and τ_{rb0} is the bulk lifetime (or direct recombination lifetime outside the well, also ignored in our model). The base geometry factor, v, gives the fraction of the base charge captured in the QW and defines as (Zhang & Leburton, 2009)

$$v = (W_{qw}/W_b)(1 - x_{qw}/W_b) \tag{31}$$

where W_{qw} is the QW width, the factor we investigate here, W_b is the base width and x_{qw} is the QW location, similar but not equal to previously defined parameter of W_{EQW}. By setting all the constants, one can calculate $\tau_{B,spon}$ and then small-signal optical frequency response and bandwidth of TL for a range of QW widths. Optimization is also possible like what we did for QW location.

6. Conclusion and future prospects

An analytical simulation was performed to predict dependence of TL optoelectronic characteristics on QW position in order to find a possible optimum place for QW. Simulated base recombination lifetime of HBTL for different QW positions exhibited an increase in optical bandwidth QW moved towards the collector within the base. Further investigations of optical response prove the possibility of a maximum optical bandwidth of about 54GHz in WEQW≈730 Å. Since no resonance peak occurred in optical frequency response, the bandwidth is not limited in this method. In addition, the current gain decreased when QW moved in the direction of collector. The above mentioned gain-bandwidth trade-off between optoelectronic parameters of TL was utilized together with other experimental methods reported previously to find a QW position for more appropriate performance. The investigated transistor laser has an electrical bandwidth of more than 100GHz. Thus the structure can be modified, utilizing the displacement method reported in this paper, to equalize optical and electrical cut-off frequencies as much as possible.
In previous sections we consider the analysis of a single quantum well (SQW) where there is just one QW incorporated within the base region. This simplifies the modelling and math-related processes. In practice, SQWTL has not sufficient optical gain and may suffer thermal heating which requires additional heat sink. Modifications needed to model a multiple QW transistor laser (MQWTL). First one should rewrite the rate equations of coupled carrier and photon for separate regions between wells. Solving these equations and link them by applying initial conditions, i.e. continuity of current and carrier concentrations, is the next step. In addition to multiple capture and escape lifetime of carriers, tunnelling of the 2-dimensional carriers to the adjacent wells should be considered. For wide barriers one may use carrier transport across the barriers instead the mentioned tunnelling. Simulation results

for diode laser (Duan et al., 2010), as one of the transistor laser parents, demonstrate considerable enhancement in optical bandwidth and gain of the device when increasing the number of quantum wells (Nagarajan et al., 1992), (Bahrami and Kaatuzian, 2010). Like the well location modelled here in this chapter, there may be an optimum number of quantum wells to be incorporated within the base region. Due to its high electrical bandwidth (≥100 GHz), it is needed to increase the optical modulation bandwidth of the TL. Base region plays the key role in all BJT transistors, especially in Transistor Lasers.

Like Quantum-Well, base structural parameters have significant effects on optoelectronic characteristics of TL which can be modelled like what performed before during this chapter. Among these parameters are base width (Zhang et al., 2009), material, doping (Chu-Kung et al., 2006), etc. For example, a graded base region can cause an internal field which accelerates the carrier transport across the base thus alters both the optical bandwidth and the current gain considerably.

7. References

Basu, R., Mukhopadhyay, B. & Basu, P.K. (2009). Gain Spectra and Characteristics of a Transistor Laser with InGaAs Quantum Well in the Base. *Proceeding of International Conference on Computers and Devices for Communication,* India, Dec. 2009

Bahrami Yekta, V. & Kaatuzian, H. (2010). Design considerations to improve high temperature characteristics of 1.3μm AlGaInAs-InP uncooled multiple quantum well lasers: Strain in barriers. *Optik, Elsevier,* doi:10.1016/j.ijileo.2010.03.016

Chan, R., Feng, M., Holonyak, N. Jr., James, A. & Walter, G. (2006). Collector current map of gain and stimulated recombination on the base quantum well transitions of a transistor laser. *Appl. Phys. Lett.,* Vol. 88, No. 143508

Chu-kung, B.F., Feng, M., Walter, G., Holonyak, N. Jr., Chung, T., Ryou, J.-H., Limb, J., Yoo, D., Shen, S.C. & Dupuis, R.D. (2006). Graded-base InGaN/GaN heterojunction bipolar light-emitting transistors. *Appl. Phys. Lett.* Vol. 89, No. 082108

Dixon, F., Feng, M. & Holonyak, N. Jr. (2010). Distributed feedback transistor laser. *Appl. Phys. Lett.* Vol. 96, No. 241103

Dixon, F., Chan, R., Walter, G., Holonyak, N. Jr. & Feng, M. (2006). Visible spectrum light-emitting transistors. *Appl. Phys. Lett.* Vol. 88, No. 012108

Duan, Z., Shi, W., Chrostowski, L., Huang, X., Zhou, N., & Chai, G. (2010). Design and epitaxy of 1.5 μm InGaAsP-InP MQW material for a transistor laser. *Optics Express,* Vol. 18, Issue 2, pp. (1501-1509), doi:10.1364/OE.18.001501

Faraji, B., Pulfrey, D.L. & Chrostowski, L. (2008). Small-signal modeling of the transistor laser including the quantum capture and escape lifetimes. *Appl. Phys. Lett.,* Vol. 93, No. 103509

Faraji, B., Shi, W., Pulfrey, D.L. & Chrostowski, L. (2009). Analytical modeling of the transistor laser. *IEEE Journal of Quantum Electron.,* Vol. 15, No. 3, pp (594-603)

Feng, M., Holonyak, N. Jr. & Hafez, W. (2004a). Light-emitting transistor: light emission from InGaP/GaAs heterojunction bipolar transistors. *Appl. Phys. Lett.* Vol. 84, No. 151

Feng, M., Holonyak, N. Jr. & Chan, R. (2004b). Quantum-well-base heterojunction bipolar light-emitting transistor. *Appl. Phys. Lett.,* Vol. 84, No. 11

Feng, M., Holonyak, N. Jr., Walter, G. & Chan, R. (2005). Room temperature continuous wave operation of a heterojunction bipolar transistor laser. *Appl. Phys. Lett.* Vol. 87, No. 131103

Feng, M., Holonyak, N. Jr., Chan, R., James, A. & Walter, G. (2006a). Signal mixing in a multiple input transistor laser near threshold. *Appl. Phys. Lett.* Vol. 88, No. 063509

Feng, M., Holonyak, N. Jr., James, A., Cimino, K., Walter, G. & Chan, R. (2006b). Carrier lifetime and modulation bandwidth of a quantum well AlGaAs/ InGaP/GaAs/InGaAs transistor laser. *Appl. Phys. Lett.* Vol. 89, No. 113504

Feng, M., Holonyak, N. Jr., Then, H.W. & Walter, G. (2007). Charge control analysis of transistor laser operation. *Appl. Phys. Lett.* Vol. 91, No. 053501

Feng, M., Holonyak, N. Jr., Then, H.W., Wu, C.H. & Walter, G. (2009). Tunnel junction transistor laser. *Appl. Phys. Lett.* Vol. 94, No. 041118

Kaatuzian, H. (2005). *Photonics, Vol. 1*, AmirKabir University of Technology press, , Tehran, Iran

Kaatuzian, H. & Taghavi,I. (2009). Simulation of quantum-well slipping effect on optical bandwidth in transistor laser. *Chinese optics letters.* doi:10.3788/COL20090705.0435, pp. 435–436

Nagarajan, R., Ishikawa, M., Fukushima, T., Geels, R. & Bowers, E. (1992). High speed quantum-well lasers and carrier transport effects. *IEEE Journal of Quantum Electron,* Vol. 28, No. 10, pp (1990-2008)

Shi, W., Chrostowski, L & Faraji, B. (2008). Numerical Study of the Optical Saturation and Voltage Control of a Transistor Vertical-Cavity Surface-Emitting Laser. *IEEE Photonics Technology Letters,* Vol. 20, No. 24

Suzuki, Y., Yajima, H., Shimoyama, K., Inoue, Y., Katoh, M. & Gotoh, H. (1990). (Heterojunction field effect transistor laser). *Ellectronics Letters,* Vol. 26, No. 19

Taghavi,I & Kaatuzian, H. (2010). Gain-Bandwidth trade-off in a transistor laser : quantum well dislocation effect. *Springer, Opt Quant Electron.* doi: 10.1007/s11082-010-9384-0, pp. 481–488

Then, H.W., Walter, G., Feng, M. & Holonyak, N. Jr. (2007a). Collector characteristics and the differential optical gain of a quantum-well transistor laser. *Appl. Phys. Lett* Vol. 91, No. 243508

Then, H.W., Feng, M. & Holonyak, N. Jr. (2007b). Optical bandwidth enhancement by operation and modulation of the first excited state of a transistor laser. *Appl. Phys. Lett.* Vol. 91, No. 183505

Then, H.W., Feng, M., Holonyak, N. Jr. & Wu, C.H. (2007c). Experimental determination of the effective minority carrier lifetime in the operation of a quantum-well n-p-n heterojunction bipolar light-emitting transistor of varying base quantum-well design and doping. *Appl. Phys. Lett.* Vol. 91, No. 033505

Then, H.W., Walter, G., Feng, M. & Holonyak, N. Jr. (2008). Optical bandwidth enhancement of heterojunction bipolar transistor laser operation with an auxiliary base signal. *Appl. Phys. Lett.* Vol. 93, No. 163504

Then, H.W., Feng, M. & Holonyak, N. Jr. (2009). Bandwidth extension by trade-off of electrical and optical gain in a transistor laser: three-terminal control. *Appl. Phys. Lett.* Vol. 94, No. 013509

Then, H.W., Feng, M. & Holonyak, N. Jr. (2010). Microwave circuit model of the three-port transistor laser. *Appl. Phys. Lett.* Vol. 107, No. 094509

Walter, G., Holonyak, N. Jr., Feng, M. & Chan, M. (2004). Laser operation of a heterojunction bipolar light-emitting transistor. *Appl. Phys. Lett.* Vol. 85, No. 4768

Walter, G., James, A., Holonyak, N. Jr., Feng, M. & Chan, R. (2006). Collector breakdown in the heterojunction bipolar transistor laser. *Appl. Phys. Lett.* Vol. 88, No. 232105

Zhang, L. & Leburton, J.-P. (2009). Modeling of the transient characteristics of heterojunction bipolar transistor lasers. *IEEE Journal of Quantum Electron.* doi:10.1109/JQE.2009.2013215, pp. (359–366)

2

Energy Efficient Semiconductor Optical Switch

Liping Sun and Michel Savoie
Communications Research Centre
Canada

1. Introduction

Energy-saving technology that reduces power consumption is of increasing importance due to the ever-increasing demand for Internet services. To prevent the traffic growth from being strangled by energy bottlenecks, novel architectural and technological solutions are indispensable. The most obvious way to cope with the issue is to reduce the energy consumed by the network elements.

Fast optical switching is an important enabler of advanced optical networks, in particular such functions as routing burst and packet optical signals, optical path provisioning and fault restoration. Semiconductor Digital Optical Switches (DOSs) can fulfill such high speed applications due to their nanosecond switching times, step-like switching responses, and immunity to variations in temperature, wavelength, polarization, refractive index and device fabrication tolerances. Moreover, semiconductor DOSs offer the potential for integration with other semiconductor optoelectronic components and thus promise considerable reductions in the size, complexity and cost of an overall optical system.

For optical waveguide switches, fast optical switching may be achieved by a refractive index change, induced either by carrier injection (Zegaoui et al., 2009; Bennett et al., 1990) or by the electro-optic effect (Cao et al., 2009; Agrawal et al., 1995), within III-V semiconductors, such as GaAs-based and InP-based. Compared to carrier-injection switches, electro-optic switches have faster switching speeds but larger switching voltages since the refractive index change induced by an electro-optic effect is about two orders of magnitude smaller than that by carrier injection.
Therefore, until now, most of the commercially available semiconductor DOS products have been based on carrier-injection (Ikezawa et al., 2008). These devices typically utilize carrier-induced Total Internal Reflection (TIR) at a waveguide branching or crossing point to switch the light path from one waveguide to another. Such TIR-based semiconductor switches typically require a large index modulation, e.g. in the order of 0.01, with the region of changed index having a well-defined boundary. Accordingly, efforts have been made to restrict current spreading and to confine the injected carriers to the desired region. Typically, achieving such carrier confinement involves using relatively complex semiconductor device technologies, such as ion implantation (Abdalla et al., 2004; Zhuang et al., 1996), electron-beam lithography (Shimomura et al., 1992), Zn diffusion (Yanagawa et al., 1990), and epitaxial regrowth (Thomson et al., 2008).

A plane view of a typical conventional semiconductor DOS is illustrated in Fig. 1(a), wherein two waveguides intersect at an angle θ forming an X-like waveguide structure. An electrode is provided over a common waveguide region where the waveguides intersect for injecting carriers into a portion of the common region to decrease its refractive index n by an amount Δn sufficient to induce TIR for the input beam and to cause it to turn by the angle θ to form the switched light. In the absence of the carrier injection, i.e. when there is no current flowing through the electrode, the input light continues its propagation along the waveguide past the common region to form the transmitted light. In order to switch the input light, the electrode is positioned so that its edge, which faces the input light, crosses the input waveguide at an angle θ/2. One drawback of the design in Fig. 1 is that, for optimal performance, the electrode should be positioned with its 'receiving' edge centered in the common waveguide region. This makes the structure asymmetrical with respect to the two input waveguides, so that the switch is, essentially, a 'Y' switch that can only function as a 1×2 switch and not as a 2×2 switch.

A typical 2×2 semiconductor TIR DOS with the conventional electrode design is shown in Fig. 1(b). In order to reduce an optical axis misalignment of the reflected light for both input ports, the electrode has to be narrow; disadvantageously, a narrow electrode will generally result in poor reflectivity induced by the current passing through the electrode, leading to a degradation of the switch extinction ratio, i.e. the optical power ratio for the output ports.

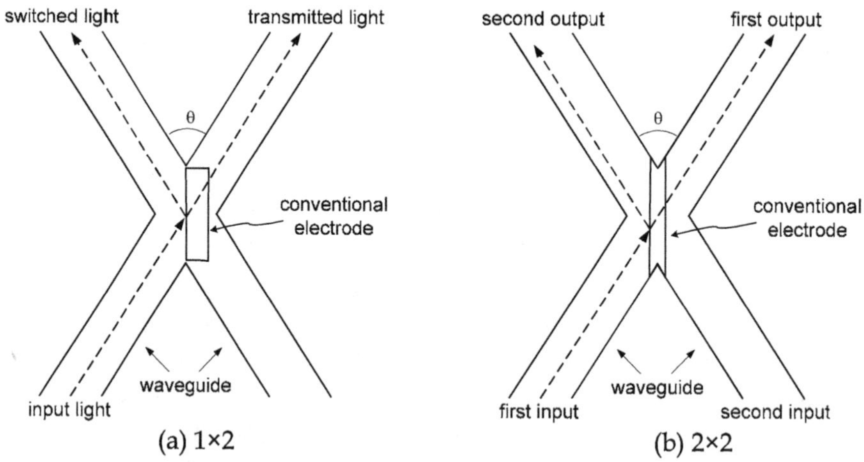

Fig. 1. Schematic diagrams of TIR optical switches with conventional electrodes.

Fig. 2 is a schematic diagram of a 2×2 switch which has a similar waveguide topology as Fig. 1(b), but utilizes the bow-tie electrode in place of the straight electrode of Fig. 1(b) (Li et al., 2001). The bow-tie electrode has a narrow part located at the central point of the waveguide intersection, and the electrode is symmetrical with respect to all four waveguide ends. The switch has a switching angle θ = 2°, and an electrode bow-tie angle of 1.5 degrees. The switch utilizes ion implantation about the electrode to decrease current spreading and confine the injected carriers. One disadvantage of the prior art bow-tie electrode design is that its electrode is still narrow at the waveguide crossing point, potentially enabling the light to leak therethrough under the carrier injection condition. Furthermore, a part of the incoming light that incidents on the second electrode segment may be refracted rather than reflected because of the increased grazing angle. All these limit the waveguide crossing angle θ of the bow-tie type 2×2 switches to about 2° in real device applications. The small switching angle increases the size of the device and limits the integration density.

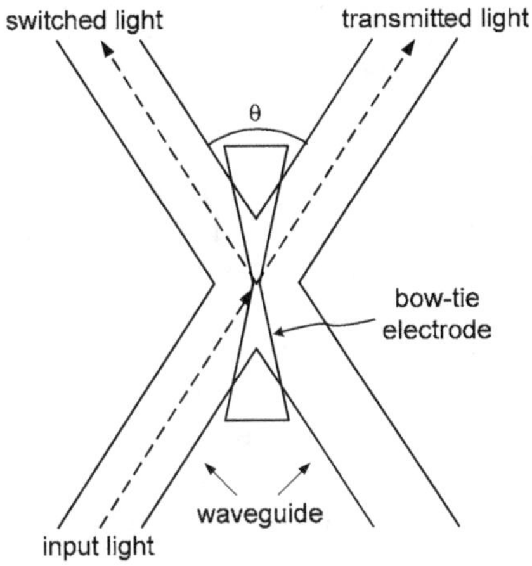

Fig. 2. Schematic diagram of a 2×2 TIR optical switch with bow-tie electrode.

Another disadvantage of the prior art TIR semiconductor switches is their relatively large power consumption, especially for devices with the switching angle at the higher end of the range, as such devices require larger electrical currents. Reconfigurable waveguide DOS devices with small switching current and low crosstalk have been reported recently (Zegaoui et al., 2009; Ng et al., 2007), but their branch angle is only 0.9°. It is therefore desirable to develop TIR switches with greater deflection angles and reduced power consumption.

This chapter presents our approaches to improve the energy efficiency of semiconductor DOSs for applications in next generation green optical networks. In the following sections, we first propose and demonstrate a novel double-reflection switch design that can reduce the power consumption and increase the switch deflection angle. Then, in the next section, an ultra-low-power 2×2 carrier-injection optical switch based on a Mach-Zehnder Interferometer (MZI) formed by Multi-Mode Interference (MMI) couplers is introduced and investigated. Conclusions and future applications are provided in the last section.

2. Double-reflection TIR DOS

This section discusses the design, fabrication and characterization of a novel double-reflection technique to fabricate 1×2 and 2×2 wide-angle AlGaAs/GaAs carrier-injection TIR optical switches. This new development uses only conventional semiconductor fabrication technologies to reduce the switching current and relax the tight restriction on the width of the electrodes located at the switch centre. To compensate for the increased switching current, compositionally graded interfaces are added at the heterojunctions to reduce the switching voltage and thus the power consumption. Switching performance is further improved by applying curved electrodes and carrier-restriction gaps.

2.1 Double-reflection structure

Our approach to improve the performance and reliability of 1×2 and 2×2 TIR optical switches with a double-reflection electrode geometry is illustrated in the two pictures shown in Fig. 3 below. The electrode in Fig. 3(a), which utilizes a well known 'Y'-shaped waveguide structure, is shaped to induce double reflection of the input light. Its first edge faces the input waveguide and is positioned for turning, in the presence of the carrier injection, the input light by a first deflection angle θ_1, so as to form the first reflected light that propagates generally towards the second edge. The second edge is positioned for turning the first reflected light by a second angle θ_2 towards the switched output waveguide for coupling thereinto as the switched light. This electrode is also shaped and positioned so that, in the absence of the carrier injection, substantially all or at least most of the input light passes under the first edge of the electrode. This design also results in that in the presence of the carrier injection through the electrode, substantially all or at least most of the input light is reflected at the first edge. Note that this feature differentiates the electrode of Fig. 3(a) from the prior art bow-tie electrode of Fig. 2, wherein the input light impinges upon two inclined edges of the bow-tie in substantially equal portions, so that only about half of the input light passes under each bow-tie edge, resulting in a loss of input light since for a large switch branching angle the part of input light that directly incidents on the second edge of the electrode will not be reflected under TIR due to their increased grazing angle.

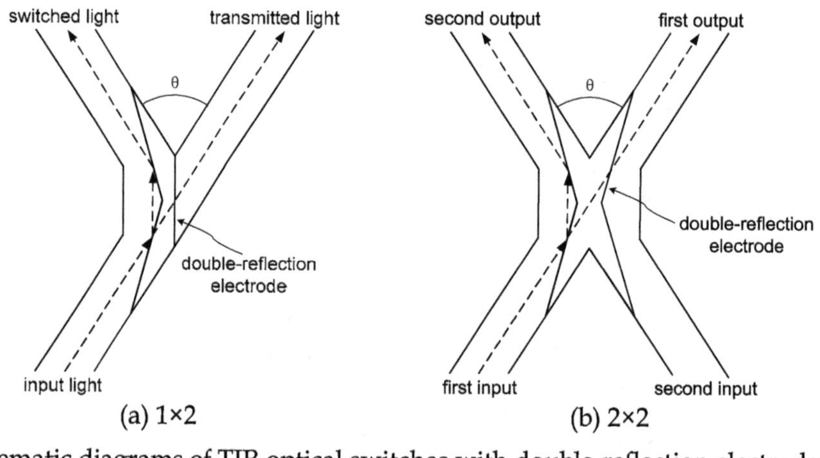

Fig. 3. Schematic diagrams of TIR optical switches with double-reflection electrode.

Continuing to refer to Fig. 3(a), in a further advantage each of the first and second deflection angles is less than the light switching angle θ, and therefore the switching of the input light's direction by the switching angle θ may be accomplished by a smaller refractive index change Δn under the electrode, and thus using a smaller injection current than would have been required for the conventional single-reflection electrode design such as that of Fig. 1. In one currently preferred electrode geometry design, the first and second angles θ_1, θ_2 are both equal to one half of the branching angle θ, i.e. $\theta_1 = \theta_2 = \theta/2$.

Advantageously, this double-reflection structure enables to have the switching angle θ twice as large as in the conventional single-reflection design for the same refractive index change Δn in the index change region, or, alternatively, to turn the light by the same angle θ using only half of the electrical current. Indeed, the reflective index change Δn that is required to induce the TIR at the first and second edges may be estimated from an approximate TIR condition,

$$\theta_i \leq \Delta n / n, \ i = 1, 2, \tag{1}$$

which approximately holds for $\Delta n/n << 1$ and the angles θ_i measured in radians, so that reducing the turning angle by a factor of 2 reduces the required index change by the same factor.

The double-reflection structure that has been described above for 1×2 optical switches, can be easily adopted for 2×2 switches, such as schematically shown in Fig. 3(b), by adding a second input waveguide and by symmetrically extending the double-reflection electrode so as to provide a second pair of electrode edges for switching the light from the second input waveguide into the first output waveguide by means of double reflection at the third and fourth electrode edges. In the 2×2 switch, the second input waveguide is optically aligned with the second output waveguide so that, in the absence of the carrier injection, input light from the second input waveguide will be transmitted by the branching section into the second output waveguide. In the presence of the carrier injection, light that enters from the second input waveguide experiences two consecutive reflections at the third and fourth edges and is directed into the first output waveguide, while light that enters from the first input waveguide also experiences two consecutive reflections at the first and second edges as described before, and is directed into the second output waveguide. The waveguide and electrode edge structure of the 2×2 switch is substantially symmetrical with respect to an axis of symmetry which bisects the waveguide branching angle. In Fig. 3(b), the electrode is shaped as an 'X', with two cut-outs around the symmetry line in the input and output portions of the structure, so as to reduce the scattering optical loss, the total electrical current required for the switching, and therefore to reduce the power consumption and device heating.

The aforedescribed double-reflection switching addresses important drawbacks of the prior art TIR-based optical switches. It eliminates the optical misalignment problem and electrode width problems of the straight and bow-tie electrodes in the 2×2 switching configurations. As a result, at the same switch crossing angle, the index change, and hence the injection current density, that is required by the TIR condition needs to be only half of that of the prior art "single reflection" design, thereby reducing the injection current density by half for the same switch crossing angle θ. On the other hand, if the same switching current density is to be used as in the prior art TIR-based switches, the switch crossing angle can be doubled, making the overall device more compact and thus allowing more devices to be integrated on a wafer. This newly developed technique enables optical switching at greater waveguide crossing angles θ and/or smaller induced refractive index changes, and thus smaller power consumption.

One potential disadvantage of the aforedescribed multi-reflection switch configuration is that, for the same switch crossing angle θ, the overall area of the switch electrode may be as much as two times larger than that of the prior art single-reflection electrodes, which may partially negate the effect of the reduced current density upon the total power consumption of the device; the larger electrode area and a relatively more complex waveguide branching area may also lead to an increase in the scattering optical loss in the waveguide branching region. Advantageously, these two potentially deleterious effects decrease dramatically by increasing the switch crossing angle θ, and may be reduced to an acceptable level at least for θ greater than about 3-4°.

2.2 Curved electrodes

A schematic layout of our proposed 2×2 switch with double-reflection structure is shown in Fig. 4. Each of the four segments of the switch electrode is curved in a logarithmic spiral shape instead of the conventional straight electrode. This electrode-curvature has been theoretically studied and reported that it could provide high power reflectivity, high extinction ratio, and low scattering loss (Nayyer et al., 2000).

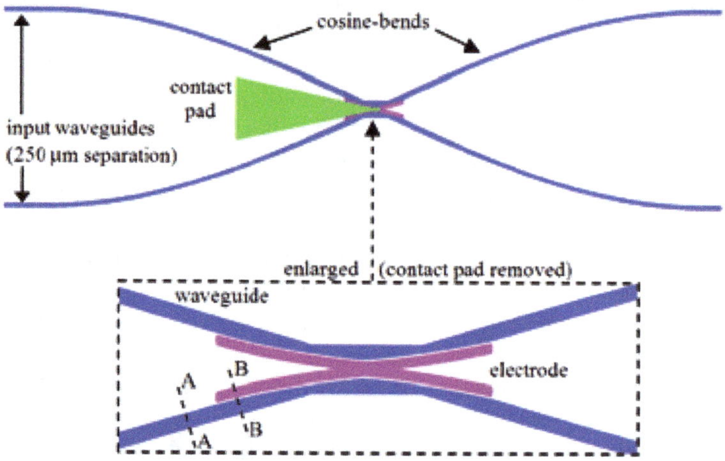

Fig. 4. Schematic layout of a proposed 2×2 double-reflection TIR optical switch.

The fabricated optical switch has the two input waveguides and two output waveguides extended with curved waveguide sections embodied as cosine bends, with a functional form $[h/2-(h/2)\cos(\pi\, x/l)]$, where h is the height of the bend, l is the length of the bend, and x is the horizontal propagation length. The device further included a top contact pad to provide an electrical contact to the switching electrode for delivering current to the switch electrode. To reduce the crosstalk at the waveguide crossing, adiabatic tapers are used in the waveguide branching region. This waveguide widening can also provide more fabrication freedom for the electrode. As shown in the figure, the double-reflection design enables us to set the electrode width at several times larger than the light penetrating depth and hence provide a sufficient index change for TIR and large extinction ratio.

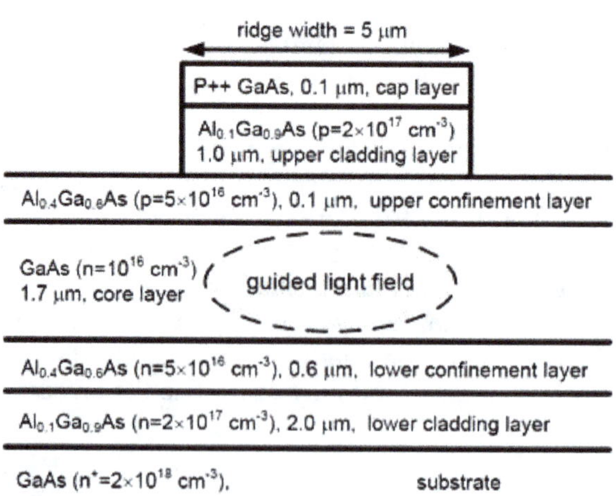

Fig. 5. The waveguide layer structure which specifies the composition ($Al_x\, Ga_{1-x}\, As$), doping profile, and layer thickness for each layer.

Fig. 5 is a schematic diagram of the vertical cross-section of the waveguide structure that we used for the 2×2 switch along the line 'A-A' shown in Fig. 4. The switch's strip-loaded waveguide design has a multilayer heterostructure with a W-shaped index profile. The waveguide core layer of GaAs is 1.7 μm thick, and is, to the best of our knowledge, the largest ever reported for a single-mode semiconductor DOS. It provides high coupling efficiency to a single mode fibre.

2.3 Current restriction gap

At the TIR interface inside our switch design, a small isolation gap is introduced between the electrode and waveguide to restrict the current spread into the waveguide regions not covered by the electrode, and to obtain a sharp carrier gradient profile for improved switching efficiency. This current restriction effect of the isolation gap is schematically illustrateed in Fig. 6. It also schematically shows an effective reflection interface wherein the reflection of the guided light occurs in the presence of the carrier injection. When the switching voltage V_b is applied between the switching electrode and a bottom contact, the switching electrical current flows through the current flow region indicated by cross-hatching. Although the gap substantially reduces the current spread into the regions not directly under the switching electrode, it does not eliminate the spreading completely, resulting in the possibility of a small lateral offset of the effective reflection interface from the reflection edge of the electrode. This offset however may be rather small, for example on the order of less than 1 μm, and is smaller than the size of the waveguide branching region.

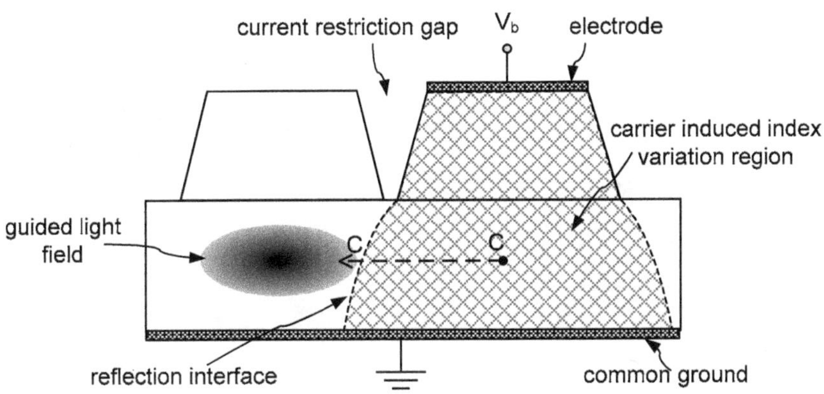

Fig. 6. Schematic diagram of the optical mode, the current flow region and the reflective interface induced by the carrier injection in the cross-sectional view of Fig. 4 taken along the line 'B-B'.

The carrier-induced index variation at the reflection interface is gradual due to the spatial spreading of injection current flow. If an abrupt spatial variation in the carrier concentration or current flow can be formed along the reflection interface, more light can be reflected to the other port. If the waveguide core layer is very thick, it is important to check how the current restriction gap affects the spreading carrier-induced index change profile at the reflection interface.

Fig. 7 is the calculated carrier concentration along line 'C-C', the middle of the core layer, in the cross-section as shown in Fig. 6 with or without a separation gap inside the splitting region, respectively. Fig. 7 was generated based on a two-dimensional simulation of Δn by using ApSys, a 2D/3D FEM optoelectronic simulation software from Crosslight Software Inc., and is based on a top electrode that is 200 μm long and with a bias current of 200 mA. Only the structure from the centerline of the biased waveguide is shown in Fig. 7. The dotted line in Fig. 7 is the edge of the top electrode. As illustrated in Fig. 7, an isolation gap in the splitting region is needed to prevent carriers from spreading beyond the electrode region. The self-heating effect was also included in the 2D FEM simulation. In the simulation, the room temperature is set at 300 K and the heat generated from contact resistance was not included. Calculation shows that the guiding of light to the destination port via modal evolution is unchanged with or without the self-heating temperature change. Fig. 8 is the corresponding carrier-induced refractive index change Δn along the middle of the core layer with or without a separation gap in the splitting region, respectively. The calculation involves a combination of bandfilling, band-gap shrinkage, and plasma effect (Bennett et al., 1990). The maximum index change is calculated to be -0.037 at the midpoint of the biasing electrode. The TIR interface located under the electrode edge is 2.5 μm apart from the biased waveguide centre. By introducing a current restriction gap inside the switch structure, the index change profile at the reflection interface is much more sharp than a switch without a gap inside. Clearly, more light can be steered into the higher-index waveguide by introducing a small gap in the splitting region.

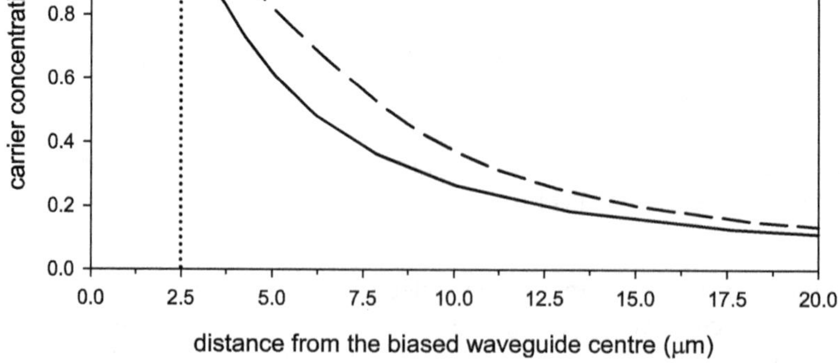

Fig. 7. Carrier concentration along the middle of core layer in the cross section at the junction branching point. The dotted line is the edge of the top electrode.

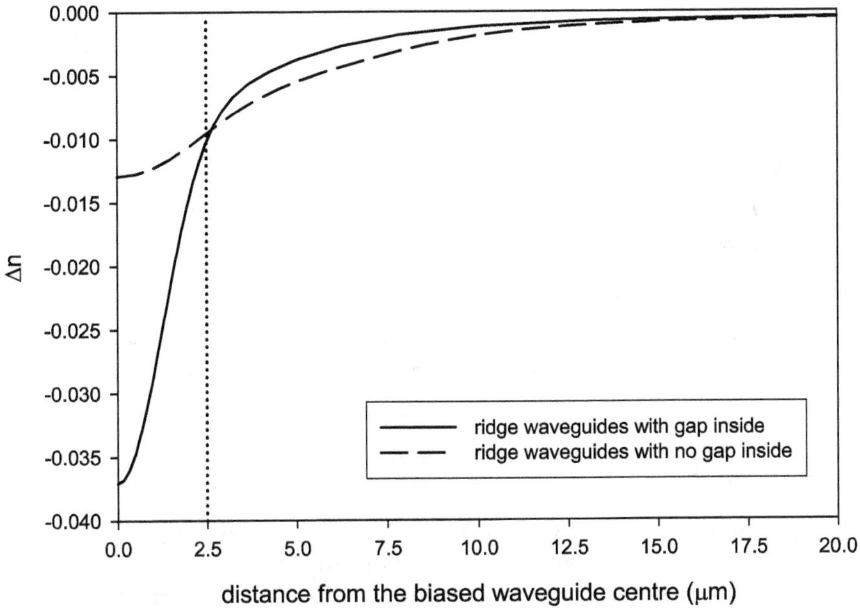

Fig. 8. Carrier-induced refractive index change Δn along the middle of core layer in the cross section at the junction branching point. The dotted line is the edge of the top electrode.

2.4 Graded heterojunction interfaces

The up-to-date growth technology of Metal-Organic Chemical Vapor Deposition (MOCVD) enables us to build semiconductor materials with precise control of thickness and composition at the atomic level. If no grading of the chemical composition of the semiconductor is applied in the vicinity of the heterostructure, all the heterojunction interfaces of the fabricated waveguide using the layer structure shown in Fig. 5 are sharply defined at atomic-level precision. One of the problems introduced by abrupt heterostructure is the band gap discontinuities caused by free charge transferring at the heterojunction between two semiconductors with different bandgap energy (Capasso et al., 1987). Carriers in the large-bandgap material will diffuse over to the small-bandgap material where they occupy band states of lower energy. As a result of the carrier transfer, and electrostatic dipole forms and leads to the band bending. For example, most published results for the $Al_xGa_{1-x}As/GaAs$ heterojunction indicate that the conduction band discontinuity $\Delta E_C \approx 0.24$ eV for x=0.30 (Wang, 1986).

Fig. 9 is the calculated band diagram of the zero-biased waveguide using the abrupt layer and doping structure shown in Fig. 5. The numerical simulation software (the APSYS of Crosslight Software) uses the FEM method to solve a 2D Poisson's equation for potential voltage, continuity equation for carriers, and thermionic emission theory for a bandgap barrier at the heterojunction interface. The adaptive mesh points were set to be very dense near the heterojunction in order to calculate the bandgap barrier correctly. The band spikes shown in Fig. 9 at the double-heterostructure interfaces near the lower confinement layer will suppress the flow of charge carriers across the junction. This band-barrier rectification effect reduces the carrier injection efficiency of the switch. It is therefore desirable to eliminate such spikes in the conduction or valence band.

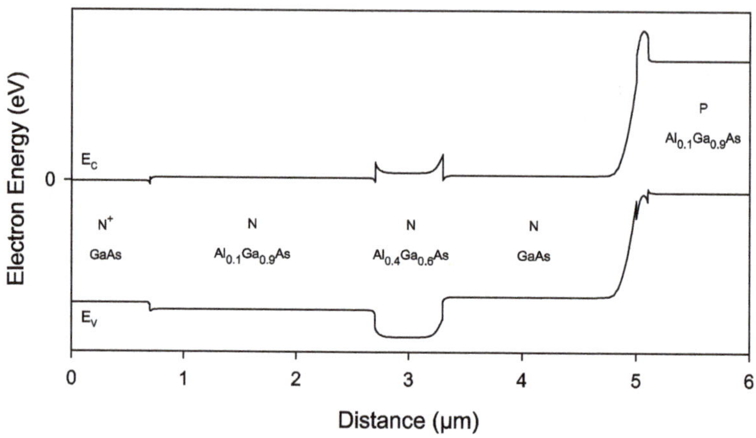

Fig. 9. Band diagram of the zero-biased waveguide using the abrupt layer and doping structure shown in Fig. 5.

Composition grading is a common technique in the design of quantum-well lasers, Heterojunction Bipolar Transistors (HBTs) and modulation-doped transistors. The compositional grading at the junction interface can smooth out a large part of the band discontinuity so that the carrier injection is increased. Usually a graded barrier with a grading length of a few hundred angstroms can effectively minimize the inference of the band spike at the abrupt interface. The smaller the carrier concentration on either side of the heterojunction, the wider the graded region has to be in order to reduce the spike height. The composition grading effects on band spikes are clearly demonstrated in Fig. 10, which shows the calculated conduction band profiles for the double-heterostructure interfaces at the lower confinement layer. Three cases were studied, varying the length of grading width L. The three simulations are for L=0, L=200 Å and L=400 Å using the layer structure shown in Fig. 5. The metallurgical grading of $Al_xGa_{1-x}As$ was set linear in the grading region.

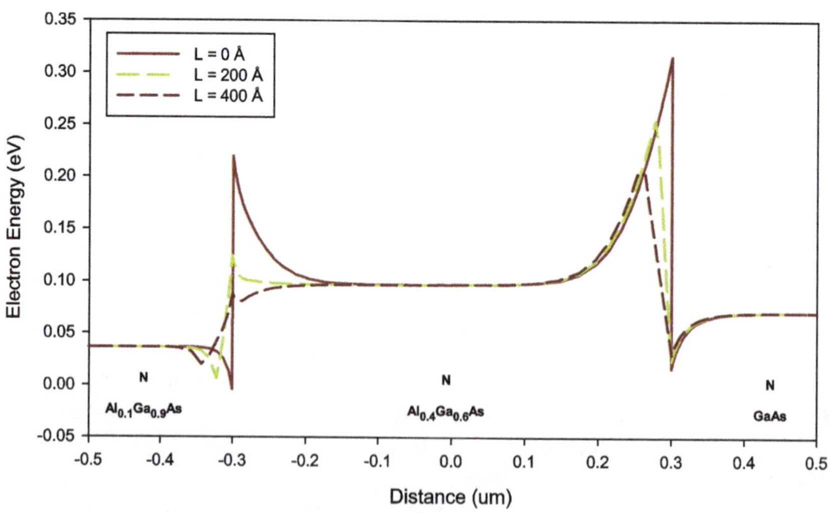

Fig. 10. Conduction band profiles for the double-heterostructure near the lower confinement layer at different grading width L.

The most important advantage of applying graded heterojunctions in the switches is that the reverse bias problem in a multilayer switch structure can be significantly reduced. The n-n heterojunction between $Al_{0.4}Ga_{0.6}As$ and $Al_{0.1}Ga_{0.9}As$ at the bottom of the lower confinement layer (see Fig. 5) would be reverse biased when a voltage is applied to operate the switch. From our simulations, over 80% of the switch driving voltage is applied across the reverse biased n-n heterojunction when all heterojunctions are abrupt. As shown in Fig. 10, this abrupt n-n heterojunction barrier can be effectively reduced or completely eliminated by over 400 Å-length grading. As a result, the switching voltage is further reduced.

2.5 Exemplary demonstrations

A prototype 2×2 carrier-injection optical switch with double-reflection electrode has been developed and fabricated. The waveguide structure, as shown in Fig. 5, has six AlGaAs/GaAs layers grown by MOCVD on an n+ GaAs substrate. For a single-mode waveguide at a wavelength of the input light of 1.55 μm, the ridge width of 5.0 μm was used, with the ridge height of about 1 μm. The GaAs waveguide core layer is lightly doped, $n=10^{16}$ cm^{-3}, to provide an optimized trade-off between low propagation loss and low switching voltage. In order to reduce energy band discontinuities at the heterojunctions that would impede the current flow therethrough, and to reduce the switching voltage, compositionally graded interfaces 20 nm - 40 nm thick were used at all the layer interfaces. The width of the current restriction gap, as illustrated in Fig. 6, is about 1 μm. These gaps may be formed using a double mask technique (van der Tol et al., 1994) to overcome photolithography problems when forming small gaps. With this technique, two masking stripes, which are independently defined in separate lithographic steps, can also be used as masks in a dry etch process to obtain ridge waveguides with small isolation gaps.

Fig. 11 is a Scanning Electron Microscope (SEM) image of the fabricated switch with the input waveguides on the left, viewed from an angle. It should be noted that the fabrication processes only employ conventional technologies, such as UV photolithography, e-beam evaporation and plasma etching. Independently formed electrode and photoresist layers were used as a double-mask in a single dry etch process to form the ridge waveguides and gaps. Uniform gap spacing was obtained using vernier alignment marks with ± 0.1 μm alignment accuracy. Fig. 12 is a SEM picture of the fabricated carrier-restriction gap, with the sample cleaved at a location close to the branching vertex of output waveguides.

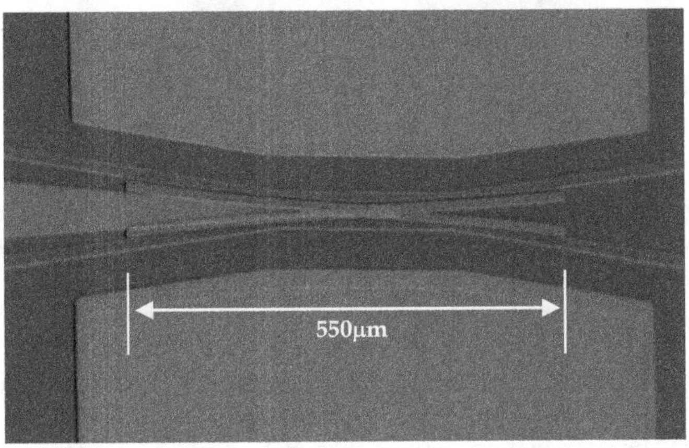

Fig. 11. SEM image of a fabricated 2×2 double-reflection TIR optical switch.

The fabricated switch has an electrode length of 550 μm, a waveguide crossing angle of 7°, and a cosine-bend section of 6.4 mm. The measured propagation losses of straight waveguides with undoped and lightly doped core layers are 0.3 dB/cm and 1.5 dB/cm, respectively, for a wavelength of 1.55 μm. Fig. 13 is the measured switching curves for light coupled into the top left input port. The switched state was obtained for an injection current of 220 mA. At this state, the driving voltage was 2.89 V and the measured device series resistance was 9.8 Ω. So the total power dissipation was at a manageable level of 0.6 W. With the area of the fabricated electrode being 3400 μm^2, the corresponding switching current density was 6.5 kA/cm^2. By using this current density and assuming a 10 ns carrier lifetime, the estimated carrier density and Δn are about 2×10^{18} cm^{-3} and -0.02, respectively. Such a high index change may be the result of two facts: one is that the carrier lifetime is quite long in the thick core layer; the other is that the current spreading decreases at very high current densities which results in current "bunching" in the core below the isolation gap and hence yields better power reflectivity. The measured switching time is 15 ns. On-off extinction ratio is 13 dB for the through port and 14 dB for the switched port. The switch is wavelength independent for a wavelength range from 1540 nm to 1570 nm. A very small polarization dependence of TE and TM modes has been observed (<0.5 dB). The measured extra insertion loss of the 2×2 switch compared with the straight waveguide fabricated on the same chip is 3 dB.

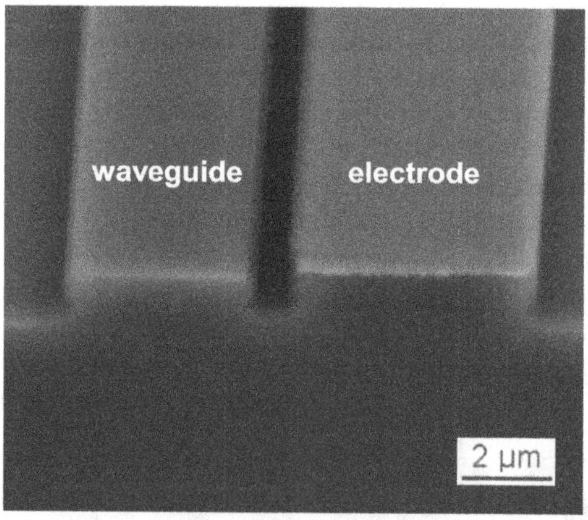

Fig. 12. SEM image of the carrier-restriction gap in an angled view of Fig. 4 taken along the line B-B.

Similarly, we also developed a 1×2 double-reflection optical switch which has a measured total power dissipation of 0.24 W with no extra cooling required to keep the switched state. As shown in Fig. 14, the fabricated switch has an electrode length of 480 μm, and a waveguide crossing angle of 6°. Based on the measured switching curve shown in Fig. 15, the switched state was obtained for an injection current of 105 mA at a bias of 2.30 V. The on-off extinction ratio is 15 dB for the through port and 14 dB for the switched port. The measured extra insertion loss of the 1×2 switch is 2 dB compared with the straight waveguide fabricated on the same chip. The switch is wavelength independent for the measured C-band. The polarization dependence of TE and TM modes input lights has been found to be less than 0.5 dB. The measured switching time is 11 ns which is close to the carrier lifetime in bulk materials. Reducing the guiding layer thickness would increase the switching speed but at the cost of decreasing the fibre coupling efficiency.

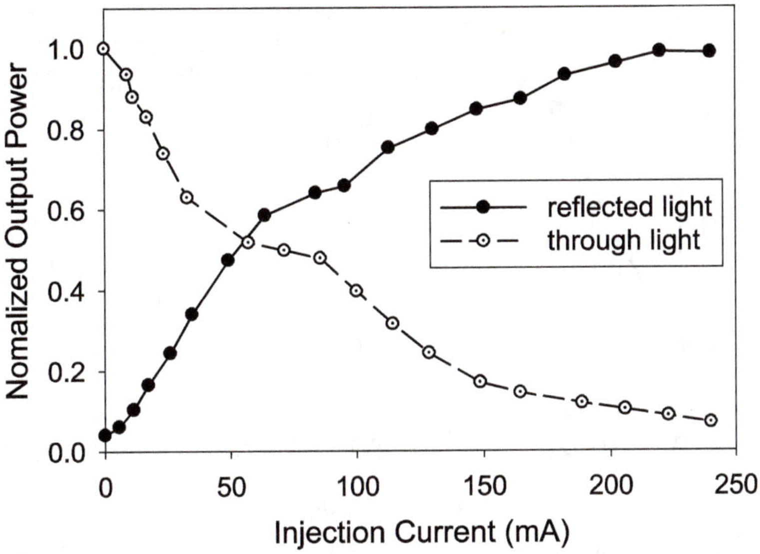

Fig. 13. Measured switching characteristics of a fabricated 2×2 double-reflection TIR optical switch.

(a) Layout of the curved double-reflection electrode design

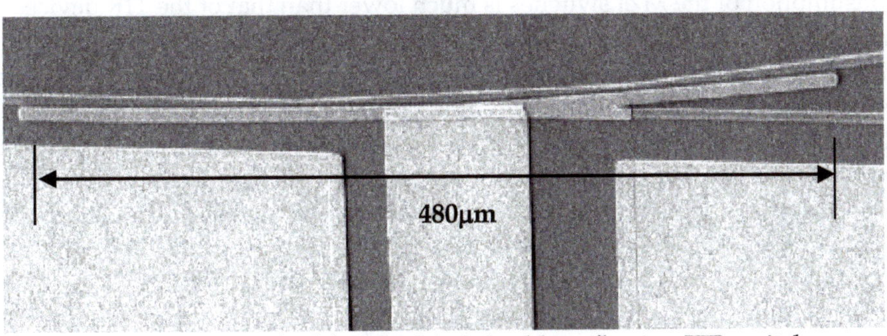

(b) SEM image of a fabricated 1×2 double-reflection TIR switch

Fig. 14. 1×2 TIR optical switch with curved double-reflection electrode.

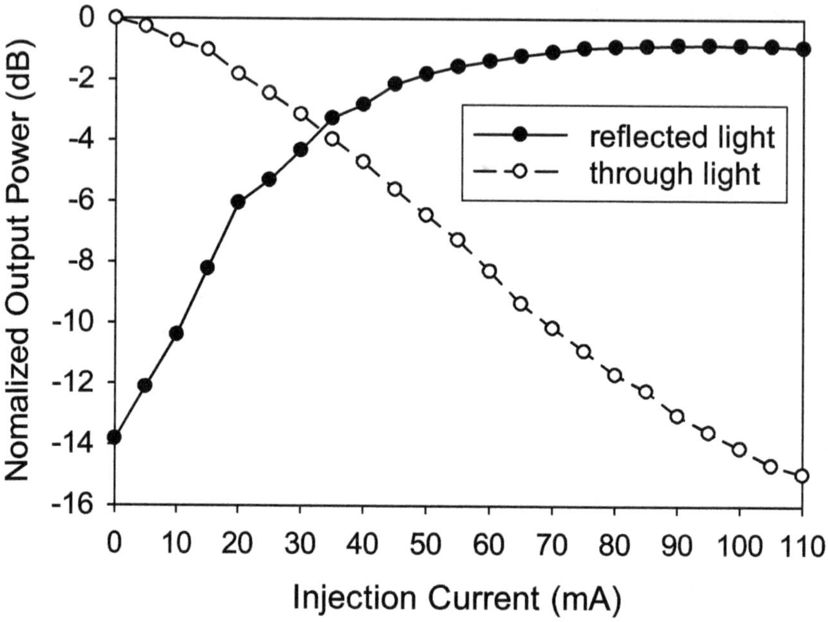

Fig. 15. Measured switching curve of a fabricated 1×2 double-reflection TIR optical switch.

3. 2×2 MMI-MZI DOS

To further reduce the switch power consumption, we have developed a 2×2 GaAs-GaAlAs carrier-injection switch based on a Mach-Zehnder Interferometer (MZI) formed by two Multi-Mode Interference (MMI) couplers. In an MZI-based switch (Cao et al., 2009; Wong et al., 2005), the two MZI arms are formed by two identical single-mode waveguides with electrodes, fabricated on the top of the waveguides, which allow the variation of the phase difference between them, based on the carrier-injection effect. Compared to the TIR switches, MZI-based switches can offer a high on-off extinction ratio and avoid the electrode fabrication complexity. More importantly, the injected current required to achieve π phase difference between the MZI arms is much smaller than that to achieve TIR, and therefore the power consumption of the MZI switches is much lower than that of the TIR devices.

A schematic of the fabricated 2×2 MMI-MZI carrier-injection switch is illustrated in Fig. 16, the device consists of an MZI configuration in which two 2×2 identical MMI couplers are connected by two identical parallel single-mode waveguides, that act as the two MZI arms. Metal electrodes are fabricated on the top of each MZI arm. Each electrode is 250 μm long and 5 μm wide. The MMI couplers are used as the 3-dB optical power splitter and combiner because of their polarization insensitive, fabrication and wavelength-tolerant properties (Besse et al., 1994). The MMI width is 20 μm and the MMI length is 1983 μm. The access (input/output) waveguides are single-mode waveguides with a width of 5 μm. Each access waveguide has two short straight sections sandwiching a long s-bend section to facilitate mode relaxation. The s-bend is designed in such a way that the additional loss caused by bending in this section is negligible.

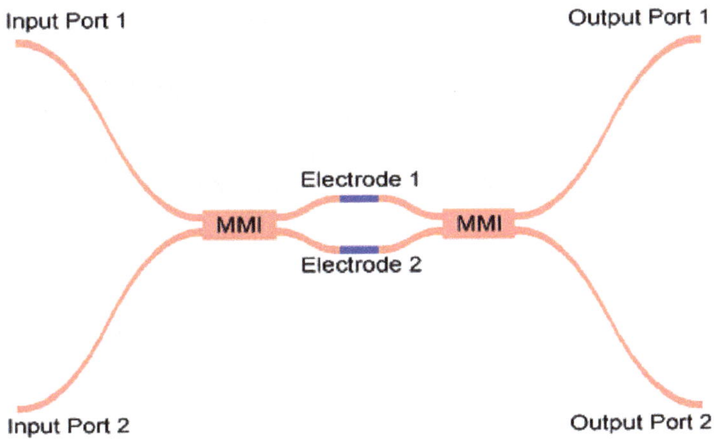

Fig. 16. Schematic diagram of a 2×2 MMI-MZI carrier-injection optical switch.

The separation between the input (output) ports is 250 μm and that between two electrodes is 35 μm. The total length of the device is about 1.4 cm. The operational principle of the 2×2 MMI-MZI switch is as follows. If two identical 2×2 MMI couplers are connected together, the two light beams, equally split from the input light by the first MMI coupler, will merge again after passing through the second MMI coupler but in the cross state. When sufficient carriers are injected into one of the electrodes to produce a π phase difference between the MZI arms, the device changes to the bar state.

The 2×2 MMI-MZI carrier-injection switch is fabricated on the epitaxial layer structure similar to the ridge waveguide structure shown in Fig. 5, with the core layer and two nearby confinement layers undoped. Compositionally graded interfaces 20 nm - 40 nm thick were used at all the layer interfaces to reduce energy band discontinuities at the heterojunctions. The calculated propagation loss of this single-mode waveguide is 0.19 dB/cm for the TE mode and 0.21 dB/cm for the TM mode.

The measured normalized output power at Output Ports 1 and 2 as a function of the injected current of Electrode 1 is shown in Fig. 17. The input light at wavelength 1.55 μm is coupled into the device at Input Port 2. As shown in Fig. 17, the first cross state (Output Ports 1 and 2 are on and off, respectively) occurs at 15.8 mA, and the first bar state (Output Ports 1 and 2 are off and on, respectively) occurs at 44.3 mA. The measured on-off extinction ratio is 20.6 dB for the cross state and 21.5 dB for the bar state. The measured voltage applied to the electrode is 1.35 V for the cross state and 1.48 V for the bar state. Therefore, the corresponding power consumption is 21.3 mW for the cross state and 65.6 mW for the bar state. Because of this very low power consumption, no extra cooling is required to maintain stable switch operation, which has been observed in our measurements. On the other hand, such low power consumption is very important for building the future optical Green Internet. From Fig. 17, we can obtain that the injected current required to achieve π phase difference between the MZI arms is 28.5 mA, which is very close to the theoretically estimated value of 29.4 mA by considering the guiding layer as an intrinsic semiconductor bulk material. Compared to the straight waveguides, the fabricated MMI-MZI devices have almost no measurable additional loss due to the MMI couplers and bend waveguides. The measured propagation loss of straight waveguides is typically around 0.5 dB/cm for a wavelength of 1.55 μm and has very small polarization dependence (< 0.5 dB). Based on these observations, the estimated total on-chip insertion loss would be about 2 dB. The measured switching time of 15 ns is close to the carrier lifetime in the GaAs bulk material. Reducing the guiding layer thickness below 0.5 μm would increase the switching speed to a few ns but at the cost of decreasing the fibre coupling efficiency. It is noted that the switch should be at the cross state when no current is injected but this is not the case as shown in Fig. 17 due to the asymmetrical structure of the device because of fabrication errors.

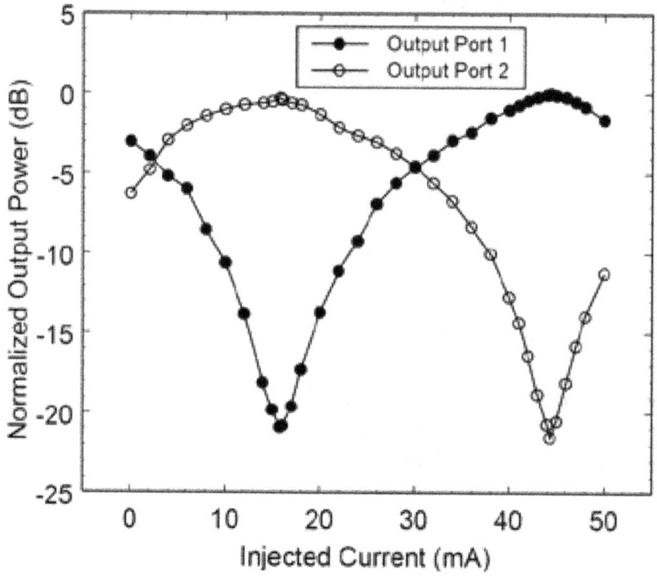

Fig. 17. Switching curves of the 2×2 MMI-MZI optical switch as a function of the injected current of Electrode 1. A light at wavelength 1.55 μm is coupled into the device at Input Port 2. The output optical power is measured at Output Ports 1 and 2.

4. Conclusions

Wide-angle 1×2 and 2×2 AlGaAs/GaAs TIR DOSs were developed by using only conventional semiconductor device fabrication technologies, suitable for monolithic circuit integration due to their small size. The developed 2×2 switch has a 7° waveguide crossing angle and a manageable maximum 0.6W dc power consumption. The 1×2 switch has a 6° branching angle and maximum quarter Watt dc power consumption. They can serve as the basic building blocks for high dense switching matrix. The techniques that we used, such as carrier restriction gap, curved electrodes, double-reflection interfaces and graded heterojunctions, can be easily applied to other switches made with other materials such as InP based, SiGe heterostructures and alike.

An ultra-low-power 2×2 MMI-MZI DOS with maximum 66 mW power consumption has also been demonstrated. Such an optical switch has a low insertion loss, high on-off extinction ratio, and very low power consumption. This device, together with the above TIR DOSs, will have significant importance in energy savings for future optical Green Internet applications, when the traffic moves down from the packet switching/routing layer to the circuit switching layer. The use of dynamic optical path-switching networks will help to accommodate users with massive video-based service demands in the future.

5. References

Abdalla, S., Ng, S., Barrios, P., Celo, D., Delage, A., El-Mougy, S., Golub, I., He, J-J, Janz, S., McKinnon, R., Poole, P., Raymond, S., Smy, T.J. & Syrett, B. (2004). Carrier Injection-Based Digital Optical Switch With Reconfigurable Output Waveguide Arms, *IEEE Photonics Technology Letters*, Vol.16, pp. 1038-1040, ISSN 1041-1135.

Agrawal, N., Weinert, C. M., Ehrke, H.J., Mekonnen, G.G., Franke, D., Bornholdt, C. & Langenhorst, R. (1995). Fast 2x2 Mach-Zehnder Optical Space Switches Using InGaAsP-InP Multiquantum-Well Structures, *IEEE Photonics Technology Letters*, Vol.7, pp. 644-645, ISSN 1041-1135.

Bennett, B. R., Soref, R.A. & Alamo, J.A.D. (1990). Carrier-Induced Change in Refractive Index of InP, GaAs, and InGaAsP, *IEEE Journal of Quantum Electronics*, Vol.26, pp. 113–122, ISSN 0018-9197.

Besse, P.A., Bachmann, M., Melchior, H., Soldano, L.B. & Smit, M.K. (1994). Optical Bandwidth and Fabrication Tolerances of Multimode Interference Couplers, *IEEE Journal of Lightwave Technology*, Vol.12, pp. 1004-1009, ISSN 0733-8724.

Cao, S., Noad, J., Sun, L., James, R., Coulas, D., Lovell, G. & Higgins, E. (2009). Small AC Driving Voltage for MZI-Based GaAs-GaAlAs Electrooptic Modulators/Switches with Coplanar Electrodes," *IEEE Photonics Technology Letters*, Vol.21, pp. 584-586, ISSN 1041-1135.

Capasso F. & Margaritondo, G. (Eds.) (1987). *Heterojunction Band Discontinuities*, Elsevier Science Ltd, ISBN 978-0444870605, Amsterdam.

Ikezawa, K., Iio, S., Suehiro, M., Suzuki, T. & Uneme, S. (2008). Demonstration of Modulation Format Free and Bit Rate Free Characteristics of 2 ns Optical Switch for Optical Routers, *Proceedings of Optical Fiber Communication Conference and Exposition*, no. JWA31, ISBN: 978-1-55752-856-8, San Diego, CA, USA, Feb 2008.

Li, B. & Chua, S.J. (2001). 2x2 Optical Waveguide Switch with Bow-Tie Electrode Based on Carrier-Injection Total Internal Reflection in SiGe Alloy, *IEEE Photonics Technology Letters*, Vol.13, pp. 206-208, ISSN 1041-1135.

Nayyer, J., Niayesh, K. & Yamada, M. (2000). Dynamic Characteristics of Optical Intersecting-Waveguide Modulators/Switches with Curved Electrodes, *IEEE Journal of Lightwave Technology*, Vol.18, pp. 693-699, ISSN 0733-8724.

Ng, S., Janz, S., Mckinnon, W.R., Barrios, P., Delage, A. & Syrett, B.A. (2007). Performance Optimization of a Reconfigurable Waveguide Digital Optical Switch on InGaAsP–InP, *IEEE Journal of Quantum Electronics*, Vol.43, pp. 1147-1157, ISSN 0018-9197.

Shimomura, K., Tanaka, N., Aizawa, T. & Arai, S. (1992). 2V Drive-Voltage Switching Operation in 1.55μm GaInAs/InP MQW Intersectional Waveguide Optical Switch, *Electronics Letters*, Vol.28, pp. 955-956, ISSN 0013-5194.

Thomson, D., Gardes, F.Y., Mashanovich, G.Z., Knights, A.P. & Reed, G.T. (2008). Using SiO_2 Carrier Confinement in Total Internal Reflection Optical Switches to Restrict Carrier Diffusion in the Guiding Layer, *IEEE Journal of Lightwave Technology*, Vol.26, pp. 1288-1294, ISSN 0733-8724.

van der Tol, J.J.G.M., Pedersen, J.W., Metaal, E.G., Ose, Y.S., Groen, F.H. & Demeester, P. (1994). Sharp Vertices in Asymmetric Y-junctions by Double Masking, *IEEE Photonics Technology Letters*, Vol.6, pp. 249-251, ISSN 1041-1135.

Wang, W.I. (1986). On the Band Offsets of AlGaAs/GaAs and Beyond, *Solid-State Electronics*, Vol.29, pp. 133–139, ISSN 0038-1101.

Wong, H. Y., Sorel, M., Bryc, A.C., Marsh, J.H. & Arnold, J.M. (2005). Monolithically Integrated InGaAs-AlGaInAs Mach-Zehnder Interferometer Optical Switch Using Quantum-Well Intermixing, *IEEE Photonics Technology Letters*, Vol.17, pp. 783-785, ISSN 1041-1135.

Yanagawa, H., Ueki, K. & Kamata, Y. (1990). Polarization- and Wavelength-Insensitive Guided-Wave Optical Switch with Semiconductor Y Junction, *IEEE Journal of Lightwave Technology*, Vol.8, pp. 1192-1197, ISSN: 0733-8724.

Zegaoui, M., Choueib, N., Harari, J., Decoster, D., Magnin, V., Wallart, X. & Chazelas, J. (2009). 2x2 InP Optical Switching Matrix Based on Carrier-Induced Effects for 1.55-μm Applications, *IEEE Photonics Technology Letters*, Vol.21, pp. 1357-1359, ISSN 1041-1135.

Zhuang, W., Duan, J., Zou, Z., Yang, P., Shi, Z., Sun, F. & Gao, J. (1996). Total Internal Reflection Optical Switch with Injection Region Isolated by Oxygen Ion Implantation," *Fiber and Integrated Optics*, Vol.15, pp. 27-36, ISSN 0146-8030.

3

Integrated ASIC System and CMOS-MEMS Thermally Actuated Optoelectronic Switch Array for Communication Network

Jian-Chiun Liou
Industrial Technology Research Institute
Taiwan

1. Introduction

The advancement of communication technology and growth of internet traffic have continuously driven the fast evolution of networks. Compared to the traditional optoelectronic switch, all-optical switch provides high throughput, rich routing functionalities, and excellent flexibility for rapid signal exchange in fiber optical network. Among various all-optical switches, thermal actuated ring switch provides the advantages of high accuracy, easy actuation, and reasonable switching speed. However, when scale up, thermal ring switch may encounter issues related to fabrication error, non-accurate wavelength response, and large terminal numbers in the control circuit.

Planar-lightwave-circuit switch (PLC-SW), employing thermo-optic (TO) effect of silica glass for light switch, is a very promising technique for communication applications because of low insertion loss, high extinction ratio, long-term stability, and high reliability. There have been many matrix switches designed based on the TO effect with low-loss, polarization insensitive operation, and good fabrication repeatability. For example, 8×8 matrix switches were fabricated by using a single Mach-Zehnder (MZ) switching unit and demonstrated well performance in transmission systems, so as an 8×8 matrix switch and a 16×16 matrix switch by the similar MZ switching unit. 32×four-channel client reconfigurable optical add/drop multiplexer on planar lightwave circuit. However, when scaled up, thermal ring switch may encounter issues related to large terminal numbers in the control circuit, fabrication error and non-accurate wavelength response. For example, if conventional driving circuits are employed for a 16×28 or even larger switch array, 448 or more terminals will be required for control. Such a large number of terminals would complicate the module structure and occupy a large area. On the other hand, when a DC-current is applied for balancing wavelength offset from fabrication error, the input power will result in a temperature elevation of the neighboring switches, thus changes the related refractive indexes and therefore deviate wavelengths.

To solve the aforementioned issues, this chapter proposes a ring resonator with silicon nitride as the core layer and silicon dioxide as the cladding layer was designed and fabricated on silicon substrate, a novel architecture of high selection speed three dimensional (3D) data registration for driving large-array optoelectronic packet switches. The 3D driving architecture can successfully reduce the total numbers of control pads into

31 for 448 switches as well as the scanning time up to 67 % reduction with a higher signal rising speed and smaller circuit area. All the sub-circuits, including power control, digital I/O, analog-to-digital converter, power drivers were integrated into a single IC. On the other hand, instead of DC current control, wavelength lock is realized by amplitude and frequency modulated heating pulses for stabilization of temperature and fine-tune of wavelength from fabrication imperfection and environment fluctuation. This planar-lightwave-circuit has been designed, fabricated, and characterized. It is demonstrated not only the functionality in optical packet switches but also the consistency between simulation and experiment results.

In this research, we design the tunable micro ring resonators and propose the employment of an integrated ASIC system CMOS technology control circuit to compensate the fabrication error and tune as well as lock the wavelength in a thermal-actuated ring-type optical switch through a frequency modulation scheme. The use of a standard and commercial CMOS technology for designing micro resonators entails a set of limitations, such as layer thicknesses, and available materials in an inalterable process sequence. From the MEMS design point of view, those restrictions will limit the electrical properties of fabricated micro-resonators. On the other hand, MEMS integration into a CMOS technology presents unique advantages, like reduction of the parasitic capacitance due to the possibility to monolithically integrate the circuitry, in addition to an expected reduction of the overall production costs.

Additional functionalities can also be added in this circuit by tailoring externally the roundtrip loss or coupling constants of the ring. The design concept can be easily scaled up for large array optical switch system without much change in the terminal numbers thanks to the three dimensional hierarchy of control circuit design, which effectively reduces the terminal numbers into the cubic root of the total control unit numbers. The integrated circuit has been designed, simulated, as well as fabricated, and demonstrated a decent performance with Free Spectral Range (FSR) equal to 1.5nm at 1534 nm and very accurate wavelength modulation to 0.3 nm within 0.01 nm fluctuation for thermal actuated ring type optoelectronic switch.

Thermal actuated optoelectronic ring switch provides the advantages of high accuracy, easy actuation, and reasonable switching speed. However, when scale up, thermal ring switch may encounter issues related to fabrication error, non-accurate wavelength response, and large terminal numbers in the control circuit. Planar-lightwave-circuit switch (PLC-SW), employing thermo-optic (TO) effect of silica glass for light switch, is a very promising technique for communication applications because of low insertion loss, high extinction ratio, long-term stability, and high reliability[1-4]. There have been many matrix switches designed based on the TO effect with low-loss, polarization insensitive operation, and good fabrication repeatability. For example, 8×8 matrix switches were fabricated by using a single Mach-Zehnder(MZ) switching unit and demonstrated well performance in transmission systems [5, 6], so as an 8×8 matrix switch [7] and a 16×16 matrix switch [8] by the similar MZ switching unit. 32×four-channel client reconfigurable optical add/drop multiplexer on planar lightwave circuit [9]. However, when scaled up, thermal ring switch may encounter issues related to large terminal numbers in the control circuit, fabrication error and non-accurate wavelength response. For example, if conventional driving circuits are employed for a 16×28 or even larger switch array, 448 or more terminals will be required for control. Such a large number of terminals would complicate the module structure and occupy a large area. On the other hand, when a DC-current is applied for balancing wavelength offset from fabrication error, the input power will result in a temperature elevation of the neighbouring switches, thus changes the related refractive indexes and therefore deviate wavelengths.

To solve the aforementioned issues, this research proposes a ring resonator with silicon nitride as the core layer and silicon dioxide as the cladding layer was designed and fabricated on silicon substrate, a novel architecture of high selection speed three dimensional data registration for driving large-array optical packet switches. The 3D driving architecture can successfully reduce the total numbers of control pads into 31 for 448 switches as well as the scanning time up to 67 % reduction with a higher signal rising speed and smaller circuit area. All the sub-circuits, including power control, digital I/O, analog-to-digital converter, power drivers were integrated into a single IC. On the other hand, instead of DC current control, wavelength lock is realized by amplitude and frequency modulated heating pulses for stabilization of temperature and fine-tune of wavelength from fabrication imperfection and environment fluctuation. This planar-lightwave-circuit has been designed, fabricated, and characterized. It is demonstrated not only the functionality in optical packet switches but also the consistency between simulation and experiment results.

2. Planar-lightwave-circuit design and principle

2.1 Design of wavelength modulation and lock

In this research, PLC filter will usher as a model system for demonstration. A PLC filter is consisted of arrays of micro-rings (MRs) with the ring radii in the order of tens of micrometers, possessing extremely small-area and providing the characteristic of high selectivity. Based on the design of this device, a four-channel PLC filter is depicted with four MRs, as schematically shown in Fig. 1. The operation principle of this device is as the first column of the optical circuit is on each MR drops an incoming signal λx into one of the output channels Ioutx when the MR resonance frequencies matching that of the incoming signal [10-11]. In this research, the design, fabrication, and measurements of a four-channel thermally tuneable MR-based filter PLC controller is presented. Silicon dioxide and silicon nitride were selected as an optical thin film system to compose the micro-rings on top of a silicon substrate. As shown in Fig. 2, the width of straight and ring waveguide was chosen as 2.0 μm. To ensure single mode operation at TE polarization, the height of waveguides was determined to be 0.45 μm by a 3D finite-difference time-domain simulation using RSoft-FullWave(a finite-difference time-domain solver of Poisson's equations)[12-13]. The simulation flow is electro-thermal transport and optics, then according to follow electrostatic potential equation (1) and carrier continuity equations (2), electrons, holes relative derived among temperature, core index, and radius. The effective index method was employed to determine the radius of the ring as 9.965 μm.

1. Poisson's Equation (Electrostatic Potential)

$$\nabla \cdot \varepsilon \nabla \varphi + q(N_D^+ - N_A^- + p - n) = 0 \tag{1}$$

2. Carrier Continuity Equations (Electrons, Holes)

$$\frac{\partial n}{\partial t} + \nabla \bullet j_n + U = 0 \quad \text{And} \quad \frac{\partial p}{\partial t} + \nabla \bullet j_p + U = 0 \tag{2}$$

3. Lattice Heat Equation (Temperature)

$$\frac{\partial}{\partial t}(c_L + \frac{3}{2}(n+p)k)T = \nabla(K_L \nabla T - \overrightarrow{S_n} - \overrightarrow{S_p}) + \overrightarrow{j_n}\vec{E} + \overrightarrow{j_p}\vec{E} + R_{dark}E_g \tag{3}$$

4. Photon Rate Equation (Modal Photons)

$$\frac{\partial S_{m,w}}{\partial t} = (G_{m,w} - \frac{1}{\tau_{m,w}})S_{m,w} + R_{m,w}^{spon} \tag{4}$$

5. Derived Helmholtz Equation (Mode Profile)

$$\nabla^2 \phi + \varepsilon k_o^2 \phi = 0 \tag{5}$$

Where,
n, p: Electron/Hole Density
V: Electrostatic Potential
T: Lattice Temperature
G, R: Optical Gain/Recombination
f: Lasing Frequency
A: Optical Wave Amplitude
S: PhotonNumber
K: Quantum Mechanical Wavefunction
The condition for single mode operation of a Si3N4 optical rib waveguide with a refractive index 2.06 operating at 1.55μm was analyzed using beam propagation method (BPM). The full vectorial BPM was used to analyze the waveguide structure in Fig. 3 for single mode propagation and polarization independence.

Height\width(μm)	0.2-0.4	0.5-0.8	0.9-2.0
0.3	single	multi	multi
0.4	single	single	multi
0.5	single	single	single
0.6	single	single	single
0.7	single	single	single
0.8	single	single	single
0.9	single	single	single

Table 1. Summarizes the waveguide mode.

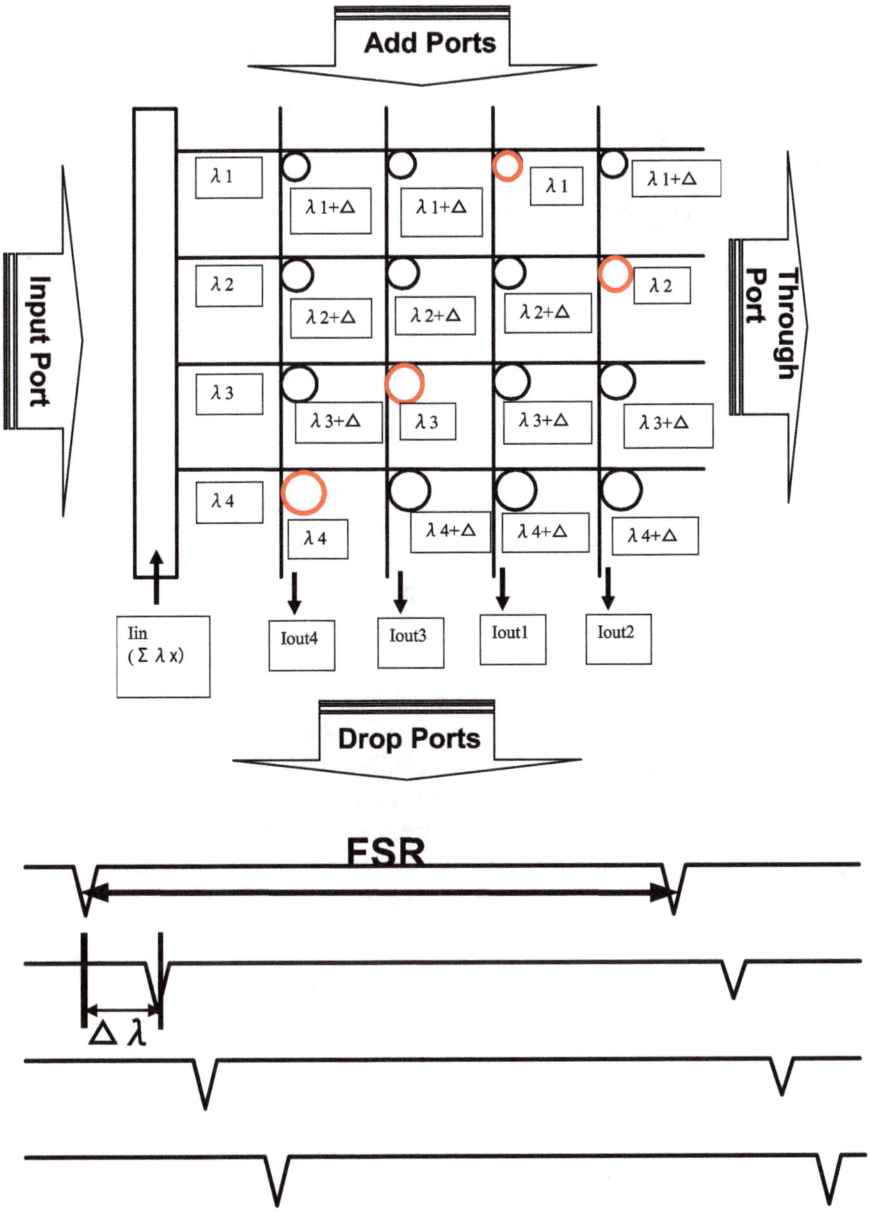

Fig. 1. Schematic of a four-channel PLC.

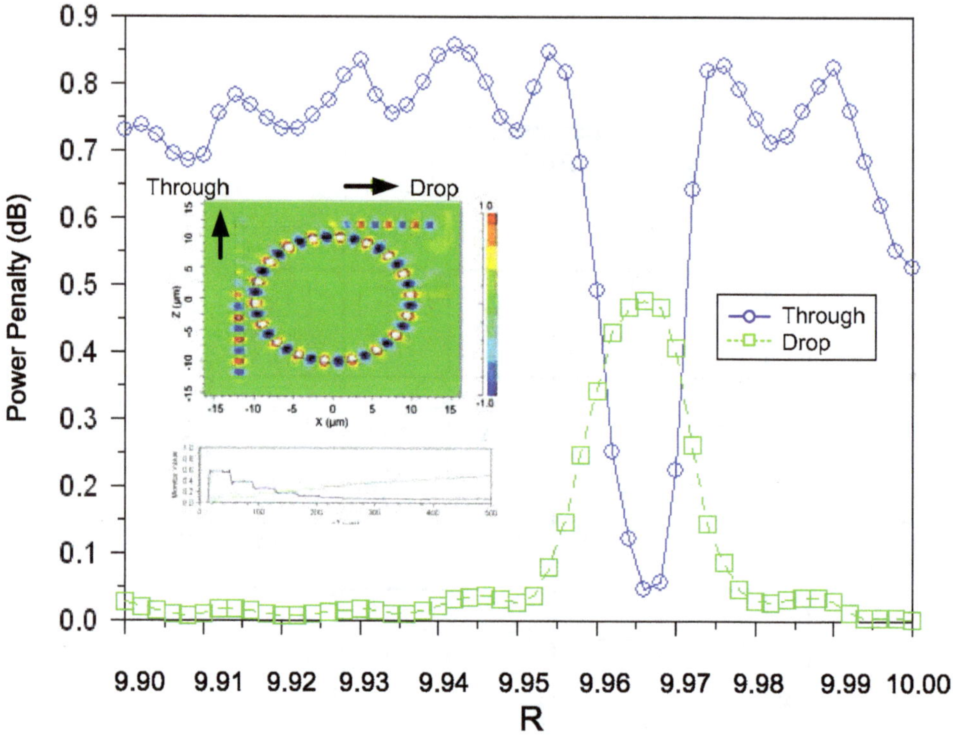

Fig. 2. Transmittance for the radius of micro ring.

According to the BPM calculations, for height = 0.4~0.5μm, width = 0.9~2.0μm enable single mode in the TE polarization. Table 1 summarizes the calculated number of waveguide modes for our studied height and width ratios. Similar single mode output intensity patterns were imaged from other fabricated control waveguide of various height and width. Fig.4(a) 、(b) and (c) depicts the imaged mode profiles have only one field maximum near the center for both the horizontal and vertical directions. Insets show the BPM calculated fundamental mode profiles from the same waveguide geometries, depicting similar Gaussian-like mode profiles.

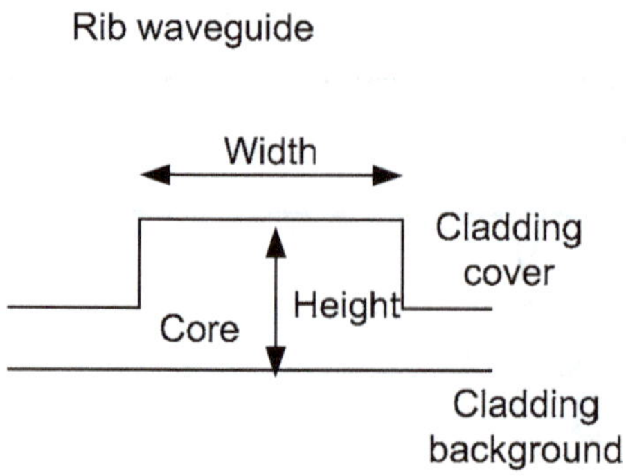

Fig. 3. Schematic cross section of a rib waveguide.

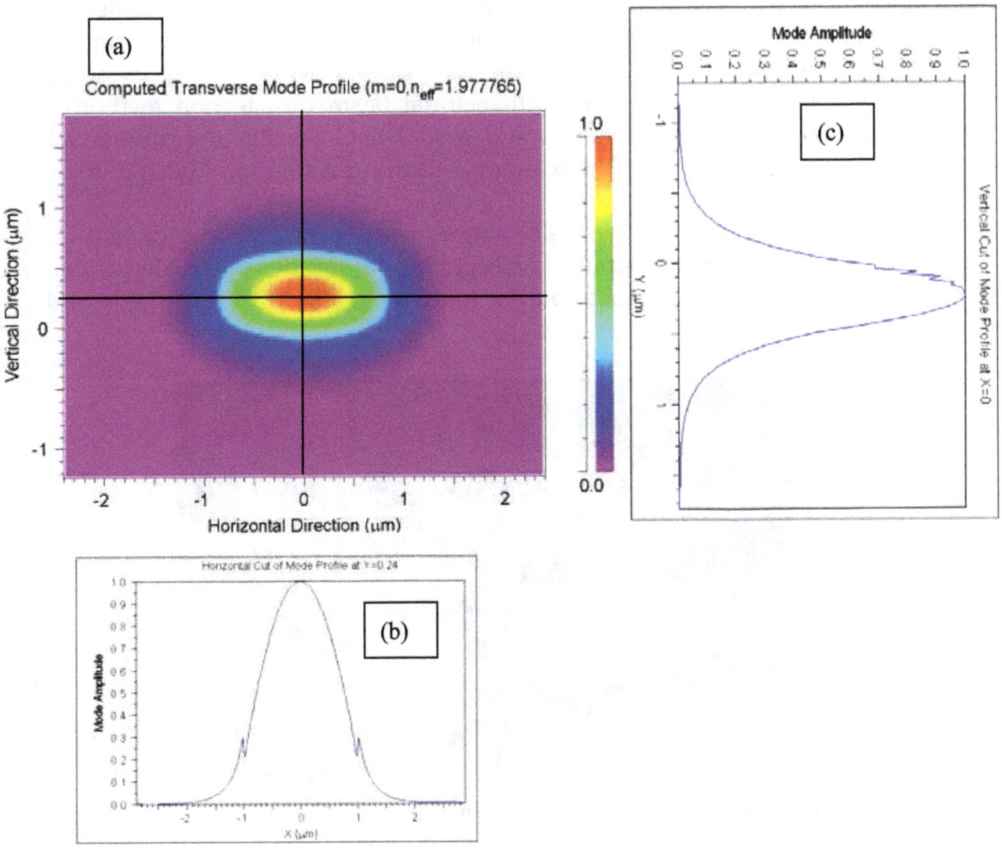

Fig. 4. Simulated fundamental mode profile from BPM-based calculations.

In the process of making the rib waveguide as shown in Table 2, we also examined four different process parameters to see how they affect the final waveguide's dimensions. The first step was to investigate the polarization characteristics of the waveguide due to its geometry. Various waveguide heights ranging from 0.3 to 1.0um and width ranging from 0.9 to 2.0um are considered. While keeping waveguide height constant during the computation, the difference in effective indices of the fundamental TE waveguide modes has been evaluated as the etch depth and waveguide width were varied.

Process number	Process step
1	Thermally grown SiO2 on Si wafer
2	LPCVD deposition of Si3N4 layer
3	Spinning of resist, patterned by photolithography(E-Beam) and structure by RIE
4	Deposition of PECVD SiO2 cladding layer and annealing of layer stack
5	Sputtering a Platinum (Pt) thin film
6	Patterned by photolithography and Pt wet-etch.

Table 2. Process flow for SiO2/Si3N4 coupled microring resonators.

Single mode propagation is an important requirement for optical waveguide devices for use with single-mode fiber; it can reduce the coupling loss. In this research, a technique is used for calculating the field distribution of the Si3N4 rib waveguide. The waveguide was modeled using the three-dimensional full-vectorial beam propagation method (BPM) to calculate the effective refractive indices and modal field profiles for various waveguide widths, heights and etch depths. The scanning electron micrograph (SEM) of Fig. 5 shows such a fabricated rib waveguide.

We experimentally verify the practically single mode nature of our deeply etched rib waveguides by imaging control straight waveguides output intensity patterns. Fig.6 shows the representative imaged waveguide output intensity profile with waveguide of the order of 1550nm.

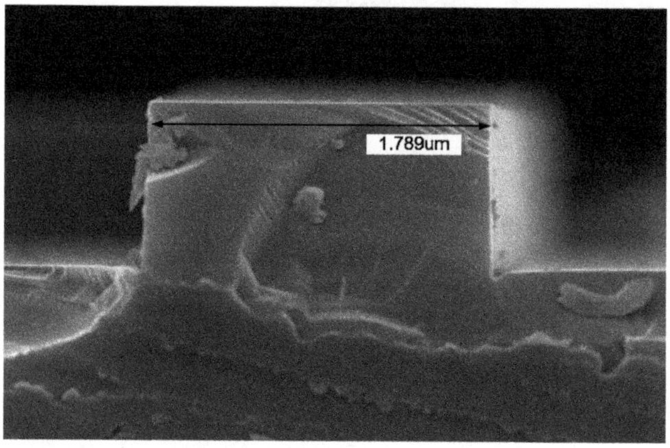

Fig. 5. SEM image of the waveguide cross section.

Fig. 6. Imaged waveguide output intensity profile.

In the wavelength modulation, thermal optic effect was employed to shift the resonance wavelength at an amount of $\triangle\lambda$ by the tuning of effective index at different temperatures. This wavelength shift can be used to tune the passband to the desired wavelength. The principle of wavelength modulation is shown in Fig. 1, illustrating that the elevation of temperature on one MR switch shifts the center wavelength by $\triangle\lambda$ but remain the same Free Spectral Range (FSR). The term FSR is borrowed from Fabry Perot interferometry, and describes the maximum spectral range one can arbitrarily resolve without the interference from the neighbouring signals. On the other hand, a high extinction ratio can be obtained through the filtering effect from the MRs with a steep wavelength response. A relationship between the radius of the ring R, the effective group index n_g, and the FSR is given by equation (6):

$$FSR = \frac{\lambda^2}{2\pi R n_g} \tag{6}$$

where λ is the wavelength [14-15].

In temperature control, frequency modulation was employed instead of voltage level modulation due to the simplicity of implementation by digital signals. Through frequency modulation, the temperature in the thermally tuneable PLC modules can be maintained almost constant and this will result in a more accurate center wavelength for the optical communication channel. It also ensures rapid response of the PLC module as the heater has been modulated on and off in a high frequency (~MHz). As a result, the PLC module at room temperature was able to achieve a very small temperature fluctuation within 0.1°C which can not be achieved by using traditional DC controls.

In order to compensate the fabrication error of the thermal ring switch, a simple and practical phase-trimming technology was employed to avoid the need of electrical biasing. The phase-trimming technique employs a local heating technology by the employment of a thin-film heater embedded under the optical ring in a feedback loop for the fine tune of the optical phase. However, if DC bias is employed in the phase control, the temperature of the neighbouring switch may encounter drift (cross talk) as well as slow response for temperature compensation. To lower down the cross talk effect, provide more accurate temperature control, and speed up thermal response of the optical ring, a frequency modulated heating scheme is employed by dynamic feedback of the frequency of heating pulses.

To achieve the above goal by frequency modulation for accurate temperature control, this study employs a selection algorithm to select a proper waveform pre-stored in the lookup table in an ASIC(Application specific integrated circuits) chip, in which all waveforms have been simulated and optimized for different temperature situations. Each drop and filter channel is assigned a temperature for the desired wavelength shift. The temperature is maintained by a corresponding waveform from the result of the sum of three signals, including data (address), select, and power.

2.2 Design of three dimensional controller

In traditional control circuit design for thermal optical type switch array, each optical ring requires one heater for wavelength adjustment. As a result, when the optical switches scale up into a large array, the numbers of input/output ports will increase enormously. To handle large array of driving circuits for such a heater array, two dimensional (2D) circuit architecture was employed by traditional driving circuits to reduce the IO number from n*n into 2n+1. However, this reduction still can not meet the requirement for high speed signal scanning with low data accessing points when switch numbers greater than 1000.

To achieve this, in this study, a three dimensional data registration scheme to reduce the number of data accessing points as well as scanning lines for large array optical packet switching chip with switch number more than 1000 is proposed. The total numbers of data accessing points will be N=3× $\sqrt[3]{Y}$ +1, which is 31 for 1000 switches by the 3D novel design, the scanning time is reduced down to 33% (The scanning speed is also increased by 3 times) thanks to the great reduction of lines for 3D scanning, instead of 2D scanning. The property comparison among 1D, 2D, and 3D architectures is listed in Table 3. As the optical switch number increases, a higher order control circuit can effectively reduce the pad number. In addition, the shape and amplitude of the driving signal can be optimized to increase the speed of the response with low driving powers [16].

X :Pads 、Y :Switches	X~Y +1	X=2×$\sqrt[2]{Y}$+1	X=3× $\sqrt[3]{Y}$+1 (X : Connection lines ,Y :Switches)
Switches	1000	1024	1000
Ring resistors	1000	1024	1000
Interconnect Pad	1001	65	31

Table 3. Performance comparison among 1D, 2D, and 3D driving schemes.

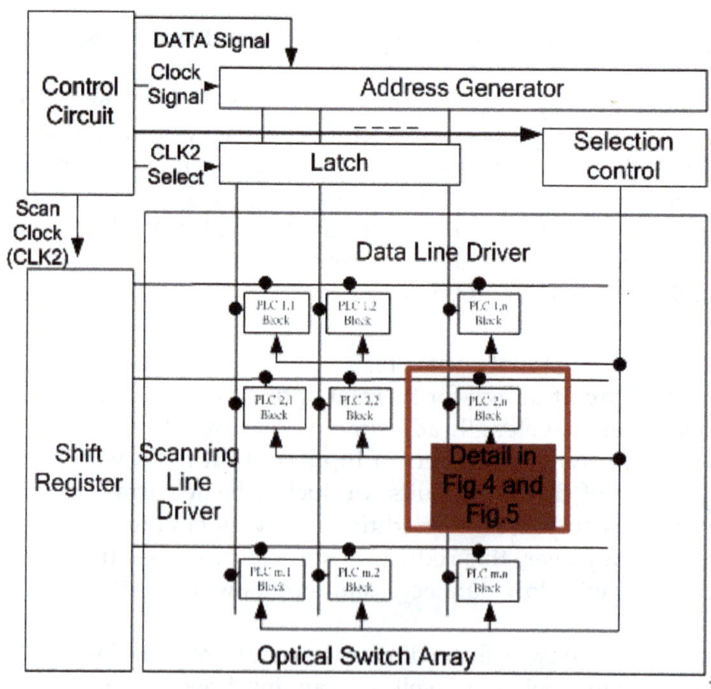

Fig. 7. Block diagram of control algorithm for micro-ring switches.

In the proposed novel 3D design, different from the 2D one, as shown in Fig. 3, the digital driver includes a clock-control circuit, a serial/parallel-conversion circuit, a latch circuit, a level shifter, a D/A converter comprised of a decoder, and an output buffer comprised of an operation amplifier. The D/A converter receive a gray-scale reference voltage from an external source [17-21]. The clock-control circuit receives control signals from an external control circuit. Based on the received control signals, the clock-control circuit attends to control of the latch circuit, the D/A converter [22-24], the output buffer by using a latch-control signal.

The general strategy that we employ is to integrate all relatively small-signal electronic functions into one ASIC to minimize the total number of the components. This strategy demonstrates that both the cost is lowered and the amount of the printed circuit board area is reduced. Based on this concept, a smart 3D multiplexed driver for optical packet switching chip with more than 1000 rings are proposed and the circuit architecture is shown in Fig. 7. Three lines are employed to control the heating of one micro-ring, including voltage, shift register, and data line. The relationship among the waveforms is shown in Fig. 8. Each heater resistor requires a voltage line for the driving current flow and shares the same ground with the other resistors. The resistors are individually addressable to provide unconstraint signal permutations by a serial data stream fed from the controller. The shift register is employed to shift a token bit from one group to another through AND gates to power the switch of a micro-ring group. The selection of a ring is thus a combined selection of the shift register for the group and the data for the specific ring. Such an arrangement allows encoding one data line from the controller to provide data to all of the rings, permitting high-speed printing by shortening the ring selection path and low IC fabrication cost from the greater reduction of circuit component numbers.

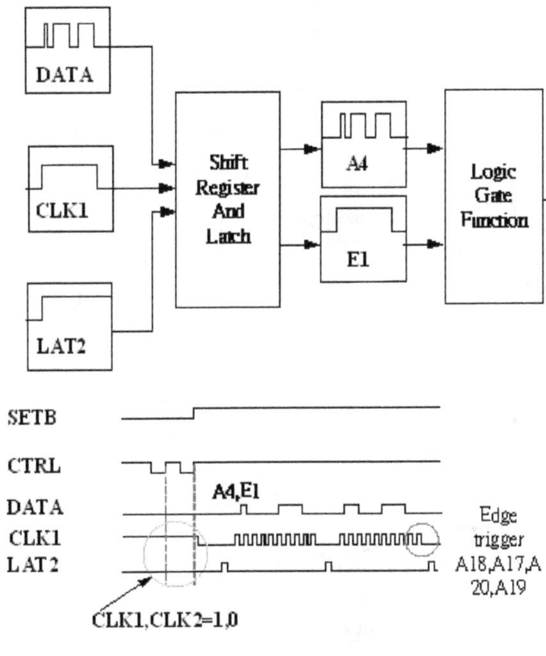

Fig. 8. Example of input waveform for controller from FPGA.

The received optical data information has to be converted into data at an optimal transfer rate (frequency) in order to conform to the ring characteristics. To the end, the clock-control circuit divides the 8-bit optical packet switching signals supplied to the data driver, as shown in Fig. 7, with an aim at lowering the operation frequency. The serial/parallel-conversion circuit converts serial signals of a plurality of channels into parallel signals, and supplies the parallel signals to the latch circuit. The latch circuit temporarily stores the received parallel signals, and supplies them to the level shifter and the D/A converter at predetermined time [25-27]. The level shifter converts a logic level ranging approximately from 3.5 V to 5 V into a ring array voltage level that ranges from 5.5V to 8.5V for various heater resistors as a result of variation processing conditions.

In the signal flow design, optical switches are usually scanned over one by one without jumping on un-activity switches. As a result, for the optical packet switching chip with 448 optical switches, a 1, 2 or 3 dimensional circuit architect will needs 448, 36, and 5 unit times for scanning over all of the switches. Therefore, the scanning time of the 3D multiplexing circuit from the first address line to the 16th, as an example, takes only 5 units of clock time from the simulation result, much faster than that of the 2D configuration with 16 units of clock time. Thus the maximum scanning time for the 3D circuit will be reduced to 30% of that in the 2D case.

To simultaneously write signals into the driving circuit, multiplexing data latches and shift registers are employed by the application of commercial available CMOS ICs. Small numbers of shift registers, control logics, and driving circuits can be electrically connected and integrated with optical packet switching using standard CMOS processes. Fig. 9 shows the driving circuit of the three-dimensional architect. The desired signal for "S" selections and "A" selections can be pre-registered and latched in the circuit for one time writing.

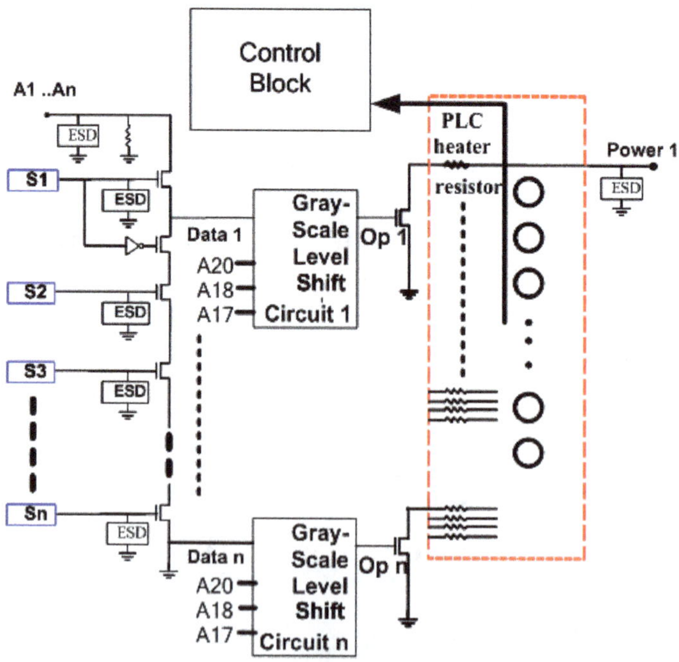

Fig. 9. Architect of three-dimensional driving circuit for micro-ring switches.

The SPICE simulation results on the relationship of input and output signal at 5μs clock time. Fig. 10 demonstrates that not only the switch speed is higher by the level shift device than that of one without level shift circuit, but also the voltage has been enhanced to 5V. An adjustable voltage pulse from 7.98V to 8.02V amplitude modulation is applied to the various heater resistors thanks to the processing condition.

The cooling down of the structure is equally important, though enhancing the speed of the cooling down process might be done by active cooling, but this would require major adaptations to the device and the low cost low power principle would not hold anymore. A much easier way to do this is biasing. In biasing, a DC-current is applied, that will result in a relatively small change in temperature, refractive index and therefore resonance wavelength. To heat the device, the wide pulse width signal or high gray scale level voltage is applied; to cool it down, a narrow pulse width signal or low-level voltage is applied. See Fig. 10 for the simulated behavior of high and low bias driving. The maximum current that can be applied is limited, due to the destruction of the heaters at high powers. The use of a bias will therefore cause a smaller modulation depth, but the modulation will be faster, since the time needed for cooling down is reduced.

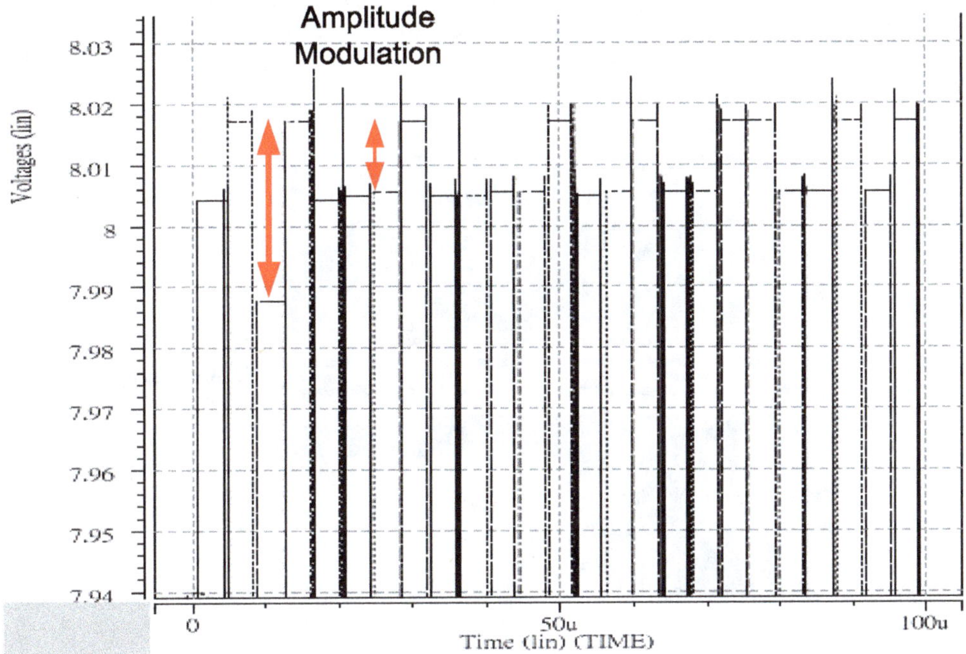

Fig. 10. Transient simulation of the input and output signals of the level shift device.

Power consumption of a narrow pulse width signal or low-level voltage applied for the thermally actuated optical switch array is very small comparing a DC-current applied. The main advancement of this new concept is that the drive signal opening the switch tracks the serial/parallel- conversion circuit, which converts serial signals of a plurality of channels into parallel signals, and supplies the parallel signals to the latch circuit.

If we analyze the total wire power, we calculate that the intermediate interconnection power is the dominant part of the total wire power. The total wire power is scaled down of the three dimensional hierarchy of high gray scale control circuit design, which effectively reduces the terminal numbers into the cubic root of the total control unit numbers.

3. Experimental and results

3.1 Wavelength modulation and lock

By using a Commercial Finite Difference Solver (CFD-RC, USA) for thermo-optical problems, the temperature profile of the MR and the relative changes of refractive indexes can be simulated, as shown in Fig.11(a) and Fig.11(b). Electro-thermal changed temperature by ASIC multiplexing data signal applying to coupled-ring-resonator for adjustment core index. Optimal tunable center wavelength 1511nm conform the shifted core indexes from 2.000 to 2.008. Although the temperature distribution on the ring is about 1 °C, the average temperature of the ring is employed as a reference for the temperature control and the tolerance is within 0.1 °C. To reduce overshooting and obtain rapid set up of ring temperature, heating pulses with amplitude modulation were employed. Through simulation, optimized driving signal can be obtained to maintain stable wavelength in 0.1 ms by accurate temperature modulation [28]. The temperature fluctuation can be controlled within 0.1°C, with a wavelength variation locked in 0.01 nm, as the measured result shown in Fig. 12.

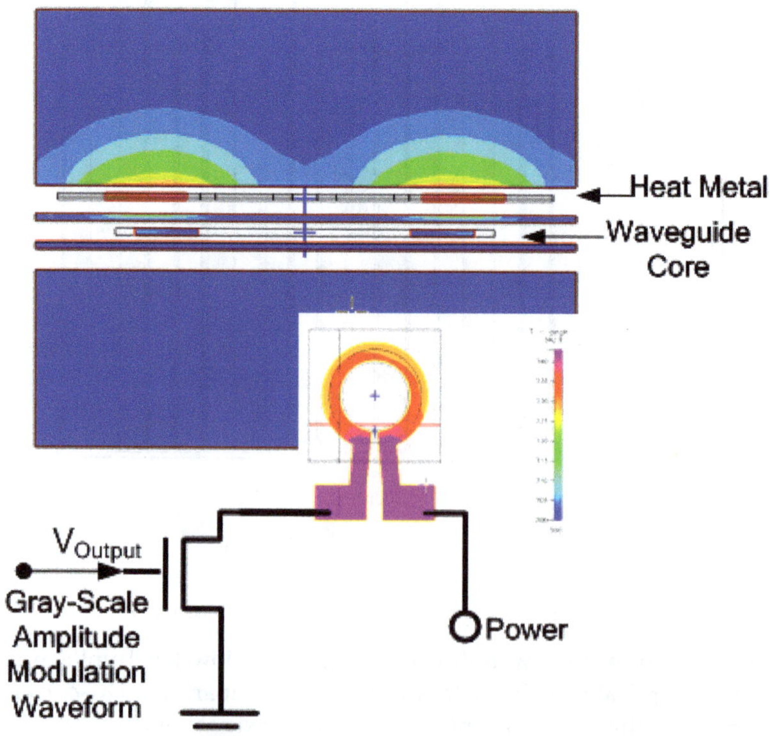

Fig. 11(a). Driving architecture of wavelength lock and simulation profile.

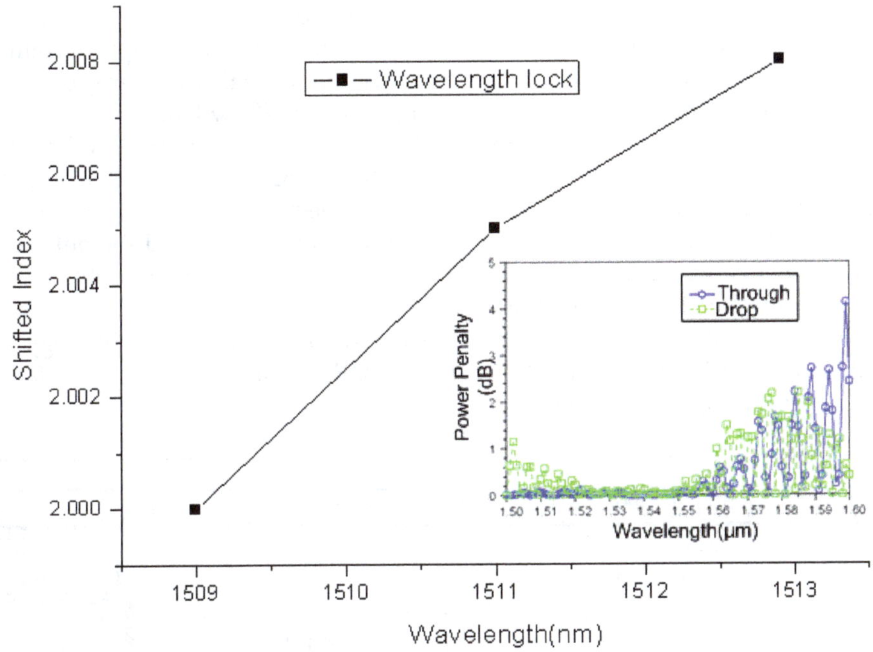

Fig. 11(b). The relative of shifted index and wavelength.

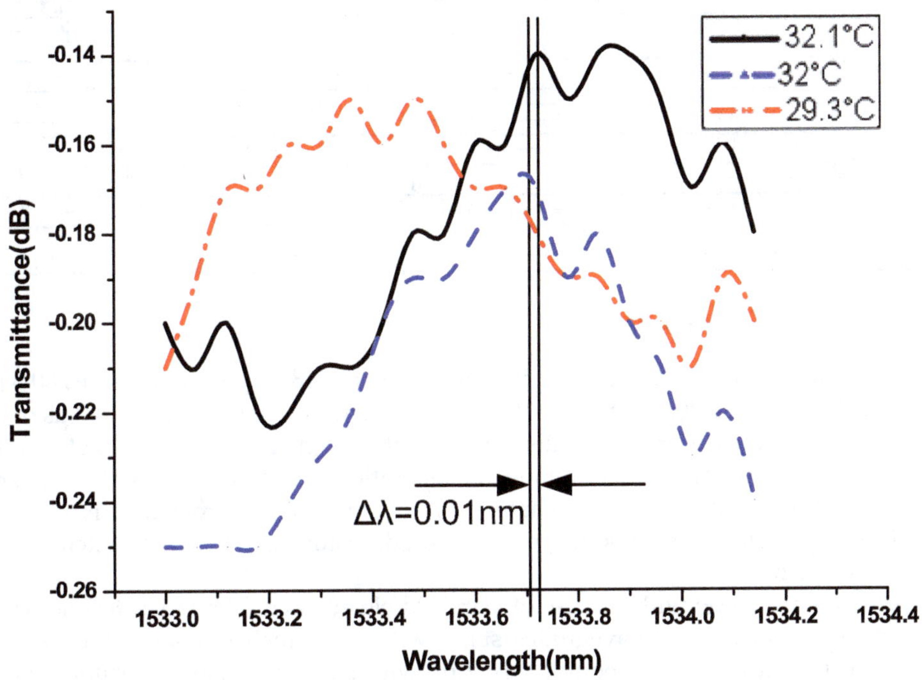

Fig. 12. Wavelength lock.

3.2 Controller and wavelength modulation result

To demonstrate basic functions of the 3D controller, as a result, we were able to reduce the number of electrical terminals to 5 control terminals and 1 power supply terminal. The controller was designed for a 0.35 μm CMOS process with a total circuit area of 2500×2500 μm^2, which is 80% of the circuit area by 2D configuration for 448 switches.

In the Logic Analysis, the relationship between the ASIC input and output is shown in Fig. 13. The input signals include DATA (signal for selected switch action), CLK1(signal to scan DATA signal), CLK2(signal to latch DATA signal or select), CTRL(signal to select enable type), as well as SETB(the time sequence to set up CTRL or power), and the output signals match the designed ASIC signals very well. Fig. 14. shows the image of a fabricated 16×28 matrix switch controller module. In this module, the chip area of 2.5×2.5 mm and was fabricated by a two-poly four-metal (2P4M) 0.35μm twin-well CMOS technology (TSMC, Taiwan Semiconductor Manufacturing Company Ltd). Each transistor is surrounded by full guard ring for preventing electrostatic shock.

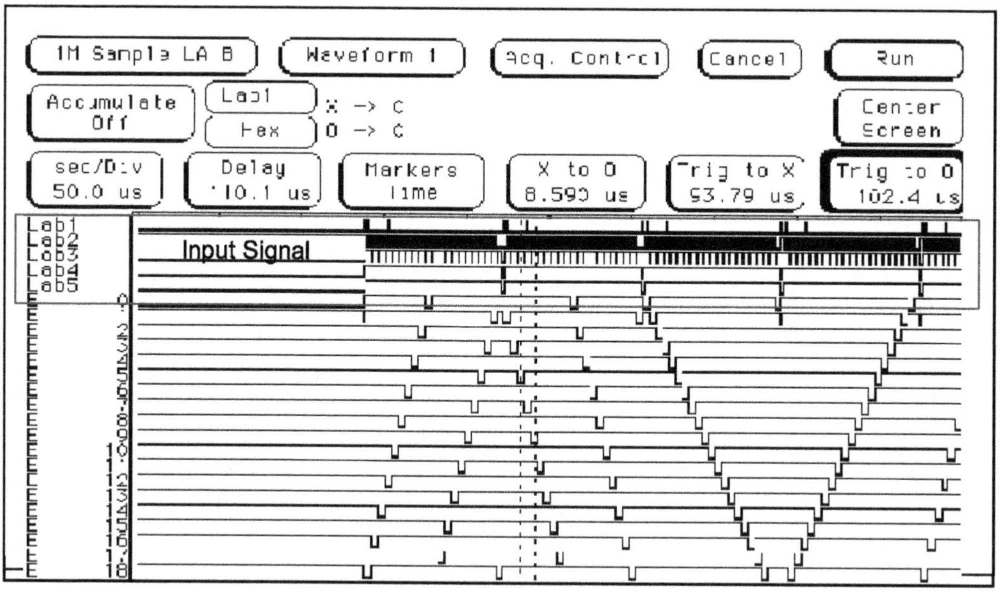

Fig. 13. FPGA verification result.

The testing result of the IC demonstrated the scanning of 448 ring switches takes 60.5 μs for 2D circuit architect while 20.5 μs for the 3D one, representing a time saving of 40 μs or a 67% time reduction. The measurement results of serial output signals for four channels, as shown in Fig. 15, demonstrated a simultaneous operation of four different temperature /wavelength modulations in each channel. By using the optimized driving signals, modulation frequencies up to 10 kHz were measured, resulting in thermal switching speeds in the order of 0.1 ms.

The micro-rings are made with the use of standard clean room fabrication technology. The fabrication of silicon nitride waveguides starts with a six inch diameter polished <100> silicon wafer. First a planar waveguide structure with a SiN(n=2.06@ λ =1550nm) core and SiO2(n=1.452@ λ =1550nm) cladding is formed. Finally the heater layer is deposited by sputtering a Platinum (Pt) thin film and patterned by photolithography and Pt wet-etch. Some results of temperature coefficient of resistance (TCR) measurements on platinum thin films. The shift in center wavelength of the ring λc is a function of the difference in effective index induced by heating the device that is given by equation (7):

$$\delta\lambda c = \frac{\lambda \Delta Neff}{Neff} \tag{7}$$

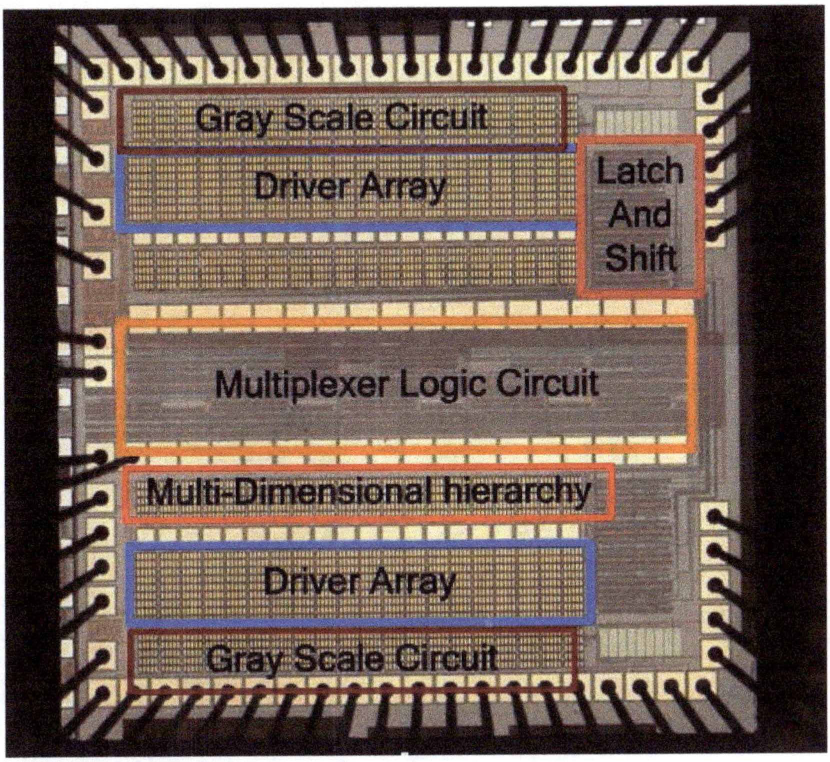

Fig. 14. Photograph of fabricated control IC chip.

In the wavelength modulation, temperature variation induced spectrum shift was measured, and the result is shown in Fig. 16, for the temperature changed from 29°C to 32°C, for which the thermal resonance shift is determined to be 0.1nm/°C. The temperature fluctuation can be controlled within 0.1°C, with a wavelength variation locked in 0.01 nm. The measured values are FSR=1.5nm and center wavelength shift λc=0.3nm at a center wavelength of λ=1532nm.

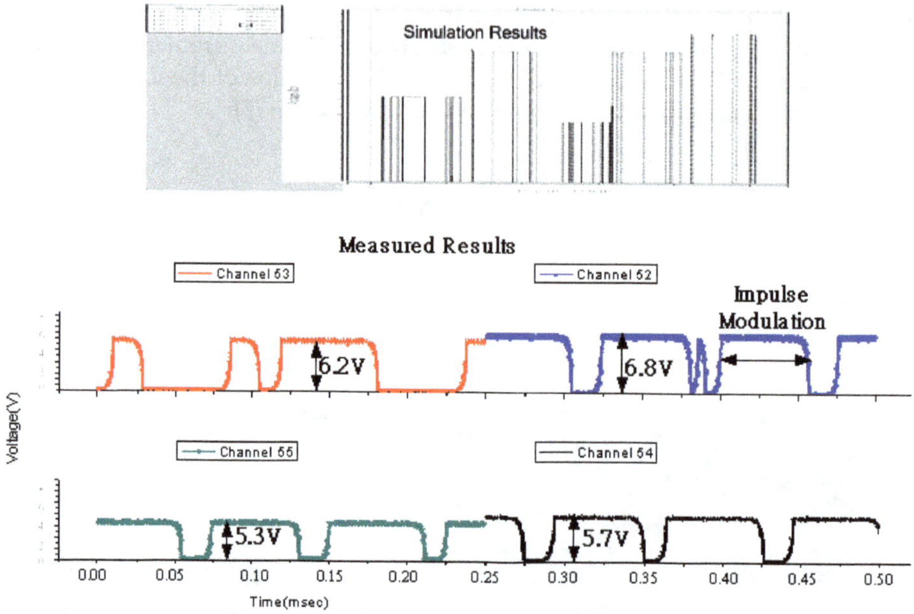

Fig. 15. Measurement result of serial outputs.

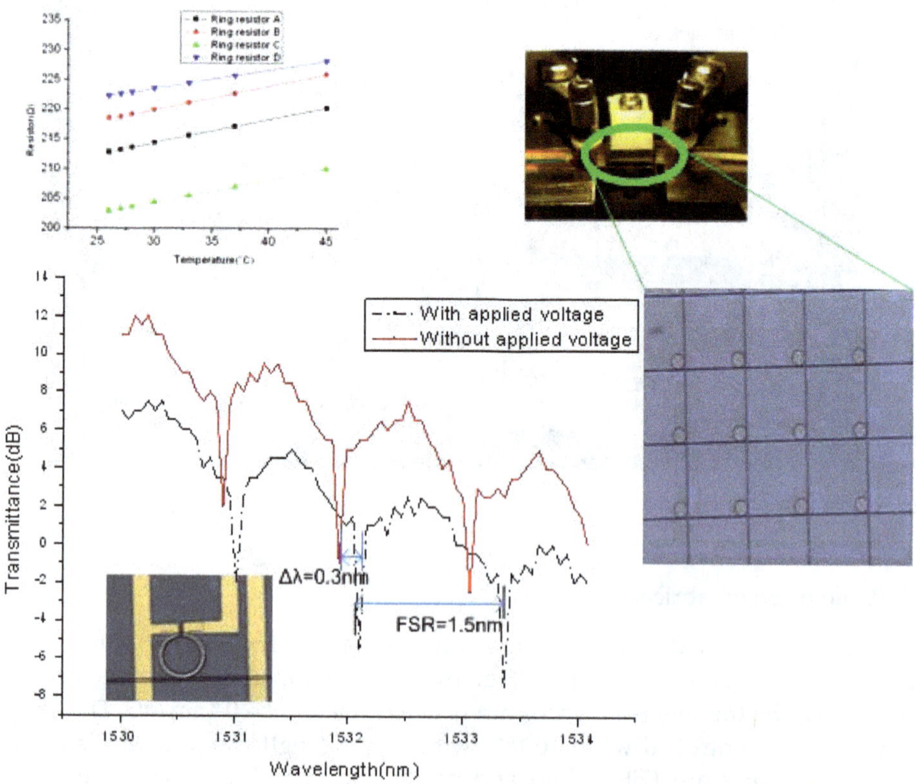

Fig. 16. Wavelength shift of transmission spectrum in coupled-ring-resonator.

4. Conclusion

The next generation of optical networking requires optical switches with complex functionality, small size and low cost. In this research, we have successfully designed and fabricated a silica-based 16×28 PLC-SW controller module in which we incorporated a switch chip based on PLC technology and new driving circuits with a serial-to-parallel signal conversion function. The new driving circuits significantly reduced the number of control terminals, and enabled us to realize a simple module structure suitable for use in a large-scale switch. It has been demonstrated that the scanning of 448 ring switches takes 20.5 μs by the novel 3D architect, representing a 67% time reduction.

On the other hand, thermal-optical effect was employed for wavelength modulation in this optical switch. To reduce overshooting and obtain rapid set up of ring temperature, heating pulses with amplitude modulation were employed. A temperature variation within 0.1°C can be maintained by this design, which can provide a very accurate wavelength modulation to 0.3 nm within 0.01 nm variation.

5. References

C. A. Brackett, et. al.(1993) "A Scalable Multi-wavelength Multihop Optical Network: A Proposal for Research on All-Optical Networks", J.Lightwave Technol., vol. 11, 736-753.

K. Okamoto, K. Takiguchi & Y. Ohmori,(1995) "16-channel optical add/drop multiplexer using silica-based arrayed-waveguide gratings", Electron. Lett., vol. 31, pp. 723-724.

H. Li, C. Lee, W. Lin, L.S. Didde, Y. J. Chen & D. Stone(1997) "8- wavelength photonics integrated 2x2 WDM cross-connect switch using 2xN phased-array waveguide grating (PAWG) multi/demultiplexers", Electron. Lett., vol. 33, pp. 592-594.

C. R. Doerr, et. al.,(1999) "40-wavelength add-drop filter", IEEE photon. Technol. Lett., vol. 11, pp. 1437-1439.

M. Okuno, A. Sugita, T. Matsunaga, M. Kawachi, Y. Ohmori, & K. Katoh,(1993) "8×8 optical matrix switch using silica-based planar lightwave circuits," IEICE Trans. Electron., vol. E76-C, no. 7, pp. 1215–1223.

M. Okuno, K. Kato, Y. Ohmori, M. Kawachi, & T. Matsunaga,(1994) "Improved 8×8 integrated optical matrix switch using silica-based planar lightwave circuits," J. Lightwave Technol., vol. 12, pp. 1597–1606.

T. Goh, A. Himeno, M. Okuno, H. Takahashi, & K. Hattori,(1998) "Highextinction ratio and low-loss silica-based 8×8 thermooptic matrix switch," IEEE Photon. Technol. Lett., vol. 10, pp. 358–360.

T. Goh, M. Yasu, K. Hattori, A. Himeno, M. Okuno, & Y. Ohmori,(1998) "Low-loss and high-extinction-ratio silica-based strictly nonblocking 16×16 thermo-optic matrix switch," IEEE Photon. Technol. Lett., vol. 10, pp. 810–812.

Wenlu Chen, Zhonghua Zhu, Yung Jui(Ray) Chen, Jacob Sun, Boris Grek, & Kevin Schmidt,(2003) "Monolithically integrated 32×four-channel client reconfigurable optical add/drop multiplexer on planar lightwave circuit", IEEE photonics technology letters,vol.15,no10,pp.1413-1415.

J.E. Ford, V. Aksyun, D. Bishop & J. Walker,(1999) "Wavelength add-drop switching using tilting mirrors", J. Lightwave Technol., vol. 17, pp. 904-911.

C. Pu, L. Lin, E. Goldstein & R. Tkach,(2000) "Client-configurable eight channel optical add/drop multiplexer using micromachining technology", IEEE photon. Technol. Lett., vol. 12, pp. 1665-1667.

William M. J. Green, Hendrik F. Hamann, Lidija Sekaric, Michael J. Rooks, & Yurii A. Vlasov,(2006) "Ultra-compact reconfigurable silicon optical devices using micron-scale localized thermal heating", Optical Society of America, pp.1-3.

Andreas Witzig, Matthias Streiff, Wolfgang Fichtner ,"Eigen-mode Analysis of Vertical-Cavity Lasers",pp.1-34.

N. A. Riza & S. Yuan,(1999) "Reconfigurable wavelength add-drop filtering based on a Banyan network topology and ferroelectric liquid crystal fiber-optic switches", J. Lightwave Technol., vol. 17, pp. 1575-1584.

Wenhua Lin, Haifeng Li, Y. J. Chen, M Dagenais & D. Stone,(1996) "Novel Dual-Channel-Spacing WDM Multi/Demultiplexers Based on Phased- Array Waveguide Grating", Photonics Tech. Lett. 8, 1501.

Jian-Chiun Liou, & Fan-Gang Tseng,(2008)" Integrated Control Circuit For Thermally Actuated Optical Packet Switch",Wireless And Optical Communications (WOC) pp.213-218.

Young-Kai Chen, Andreas Leven, Ting Hu, Nils Weimann, Rose Kopf, Al Tate,(2008)" Integrated Photonic Digital-to-Analog Converter for Arbitrary Waveform Generation", Photonics in Switching (PS), Optical Switches and Routing Devices, D-08-2, pp.1-2.

WANG Meng-Yao, WEI PAN, BIN LUO, ZHANG Wei-Li, & ZOU Xi-Hua (2007) "Optimization of gray-scale performance in pixellated-metal-mirror FLC-OASLM by equivalent circuit model" Microelectronics journal, vol. 38, no2, pp. 203-209.

Christopher M Waits, Alireza Modafe & Reza Ghodssi,(2003)"Investigation of gray-scale technology for large area 3D silicon MEMS structures", J. Micromech. Microeng. 13 170-177.

S. Mias, L.G. Manolis, N. Collings, T.D. Wikinson, W.A. Crossland,(2005) Phase-modulating bistable optically addressed spatial light modulators using wide-switching-angle ferroelectric liquid crystal, Opt. Eng. 44 (1) 014003 - 014017.

C.A.T.H. Tee, W.A. Crossland, T.D. Wilkinson, A.B. Davey,(2000) Binary phase modulation using electrically addressed transmissive and silicon backplane spatial light modulators, Opt. Eng. 39 (9) 2527-2534.

S. Mias, N. Collings, T.D. Wilkinson, S. Coomber, M. Stanley, W.A. Crossland,(2003) Spatial sampling in pixelated-metal-mirror ferroelectricliquid- crystal optically addressed spatial-light-modulator, Opt. Eng. 42 (7) 2075 - 2081.

Y. Kuno, Y. Taura, M. Yagura, S. Hamada, N.S. Takashashi, S. Kurita,(1991) Amorphous-silicon photosensor-based ferroelectric-liquidcrystal light valve, Proc. SID 32 (3), 187-190.

W. Li, R.A. Rice, G. Moddel, L.A. Pagano-Stauffer, M.A.Handschy,(1989) Hydrogenated amorphous-silicon photosensor for opticallyaddressed high-speed spatial light modulator, IEEE Trans.Electron Dev. 36 (12) 2959 - 2964.

M.Y. Wang, W. Pan, B. Luo, X.H. Zou, W.L. Zhang, (2007) The t - V min addressing mode and an improved equivalent circuit model for ferroelectric liquid crystal displays, IEE Proc.- Opto- electron , vol. 1, no1, pp. 16-22.

M.Y. Wang, W. Pan, B. Luo, (2004) Study of the respondent characteristicsin ferroelectric liquid crystal spatial light modulator, J. Funct. Mater.Dev. 10 (4) 489 - 492.

Meng-yao Wang, Wei Pan, Bin Luo, Wei-li Zhang and Xi-hua Zou,(2007)"Optimization of gray-scale performance in pixellated-metal-mirror FLC-OASLM by equivalent circuit model "Microelectronics Journal (38) 203-209.

Jian-Chiun Liou & Fan-Gang Tseng (2008) "An Intelligent Multiplexing Control Thermal Actuated Optical Packet Switch", Photonics in Switching (PS), Optical Switches and Routing Devices, D-06-2, pp.1-2.

4

Optoelectronics in Suppression Noise of Light

Jiangrui Gao, Kui Liu, Shuzhen Cui and Junxiang Zhang
State Key Laboratory of Quantum optics and Quantum optics Devices
Institute of Opto-Electronics, Shanxi University, Taiyuan
China

1. Introduction

High quality light, whenever coherent light and nonclassical light, is more and more widely used in many filed. The coherent state behave minimum classical noise and better coherence, is mostly important light in quantum information, precise measurement and quantum metrology etc. In quantum information and quantum optics, a coherent light is used to entanglement generation (Ou, Z. et al. 1992), quantum teleportation (Furusawa, A. et al. 1998), quantum computation (Menicucci, N. et al. 2006) and quantum secret sharing (Lance, A. et al. 2004). The quiet laser beam, shot noise limited laser beam is important in above application.

In the past 20 years there has been much interest in the generation and application of various nonclassical light sources, such as squeezed-state entanglement, quantum-correlated twin beams, and sub-poissonian light, for which improved signal-to-noise ratios and measurement precision beyond the quantum-noise limit are anticipated. In particular, sub-poissonian light is a promising light source in quantum information and optical communication because of its communication capacity (Tapster, P. et al. 1988; Teich, M. & Saleh, B. 1985; Machida, S. et al. 1987,1989; Richardson, W. et al. 1991; Li, R. et al. 1995).

In 1990 Mertz et al. (Mertz, J. et al. 1990) obtained sub-Poissonian light by using the feed-forward technique, and in the year of 1997, P. K. Lam et al. (Lam, P. et al. 1997) experimental demonstrate noiseless signal amplification of using positive optoelectronic feed forward. The technique of optoelectronics is applied into noise suppression of light and noiseless signal amplification, and it is extended to other displacement transformation, such as quantum teleportation (Furusawa, A. et al. 1998), quantum cloning (Andersen, U. 2005; Braunstein, S. et al. 2001); furthermore, the purification of nonclassical light recently (Hage, B. et al. 2008; Dong, R. et al. 2008).

In this paper, we introduce the application of opto-electronics in noise suppression. It includes two parts, section II is noise suppression of fiber laser to SNL by optoelectronic feed forward and section III is preparation of sub-poissonian state by opto-electronic feedforward.

2. Suppression intensity noise of fiber laser

As we all known, a general laser beam is near the coherent light, the fiber laser is the new type laser, it behave some advantageous: high output power, small size and good frequency stability, easy to coupled with the fiber, but the light is not complete coherent light, especially at the low sideband frequency, the light have higher noise than shot noise limit (SNL). It hinders fiber laser application in the quantum information, so suppression the noise of fiber laser is a significant issue. Intensity noise of fiber laser include beating noise, environmental perturbations, thermal noise, and fluctuations in pump power (Cheng, Y. et al. 1995).Various methods for partially reducing output noise have been proposed. High frequency beating noise from fiber ring laser can be reduced by intracavity spectral filtering (Sanders, S. et al.1992). Intensity noise of fiber laser at the relaxation oscillation frequency was suppressed by integrating a negative feedback circuit (Spiegelberg, C. et al. 2004; Ball, G. et al. 2008).

Optoelectronic feed forward, which is easy and practical, is a method to suppression noise of light easily. In the section we introduce an optoelectronic feed-forward loop used to suppress the intensity noise of the fiber laser. Selecting the best delay time and feed-forward gain, the intensity noise can be suppressed near to SNL at sideband frequency. The maximum noise reduction 22dB was obtained at 6.0MHz. Furthermore, altering the delay time and filter band of optoelectronic system, the light can near to shot noise at differ frequency point. After the noise suppression, the fiber laser can be widely applied into the filed of quantum information.

The experiment setup of laser noise suppression is shown in fig.1. It is based on the noise correlation between reflection light and transmission light.

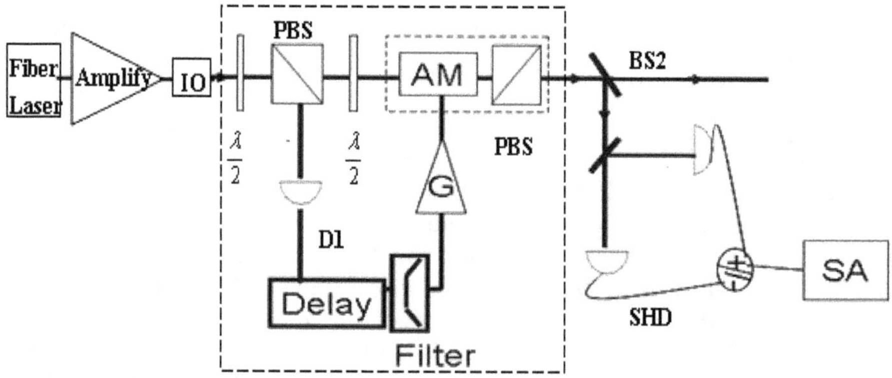

Fig. 1. The experiment setup of suppression of intensity noise of fiber laser with OptoElectronic feedforward. PBS: polarization beam splitter; BS: beam splitter; AM: amplitude modulator; D1; detector; SHD: homodyne detectors; SA: spectrum analyzer.

As shown in fig.1, the laser output from fiber laser is divided into two parts by beam splitter which is composed by polarization beam splitter (PBS) and half wave plate. The detector D1 detects the reflected light and the measurement of D1 is taken as the signal of feed forward. After passed through delay, filter and amplifier, the signal is acted on amplitude modulator (AM). As the result, the noise of transmission is suppressed, and the noise is measured by homodyne detectors.

2.1 Theory of optoelectronic feed forward in noise suppression

The output field of fiber laser can be expressed:

$$\hat{A}_{in}(t) = A_{in} + \delta\hat{A}_{in}(t) \tag{1}$$

Here, $\hat{A}_{in}(t)$ is the annihilate operator of input field, A_{in} is the average value of input field, $\delta\hat{A}_{in}(t)$ is fluctuation of input field.
After passed through AM, the light field can read as (Lam, P. et al. 1997):

$$\hat{A}_{out}(t)=\sqrt{\eta_3}[\sqrt{1-R}A_{in}+\sqrt{1-R}\delta\hat{A}_{in}(t)+\sqrt{R}\delta\hat{v}_1(t)+\delta\hat{r}(t)]+\sqrt{1-\eta_3}\delta\hat{v}_3(t) \tag{2}$$

Where $\hat{v}_1$ is vacuum noise from the first BS; $\hat{v}_3$ is vacuum noise corresponding to imperfect of homdyne detector system, η_3 is the efficiency of homdyne detectors. $\delta\hat{r}$ is fluctuation from feedback loop; η_2 is efficiency of detector D1.

$$\begin{aligned}\delta\hat{r}(t)=&\int_{-\infty}^{\infty}\kappa(\tau)\sqrt{R\eta_2}\times A_{in}[\sqrt{R\eta_2}\delta\hat{X}_{Ain}(t-\tau)\\&-\sqrt{(1-R)\eta_2}\delta\hat{X}_{v1}(t-\tau)+\sqrt{R}\delta\hat{X}_{v2}(t-\tau)]d\tau\end{aligned} \tag{3}$$

Here $\kappa(\tau)$ is response function of feedback loop; $\delta\hat{X}_{in}=\delta\hat{A}_{in}+\delta\hat{A}_{in}^{+}$, $\delta\hat{X}_{vi}=\delta\hat{A}_{vi}+\delta\hat{A}_{vi}^{+}$ are amplitude quadrature and phase quadrature, respectively.
The amplitude quadrature of output field can be deduced from ref. (1) and (2)

$$\begin{aligned}\delta\hat{X}_{out}(t)&=\delta\hat{A}_{out}(t)+\delta\hat{A}_{out}^{+}(t)\\&=\sqrt{\eta_3}[\sqrt{1-R}\delta\hat{X}_{in}(t)+\sqrt{R}\delta\hat{X}_{\hat{v}_1}(t)+\delta\hat{X}_r(t)]+\sqrt{1-\eta_3}\delta\hat{X}_{\hat{v}_3}(t)\end{aligned} \tag{4}$$

Here

$$\begin{aligned}\delta\hat{X}_r(t)&=\delta\hat{r}(t)+\delta\hat{r}^{+}(t)\\&=\int_{-\infty}^{\infty}G(\tau)[\sqrt{R\eta_2}\delta\hat{X}_{Ain}(t-\tau)-\sqrt{(1-R)\eta_2}\delta\hat{X}_{v1}(t-\tau)+\sqrt{R}\delta\hat{X}_{v2}(t-\tau)]d\tau\end{aligned}$$

$$G(\tau)=\kappa(\tau)\sqrt{R\eta_2}\times A_{in}$$

The equation (4) can be easily solved by taking the Fourier transform into the frequency domain:

$$\begin{aligned}\delta\hat{X}_{out}(\omega)=&\sqrt{\eta_3}\{\sqrt{1-R}\delta\hat{X}_{in}(\omega)+\sqrt{R}\delta\hat{X}_{\hat{v}_1}(\omega)+G(\omega)[\sqrt{R\eta_2}\delta\hat{X}_{in}(\omega)\\&-\sqrt{(1-R)\eta_2}\delta\hat{X}_{v1}(\omega)+\sqrt{R}\delta\hat{X}_{v2}(\omega)]\}+\sqrt{1-\eta_3}\delta\hat{X}_{\hat{v}_3}(\omega)\end{aligned} \tag{5}$$

Here G(ω) is the gain of feedback, $G(\omega)=\kappa(\omega)\sqrt{R\eta_2}\times A_{in}$.
We can obtain the noise spectrum of output filed:

$$\begin{aligned}V_{out}(\omega)&=\left\langle|\delta X_{Aout}|^2\right\rangle\\&=\eta_3\left|\sqrt{(1-R)}+G(\omega)\sqrt{R\eta_2}\right|^2V_{in}(\omega)+\eta_3\left|\sqrt{R}-G(\omega)\sqrt{(1-R)\eta_2}\right|^2V_1\\&+\eta_3\left|G(\omega)\sqrt{(1-\eta_2)}\right|^2V_2+(1-\eta_3)V_3\end{aligned} \tag{6}$$

Here V_1, V_2, V_3 are vacuum noise spectrum and $V_1=V_2=V_3=1$; V_{in} is the intensity noise spectrum of input field.
To infer from eq.(6), when $G(\omega)=G(\omega)_{opt}$, the intensity noise suppression is best, and then the minimum output noise $V_{out}(\omega)$ is:

$$V_{out}^{opt}(\omega) = \frac{[R\eta_2 + (1-R)\eta_3](V_{in}(\omega)-1)+1}{(V_{in}-1)R\eta_2+1}, \tag{7}$$

$$G(\omega)_{opt} = \frac{-(V_{in}-1)\sqrt{(1-R)R\eta_2}}{V_{in}R\eta_2 + \eta_2(1-R)+(1-\eta_2)}, \tag{8}$$

When η_2=0.95, η_3=0.20, the optimum output noise V_{out}^{opt} as the function of input noise V_{in} and BS reflection R is shown in fig.2. With the same R, higher the input noise is, higher the output noise is, and with the same input noise, higher the reflection R is, lower the output noise V_{out} is. Just only when $R=1$, the output noise V_{out}^{opt} reach at shot noise limit (SNL). In general, the reflection R is always smaller than 1 in order to get effective intensity of output field, the method can suppress main noises, but not all noises.

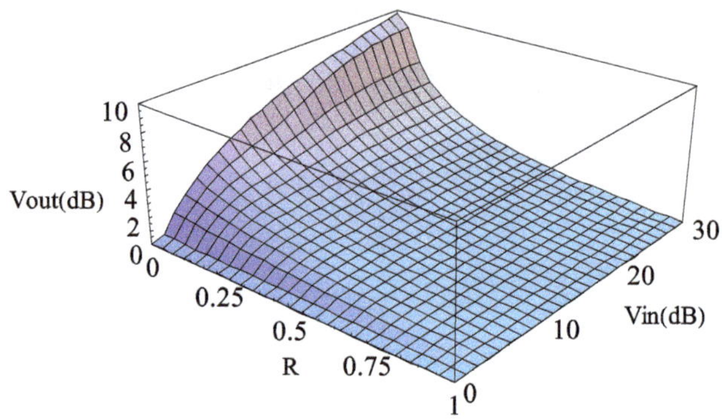

Fig. 2. The optimum output noise V_{out}^{opt} as the function of input noise V_{in} and BS reflection R, η_2=0.95, η_3=0.20.

2.2 Noise suppression

The relationship of residual output noise and reflection R is shown in fig. 3. at 6 MHz, the dot is experiment data; the dash line is theory result as eq. (7). The detector D1 can be in saturation when the R is high, we make $\eta_2 \times R = \text{constant}$ to correct the theory of dash line, and the solid line is the corrected theoretical result. In the experiment, $\eta_2 = 0.95$, C=0.24 and $V_{in} / SNL = 24dB$.

Because of the more information we can be got from laser beam when the reflection is higher, that means the signal and noise ratio is better, the noise suppression will be better as shown in fig.3. When reflection R is large enough, the increase of noise suppression efficiency become flat. In order to obtain both the output power and noise suppression efficiency simultaneously, we select the reflection $R=0.3$.

With best selecting delay time, bandwidth filter and gain of feedback loop, the noise of suppression vs analysis frequency is shown in fig. 4 when the reflection $R=0.3$. The maximum noise reduction 22dB is obtained at 6MHz, and the noise is near to SNL. The output noise only can be most suppressed at a frequency point with limited bandwidth due to limitation of bandwidth of optoelectronic feedback loop. In order to suppress noise in large range of frequency, we can use a high finesses cavity as model cleaner to suppress wideband noise effectively. We have obtained the shot noise limit laser from 3MHz by mode cleaner [Liu, K. et al. 2009].

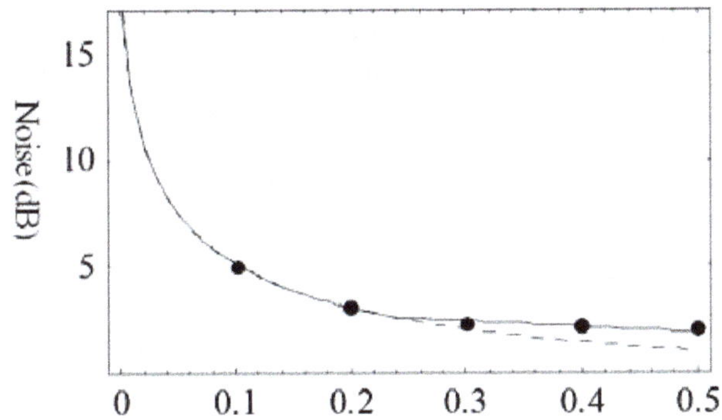

Fig. 3. Residual Noise Vs Reflectivity. • The results of experiment; --- The results of theory, $\eta_2 = 0.95$; — The results of theory, $\eta_2 \times R = 0.24$.

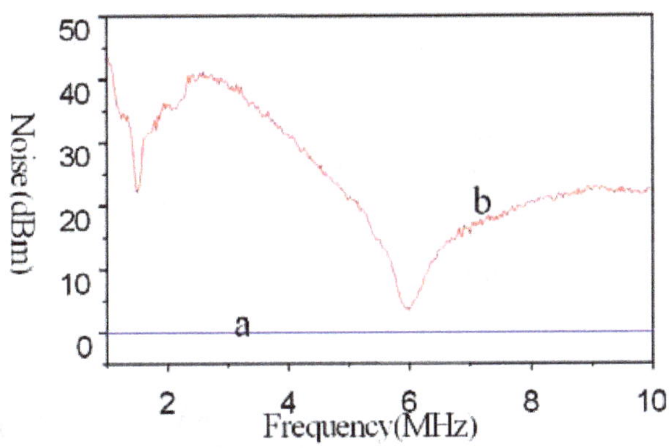

Fig. 4. The best noise suppression of Fiber Laser with Opt-Electronic feedbak at 6.2 MHz. a) the electronics noise. b) the SNL.

3. Generation of sub-poissonian state

Sub-Poissonian light is an important light source in studies of quantum information and optical communication. Sub-Poissonian light is also the light with the largest communication capacity in known light fields (Yamamoto, Y. & Haus, H. 1986). There are three techniques for generation of sub-Poissonian light: direct conversion, a feedback technique, and a feed-forward technique. In 1990 Mertz et al. (Mertz, J. et al. 1990) obtained sub-Poissonian light by using the feed-forward technique and a nondegenerate optical parametric oscillator (NOPO) pumped by the 528 nm line of an argon-ion laser; the observed noise reduction was as much as 24% below the shot-noise limit (SNL).Subsequently, Kim and Kumar (Kim, C. & Kumar, P. 1992) analyzed theoretically the generation of tunable sub-Poissonian light with an intensity feed-forward scheme for application in precision absorption spectroscopy. Then, pulses sub-Poissonian light was obtained experimentally also by traveling wave optical parametric deamplification (Li, R. et al. 1995). In 2006 Zou, H. obtained noise reduction of 45% wavelength tunable sub-Poissonian light by using the feed-forward technique (Zou, H. et al. 2006).

Zhang *et al.* (Zhang, Y. et al. 2002) presented statistics on twin beams emerging from a nondegenerate OPO. They recorded the twin beams' photocurrent difference fluctuation and demonstrated sub-Poissonian distributions of the twin beams' intensity difference. Laurat et al. (Laurat, J. et al. 2003) reported their experimental demonstration of the conditional preparation of a continuously variable nonclassical state of light from twin beams by a data-acquisition system; this is a kind of postselected sub-Poissonian light.

In this section we demonstrate the achievement of a frequency-tunable high-intensity sub-Poissonian state with optoelectronic feed-forward and quantum-correlated twin beams. A maximum noise reduction of 2.6 dB (45%) below the SNL was measured, and a wavelength tunable range of 7.4 nm was demonstrated. Sub-Poissonian distribution with twin beams was also demonstrated by direct measurement of the sub-Poissonian state.

3.1 The principle of generation sub-Poissonian light by optoelectronics

Generation of quantum-correlated twin beams is a prerequisite to our scheme. It is well known that a NOPO running above threshold is one of the best choices for intensity-correlated twin beams. Several groups of scientists have demonstrated experimentally quantum correlation of the intensity of twin beams. The intensity feed-forward scheme is shown schematically in Fig. 5.

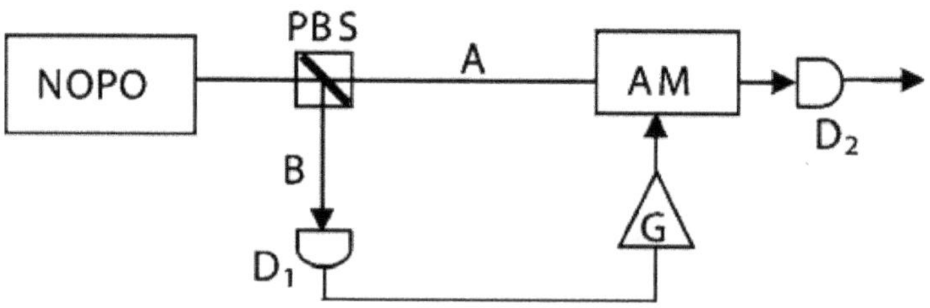

Fig. 5. Schematic of the feed-forward scheme. The quantum-correlated signal and idler beams from the NOPO are separated into beams A and B. Beam B is directly detected by detector D1 to correct beam A by modulation. PBS: polarization beam splitter; G:amplifier; AM: amplitude modulator; NOPO: nondegenerate optical parametric oscillator.

Beams A and B from the NOPO are quantum-correlated twin beams (signal and idler). Beam B is detected directly to correct beam A. After optimum correction, intensity noise S_A^{opt} of beam A at analysis frequency becomes (Mertz, J. et al. 1990)

$$S_A^{opt} = 2S_{A-B}(\Omega)[1 - \frac{S_{A-B}(\Omega)}{2S(\Omega)}] \tag{9}$$

where $S(\Omega)$ is the intensity noise spectrum of a single beam (beam A or beam B) and $S_{A-B}(\Omega)$ is the noise spectrum of the intensity difference between the twin beams, which characterizes the quantum correlation between the twin beams. Both $S(\Omega)$ and $S_{A-B}(\Omega)$ are normalized to their respective SNL. The intensity noise of a single beam is decided mainly by the OPO's state of operation. While the OPO cavity operates close to oscillator threshold, each beam generated by the OPO in general has a large amount of excess noise. Under the circumstances, the intensity noise spectrum of corrected beam A can be written approximately as

$$S_A^{opt} = 2S_{A-B}(\Omega) \tag{10}$$

It is obvious that only when the intensity difference squeezing is more than 3 dB ($S_{A-B} < 0.5$) can the noise power of the corrected signal beam be below the shot-noise level, that is, sub-Poissonian light.

3.2 Experiment and results

A schematic of the experimental setup is shown in Fig. 6. A homemade intracavity frequency-doubled and frequency-stabilized cw ring neodymium:yttrium aluminum perovskite laser serves as the light source. The output second-harmonic wave at 540 nm is used to pump the semimonolithic NOPO. The oscillation threshold is less than 120 mW, and an output power of 40 mW is obtained at a pump power of 170 mW.

The output infrared beams are reflected by dichroic mirror M and then separated by a polarizing beam splitter (PBS) into detection arms A and B. The detectors (PA_1, PA_2, PB_1, and PB_2) are Epitax 500 p-i-n photodiodes, with quantum efficiencies of about 0.94. The displacement operation is performed by use of an electro-optical modulator and a highly reflecting (99/1) mirror with the help of a local beam for reducing loss.

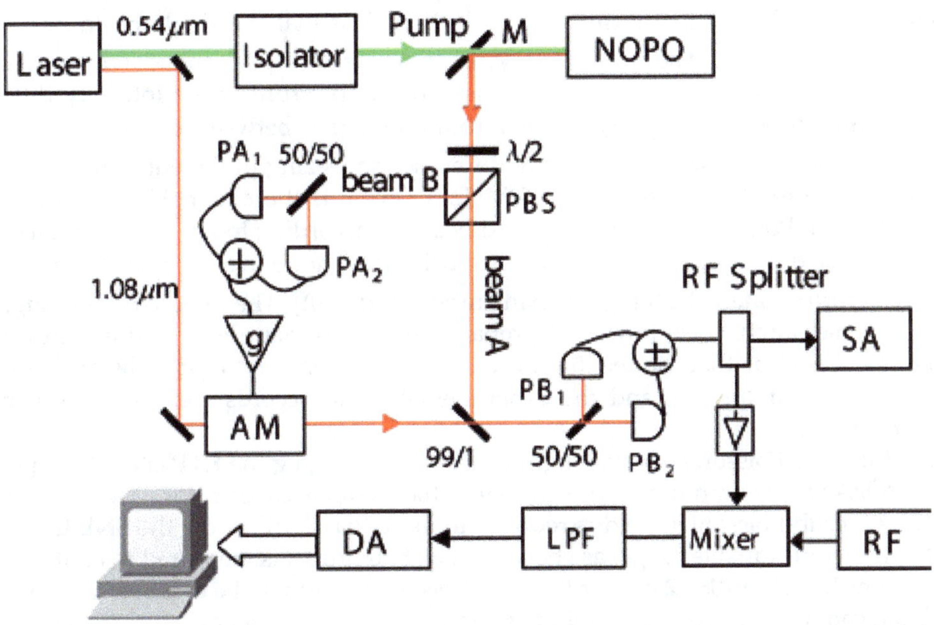

Fig. 6. (Color online) Experimental setup: PA_i, PB_i (i=1,2): photodetectors; AM: amplitude modulator; RF Splitter: radio-frequency power splitter; SA: spectrum analyzer; RF: radio-frequency signal; LPF: low-pass filter; DA: data acquisition card; PBS: polarization beam splitter; M: mirror.

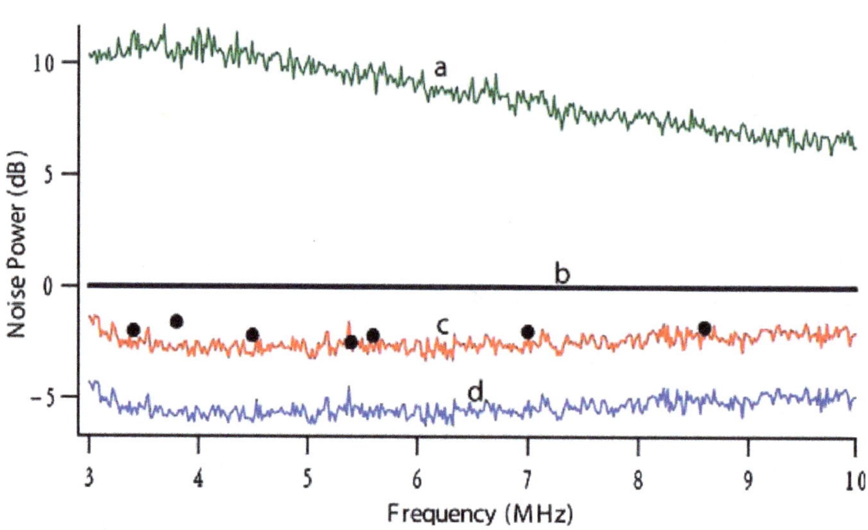

Fig. 7. (Color online) Noise spectra from 3 to 10 MHz. All noise power is normalized to the corresponding SNL. Curve *a*, noise without feed-forward correction; curve *b*, the SNL. Curve *d*, noise spectrum of the intensity difference between the signal and the idler fields. Curve *c*, obtainable noise of beam A [according to Eq. (9)] with data curve *a* and curve *d*. Filled circles, experimentally measured results.

Experimental results are shown in Fig. 7. All the noise power spectra are normalized to their respective SNL. Curve a is the noise power spectrum of beam A without feed-forward correction $S(\Omega)$; line b is the normalized SNL. It is clear that the single beam has a large mount of excess noise. The excess noise is typically 6–10 dB above the SNL in the measurement 3–10 MHz. Curve d is the noise power spectrum of the intensity difference between the twin beams $S_{A-B}(\Omega)$, the quantum correlation between beams A and B. The curve exhibits a more than 5 dB intensity difference squeezing in the entire range, and the maximum value reaches 5.7 dB near 4.5 MHz. Curve c is the obtainable minimum noise power of the sub-Poissonian beam (corrected beam A) according to Eq. (9) with data curve a $S(\Omega)$ and curve d $S_{A-B}(\Omega)$. Because there is much excess noise in each beam from the OPO, curve c is approximately 3 dB higher than curve d [Eq. (10)]. The filled circles in Fig. 7 are the directly measured noise power of prepared sub-Poissonian light at different analysis frequencies. It is clear that the maximum noise reduction is 2.6 dB (45%) below shot-noise level at 5.5 MHz frequency, and the experimental result accords well with the expected values of curve c.

We tuned the sub-Poissonian wavelength roughly by changing the KTP crystal temperature in the OPO. A wavelength range from 1078.9 to 1083.8 nm is covered for a signal beam, and the noise of a sub-Poissonian field is reduced to more than 2 dB below the SNL throughout all the wavelength-tunable range, as shown in Fig. 8. If arm A is measured to control arm B, a range from 1076.4 to 1081.2 nm for the sub-Poissonian field will be demonstrated. One can also obtain continuous frequency tuning by tuning the laser frequency

Additionally, the distribution of sub-Poissonian light is acquired as shown in Fig. 9. In the experimental setup (Fig. 6), a part of the signal after rf splitting is mixed with a sinusoidal local oscillator at 5.5 MHz. After a 100 kHz low-pass filter, the signal is collected at a sampling rate of 500 kHz by a 12 bit acquisition card. The statistical distribution of the photocurrent fluctuation is shown in Fig. 9, with 200,000 points for each curve. The points in the figure are the experimental results, and the solid curves are Gaussian fits of the probability distribution. Curve a represents the probability distribution of the prepared sub-Poissonian field, curve b corresponds to a coherent state (the SNL), and curve c corresponds to single-beam field without correction. It is shown that the sub-Poissonian distribution of light fluctuation is narrower than a standard Gaussian distribution of the coherent state. The uncorrected single-beam fluctuation distribution is a super-Poissonian and is much broader than the standard Gaussian distribution. The photocurrent fluctuation of the sub-Poissonian field can also be compared with the standard Gaussian distribution. A noise reduction of 1.2 dB below the SNL is calculated from average half-widths (Fig. 9) and does not accord well with what we observed with the spectrum analyzer because of the narrow bandwidth of the prepared sub-Poissonian field and a nonideal low-pass filter. The calculated photocurrent fluctuation of a single beam is 9 dB above the SNL, which accords well with what we observed with the spectrum analyzer.

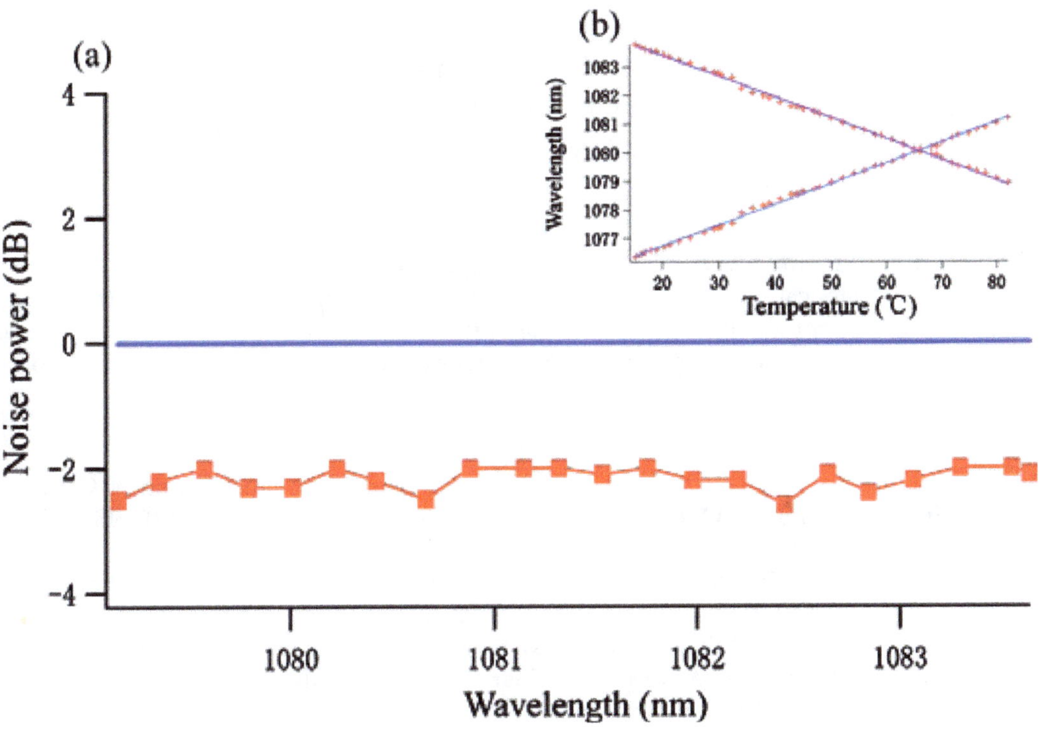

Fig. 8. (Color online) (a) Normalized sub-Poissonian light noise from 1079 to 1083.7 nm. (b) Wavelength of twin beams versus temperature of the crystal in the OPO.

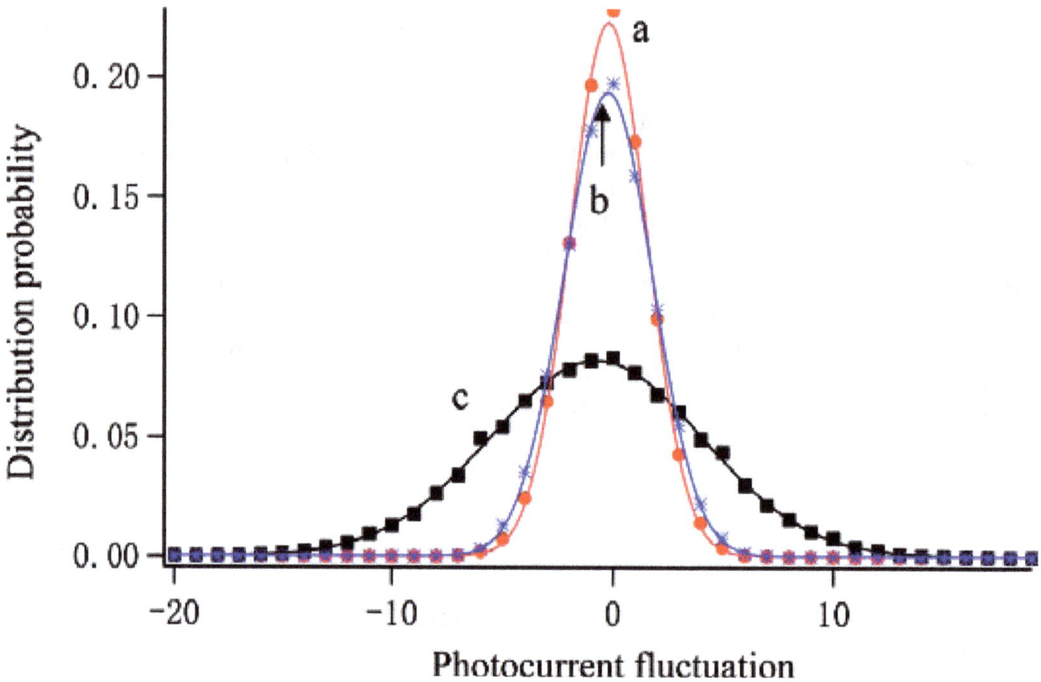

Fig. 9. (Color online) Intensity fluctuation distribution at 5.5 MHz. Curve *a*, prepared sub-Poissonian field; curve *b*, coherent light; curve *c*, single beam from the NOPO (beam A).

4. Conclusion

We introduce the application of opto-electronics feed-forward in noise suppression, including both classical noise (fiber laser noise suppression) and quantum noise (preparing sub-Poissonian) suppression. The technique of opto-electronics has been widely applied and will be more and more significant in the field of quantum optics and quantum information.

5. References

Andersen, U.; Josse,V. & Leuchs, G. (2005). Unconditional Quantum Cloning of Coherent States with Linear Optics. Phys.Rev. Lett. Vol. 94, No. 24, pp.240503, ISSN:1079-7114.

Ball, G.; Hull-Allen,G. & Holton, C. (2008). Low noise single frequency linear fiber laser. Electronics Letters. Vol. 29, No. 18, pp. 1623-1625, ISSN: 0013-5194.

Braunstein, S. Nicolas, J.; Iblisdir, S.; Loock, P. & Massar, S. (2001). Optimal Cloning of Coherent States with a Linear Amplifier and Beam Splitters. Phys.Rev. Lett. Vol. 86, No. 21, pp.4938-4941, ISSN:1079-7114.

Cheng, Y.; Kringlebotn, J.; Loh, W.; Laming, R. & Payne, D. (1995). Stable single-frequency travelling-wave fiber loop laser with integral saturable-absorber-based tracking narrow-band filter. Opt. Lett. Vol. 20, No. 8, pp. 875-877, ISSN:0146-9592.

Dong, R.; Lassen, M.; Heersink, J.; Marquardt, C.; Filip, R.; Leuchs, G. & Andersen, G. (2008). Experimental entanglement distillation of mesoscopic quantum states. Nature Phys. Vol. 4, No. 2 November, pp.919-923, ISSN:1745-2473.

Furusawa, A.; Sørensen, J.; Braunstein,S.; Fuchs, C.; Kimble, H. & Polzik, E. (1998). Unconditional Quantum teleportation. Science. Vol.282, No.5389, pp.706-709, ISSN:1095-9203.

Hage, B.; Samblowski, A.; DiGuglielmo, J.; Franzen, A.; Fiurášek, J. & Schnabel, R. (2008). Preparation of distilled and purified continuous-variable entangled states. Nature Phys. Vol. 4, No. 2 November, pp.915-918, ISSN:1745-2473.

Kim, C. & Kumar, P. 1992. Tunable sub-Poissonian light generation from a parametric amplifier using an intensity feedforward scheme. Phys.Rev. A. Vol. 45, No. 7, pp.5237-5242, ISSN:1050-2947.

Lam, P.; Ralph, T.; Huntington, E. & Bachor, H. (1997). Noiseless Signal Amplification using Positive Electro-Optic Feedforward. Phys.Rev. Lett. Vol. 79, No. 8, pp.1471-1474, ISSN:1079-7114.

Laurat, J.; Coudreau, T.; Treps, N.; Maitre, A. & Fabre, C. (2003). Conditional Peraration of a Quantum State in the Continuous Variable Regime: Generation of a sub-Poissonian State from Twin Beams. Phys.Rev. Lett. Vol. 91, No. 21, pp.213601, ISSN:1079-7114.

Li, R.; Choi, S. & Humar, P. (1995). Generation of sub-Poissonian pulses of light. Phys.Rev. A. Vol. 51, No. 22, pp.R3429-R3432, ISSN:1050-2947.

Liu, K., Cui S., Zhang, H. Zhang, J. and Gao J. R. et al. 2011. Chin. Phys. Lett. Vol. 28, No.7, pp.074211, ISSN:1741-3540.

Machida, S.; Yammmoto, Y. & Itaya, Y. (1987). Observation of amplitude squeezing in a constant- current- driven semiconductor laser. Phys.Rev. Lett. Vol. 58, No. 10, pp.1000-1003, ISSN:1079-7114.

Machida, S. & Yamamoto, Y. (1989). Observation of amplitude squeezing from semiconductor lasers by balanced direct detectors with a delay line. Opt. Lett. Vol. 14, No. 19, pp. 1045-1047, ISSN:0146-9592.

Menicucci, N.; Loock, P.; Gu, M.; Weedbrook, C.; Ralph, T. & Nielsen, M. (2006). Universal Quantum Computation with Contiunous-Varible Cluster States. Phys.Rev. Lett. Vol. 97, No.11, pp.110501, ISSN:1079-7114.

Mertz, J.; Heidmann, A.; Fabre, C.; Giaocobino, E. & Reynand, S. (1990). Observation of high-intensity sub-Poissonian light using an optical parametric oscillator. Phys.Rev. Lett. Vol. 64, No. 24, pp.2897-2900, ISSN:1079-7114

Ou, Z.; Pereira, S. F.; Kimble, H. J. & Peng, K. C. (1992). Realization of the Einstein-Podolsky-Rosen paradox for continuous variables.Phys.Rev. Lett. Vol.68,No. 25,pp.3663-3666, ISSN:1079-7114.

Richardson, W.; Machida, S. & Yamamoto, Y. (1991). Squeezing photon-number noise and sub-Poissonian electrical partition noise in a semiconductor laser. Phys.Rev. Lett.Vol. 66, No. 22, pp.2867-2870, ISSN:1079-7114.

Sanders, S.; Park, N.; Dawson, J. W. & Vahala, K. J. (1992). Reduction of the intensity noise from an erbium-doped fiber laser to the standard quantum limit by intracavity spectral filtering. Appl. Phys. Lett. Vol. 61, pp. 1889-1891, ISSN: 0003-6951.

Spiegelberg, C.; Geng, J. H. & Hu, Y. D. (2004). Low-Noise-Narrow-linewidth Fiber Laser at 1550nm. Journal of Lightwave Technology. Vol. 22, No. 1, pp.57, ISSN: 0733-8724

Tapster, P. et al. 1988. Use of parametric down-conversion to generate sub-poissonian light. Phys.Rev. A. Vol. 37, No. 8, pp.2963-2967, ISSN:1050-2947.

Teich, M. & Saleh, B. 1985. Observation of sub-Poisson Franck-Hertz light at 253.7nm. J.Opt.Soc.Am.B. Vol. 2, No. 2, pp.275- 282, ISSN:1520-8540.

Yamamoto, Y. & Haus, H. 1986. Peaparation, measurement and information capacity of optical quantum states. Rev. Mod. Phys. Vol. 58, No. 4, pp. 1001-1020, ISSN:0034-6861.

Zhang, Y.; Kasai, K. & Watanable, M. (2002). Investigation of the photon-number statistics of twin beams by direct detection. Opt. Lett. Vol. 27, No. 14, pp. 1244-1246, ISSN:0146-9592.

Zou, H.; Zhai, S.; Guo, J.; Yang, R. & Gao, J. R. (2006). Preparation and measurement of tunable high-power sub-Poissonian light using twin beams.Opt. Lett. Vol. 31, No. 11, pp. 1735-1737, ISSN:0146-9592.

5

Transfer Over of Nonequilibrium Radiation in Flames and High-Temperature Mediums

Nikolay Moskalenko, Almaz Zaripov, Nikolay Loktev, Sergei Parzhin and Rustam Zagidullin
Kazan State University of Power
Russia

1. Introduction

Throughout the XX-th century intensive development was received by the high technologies intended for maintenance of stable rates of economic development and global competitive capacity in key industries of manufacture. The contribution of scientific and technical progress in economic growth becomes solving. Now in the developed countries development of high technologies has passed to a stage of the scientific and technical policy directed on introduction of high technologies in sphere of information services, medicine, ecology, power, military-technical manufacture, control of safety of economic activities in any branches of manufacture. Thus the power remains live-providing, a key economic branch in economy of any country and its development should be carried out by advancing rates. On the other hand, the power is a branch in which new scientific and technical achievements take root with high degree of efficiency owing to high level of automation of manufacture and energy transportation.

In the present chapter of the monography basic aspects of a problem of the transfer over of radiation in high-temperature mediums and flames and their decision with reference to problems of remote diagnostics of products of combustion in atmospheric emissions and top internal devices are considered. The special attention is given the account of nonequilibrium processes of radiation which are caused by chemical reactions at burning fuels and photochemical reactions in atmosphere. Radiation of high-temperature mediums is selective in this connection the problem of numerical modeling of spectraradiometer transfer function of atmosphere for non-uniform selective sources of radiation which are flame, combustion products of fuel, torches and traces of aerocarriers, combustion products in top internal chambers is considered. Absence of sharp selection of a disperse phase creates possibility of division of radiation of disperse and gas phases and in the presence of the aprioristic information creates conditions of their remote diagnostics (Moskalenko et al., 2010). The developed measuring complexes (Moskalenko et al., 1980a, 1992b) have allowed to specify substantially the information received earlier under radiating characteristics of products of combustion (Ludwig et al. 1973) and to investigate nonequilibrium processes of radiation in strictly controllable conditions of burning (Kondratyev et al., 2006, Moskalenko et al., 2007a, 2009b, 2010c). The developed two-parametrical method of equivalent mass for functions spectral transmission gas components of atmosphere (Kondratyev, Moskalenko, 1977) has successfully been applied in calculations of radiating heat exchange in high-temperature mediums (Moskalenko, Filimonov, 2001; Moskalenko et al., 2008a, 2009b). The method of numerical modeling of functions spectral transmission on parameters of spectral lines has been used by us for calculations of the transfer over of radiation of torches and traces of aerocarriers in atmosphere and at the decision of return problems of diagnostics of products of combustion by optical methods (Moskalenko & Loktev, 2008, 2009; Moskalenko et al., 2006). Experimental researches of speed radiating cooling a flame are executed by means of calculation of structure of products of combustion (Alemasov et al., 1972) and modeling of radiating heat exchange in chambers of combustion of measuring complexes with control of temperature of a flame by optical methods (Moskalenko & Zaripov, 2008; Moskalenko & Loktev, 2009; Moskalenko et al., 2010).

Measurements of concentration of oxides of nitrogen in flames have shown that their valid concentration much lower in comparison with the data of calculations (Zel'dovich et al., 1947). There was a necessity of finding-out of the reasons causing considerable divergences of theoretical calculations and results of measurements of concentration NO in flames. The reason strong radiating cooling of flames which didn't speak only equilibrium process of their radiation demanded finding-out.

Processes of burning gaseous, liquid and firm fuel have great value in power, and also in technological processes of various industries. At present a principal view of burned fuel in the European territory is gaseous fuel. Partially it is caused by ecological norms and requirements to combustion products. Use of gaseous fuel conducts to reduction of capital expenses at building of thermal stations and boiler installations owing to an exception of expensive filters of clearing of the list of the equipment of station. High heat-creation ability of gas fuel at low operational expenses provides high efficiency of power installations as a whole. A low cost of transportation at use of gas fuel provides its competitiveness in the market. Decrease in losses of heat at its transportation demands creation of small-sized boilers with high efficiency, high thermal stress of top internal space at the raised efficiency that leads to search of optimum design decisions by working out of power installations. Development of rocket technics, creation of space vehicles of tracking their start and support, optimization of systems of detection and supervision demands the data about structural characteristics of torches both spectral and spatial distribution of their radiation which can be received by correct methods of the decision of problems of a transfer over of radiation and radiating heat exchange in the torch. All it has demanded performance of complex researches of processes of radiation at burning and its the transfer over to medium which are discussed more low.

2. Radiating characteristics gas optically active components

Experimental researches radiating optically active components in a range of temperatures $220 \geq T \geq 800K$ have been begun in 1964 for the purpose of reception of the initial data for modeling of radiating heat exchange and spectral and spatial structure of radiation natural backgrounds of the Earth and atmosphere and anthropogenous influences on climate change (Kondratyev & Moskalenko, 1977; Kondratyev et al., 1983; Kondratyev & Moskalenko, 1984). The developed measuring complexes allowed to measure spectra of molecular absorption at pressure from 10^{-3} atm. to 150 atm. That has allowed to parameterized functions of spectral transmission of atmospheric components in a spectral range 0,2÷40 μm at the average spectral permission Δv =2-10 cm^{-1}, for atmospheres of the

Earth and other planets. Other direction of researches of radiating characteristics of products of combustion fuels developed in parallel with the first and for the known reasons is poorly reflected in publications. Further we will stop on the analysis of results of researches of radiating characteristics of ingredients a gas phase of products of combustion in a range of temperatures 600÷2500K.

2.1 Measuring devices and results of experimental researches

For the decision of many applied problems connected with the transfer over of radiation of a flame in atmosphere and radiating heat exchange in power installations, data on spectral radiating ability of the various gas components which are products of combustion of flame are required. Independent interest is represented by researches of influence of temperature on formation of infra-red and ultra-violet spectra of absorption or radiation of gas components. Depending on a sort of research problems of spectra of absorption or radiation of gas mediums of measurement it is necessary to carry out or with the average permission Δ =5-20 cm^{-1}, or with the high permission $\Delta \leq$ 0,2 cm^{-1}. In the latter case it is possible to measure parameters of spectral lines and to receive the important information on the molecular constants characterizing vibrational - rotary and electronic spectra of molecules (Moskalenko et el., 1972, 1992). In a range of temperatures 295÷1300 K research of characteristics of molecular absorption it was carried out with use the warmed-up multiple-pass ditches (Moskalenko et el., 1972). Other installation (Moskalenko et el., 1980) allowed to investigate as spectra of absorption and radiation of gases in hydrogen-oxygen, hydrogen-air, the propane-butane-oxygen, the propane-butane-air, methane-oxygen, methane-air, acetylene-oxygen, acetylene-air flames in the field of a spectrum 0,2÷25 μm at temperatures 600÷2500 K, and also to investigate characteristics of absorption of selective radiation of a flame modeled atmosphere of the set chemical composition. Besides, any other component can be entered into a flame, of interest for research.

The Block diagram of experimental installation and design of a high-temperature gas radiator is described (Moskalenko et el., 1972). It includes the lighter, high-temperature absorbing (radiating) to a ditch, system of input of investigated gas and control of their expense, optical system of repeated passage of radiation in a ditch under White's scheme, the block of the gas torches forming two counter streams of a flame in quartz ditch with the heat exchanger for decrease radiating cooling of a flame, coordinating optical prefixes for radiation designing on an entrance crack of spectrometers of reception-registering system with replaceable receivers of radiation PEA - 39A, PEA - 62, BSG - 2, cooled photodetectors with sensitive elements PbS, PbTe, GeCu, GeZn, GeAu, GeAg, germanium bolometer. The spectrum of radiating ability of the high-temperature gas medium is defined by tariroving of a spectrometer on radiation of absolutely black body or normalizing radiation sources. Radiation falling on a reception platform is modulated by the electromechanical modulator with frequency of 11 or 400 Hz (in case of work with PEA and photodetectors). Registration of spectra of radiation was made by spectrometer IRS - 21 or the spectrometers of the high permission collected on the basis of monochromators MDR - 2, DPS - 24, SDL - 1. The last are completed with replaceable diffraction lattices with number of strokes 1200, 600, 300, 150, 75 and the cutting off interferential optical filters providing a working spectral range 0,2 $<\lambda<$25 μm. The limit of the spectral permission of spectrometers made 0,1÷0,2 cm^{-1}. Spectral radiating ability of the gas medium

$$\varepsilon(\nu,T)=G(\nu)\cdot\frac{N^0(\nu,T_\beta)}{B(\nu)\cdot N^0(\nu,T)}, \tag{1}$$

where T – temperature of the investigated gas medium; G (v), B (v) – recorder indications at registration of radiation from the gas medium (flame) and absolutely black body (ABB); N^0(v, T_v) and N^0(v, T) – spectral brightness ABB at temperatures T_v ABB and T the investigated gas medium.

At work in a mode of absorption of not selective radiation by a flame the radiation modulated by the electromechanical modulator from the lighter is registered. Not modulated radiation of the flame by reception system isn't registered. In the lighter as radiation sources SI lamps - 6 - 100, DVS - 25, globar and ABB with temperature 2500K are used. Radiation from these sources, promodulated by the electromechanical modulator, by means of optical system of the lighter goes in high-temperature absorbing gas to a cell which optical part is collected under White's scheme. The thickness of the absorbing component can change by increase in an optical way at the expense of repeated passage of a beam of radiation between mirrors of system of White. The maximum thickness of the absorbing medium can reach 16 m.

Absorbing (radiating) a cell represents the device executed in the form of established in heat exchanger along an optical axis of the cell two mobile pipes, made of quartz. On a circle of entrance cavities from end faces quartz ditches are located two systems of gas torches (on 6 pieces in everyone) for reception of the hot absorbing (radiating) medium. The internal cavity is filled with two counter streams of a flame. Combustion products leave through a backlash between mobile quartz pipes, the heat exchanger and two unions, located at its opposite ends. Investigated gases can be both combustion products, and other gases entered in a cell and warm flame. For flame creation two various systems of torches are used.

At work about hydrogen-oxygen (hydrogen-air) a flame are used torches of Britske, each of which allows to receive a flame of diffusion type. We will remind that under diffusion flame such flame for which fuel and an oxidizer are originally divided is understood. Fuel and an oxidizer mix up or by only diffusion, or partially by diffusion and partially as a result of turbulent diffusion. For reception the propane-butane-oxygen, the propane-butane-air flame hot-water bottles have been designed and made, each of which allows receiving a flame of Bunsen's type. The flame of Bunsen's type is understood as a flame of preliminary mixed oxidizer and fuel.

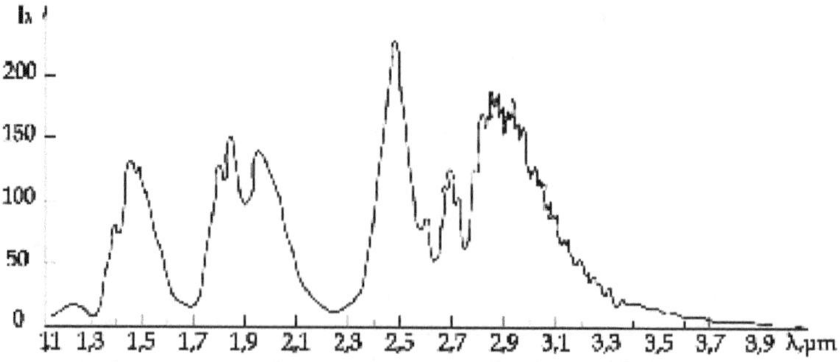

Fig. 1. Radiative spectrum of the hydrogen – oxygen flame at temperature T≈2300K in the range 1,1-4 μm.

Each torch has an adjustable angle of slope of an axis of a torch to an axis of the cell quartz in limits from 20 to 70°. Combustible gases are set fire by a spark. Change of temperature of a flame is reached by change stehiometrical parities of combustible gas and an oxidizer, and also change of combustible gas and oxidizer diluting by buffer gas. Temperature measurement is carried out W – Re and Pt – Po by thermocouples and optical methods.

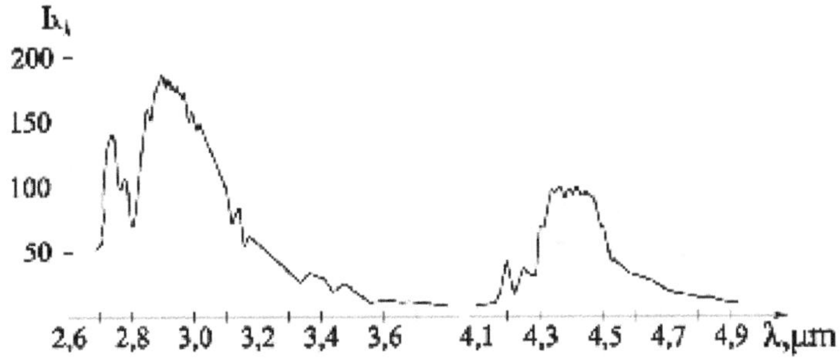

Fig. 2. Radiative spectrum of the hydrogen – oxygen flame in range 2,7 - 5 µm with addition CO_2 in quality of the research gas.

On fig. 1, 2 examples of records of spectra the radiations which have been written down by means of spectrometer IRS–21 are resulted at the average spectral permission at temperature T ≈ 2300K. For oxygen-hydrogen flame radiation bands only water vapor in a vicinity of bands 0,87; 1,1; 1,37; 1,87; 2,7 and 6,3 µm are observed. In ultra-violet spectrum areas are observed electronic spectra of radiation of a hydroxyl OH. With temperature growth considerable expansion of bands and displacement of their centers in red area is observed. At temperatures more 2000K in a flame absence of "windows" of a transparency of a flame, spectral intervals with radiating ability close to zero is observed.

At addition in a flame of gases from a number limit hydrocarbons (methane, ethane, etc.) In radiation spectra bands of carbonic gas (2; 2,7; 4,3; 15 µm) are observed. The similar picture is observed at introduction in a flame and purely carbonic gas. At introduction in flame NO the spectrum of the basic band 5,3 µm NO and a continuous spectrum of radiation NO_2 in a range from 0,3 to 0,8 µm is observed. Data processing of measurements of spectra of radiation of a flame and restoration of a profile of temperature along an axis of an ardent radiator has shown appreciable temperature heterogeneity in zones of an input of a flame in the combustion chamber (Moskalenko & Loktev, 2009) which is necessary for considering at definition of dependence of radiating characteristics of separate components from temperature. This lack has been eliminated in working out of a measuring complex of the high spectral permission (Moskalenko et el., 1992) for research of flames. On working breadboard models of this installation and the experimental sample of this installation the most part of the spectral measurements taken as a principle of parameterization of radiating characteristics of gas components of products of combustion has been executed.

The spectral measuring complex described more low also is intended for registration of spectra of radiation of flames and spectra of absorption of radiation by a flame at the high spectral permission in controllable conditions and has full metrological maintenance. On fig. 3 the block-scheme of this installation is presented. An installation basis make: the block of a high-temperature gas radiator, blocks of optical prefixes 2D-4, intended for increase in an optical way in an ardent radiator and the coordination of fields of vision of the lighter; the block of a high-temperature radiator of sources of radiation 3 for absolute calibration of a spectrum of radiation of a flame and the Fourier spectrometer of high spectral permission FS - 01. Management of experiment and data processing of measurements by means of software on the basis of measurement-calculation complex IVK - 3. The measuring complex functions in spectral area 0,2–100 μm. Registration of spectra is carried out by means of spectrometers FS - 01, SDL - 1.

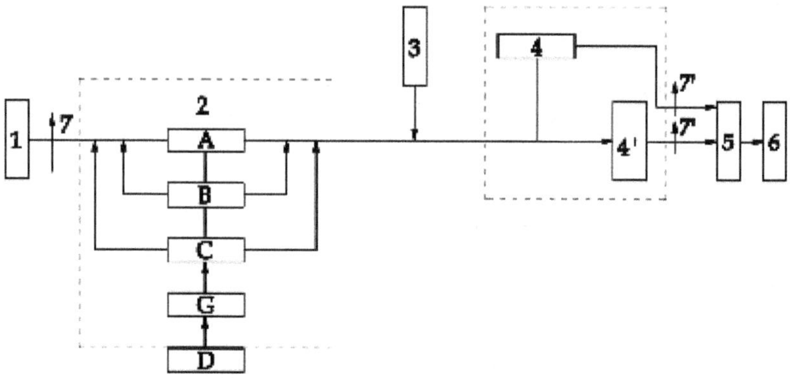

Fig. 3. The experimental installation scheme: 1 – illuminator, 2 – hightemperature gaseous radiator (A – lead – in of research gas system and contrac there expense, B – the mechanism of multiple passing ray thaw a flame, C – the gaseous burner of ascending flow of a flame, G – the gaseous provision system vacuum and control of gaseous expense, D – the system with a water circular pump); 3 – aradiative sources; 4, 4′ – optical system for agreement of in trance and exit apertures; 5 – the reception – recording system; 6 – the system of atreatment of measuring data; 7, 7′ – electrical mechanical modulators of radiation.

The high-temperature ardent radiator structurally represents the block of a gas radiator closed from above by the water cooled cap with two protective windows, stable in time. Formed at burning of gases flames have a squared shape with a size at the basis 40x20 cm^2. The torch design allows to investigate hydrogen - oxygen, hydrogen - air and hydrocarbonic flames. Measurements have shown that heterogeneity of a temperature field within a field of vision of optical system makes 3 %. Various variants of optical schemes together with system of repeated passage of radiation constructed under White's scheme, allows to investigate radiation spectra of flames and spectra of absorption of continuous radiation of a flame in a range of lengths of an optical way 0,2÷16 m. The flame temperature is measured by a method of the self-reference of spectral lines in lines of water vapor of bands 1, 38 and 1,87 μm. The average relative error of measurement of temperature of a flame makes ±2 %. Measurement of volume expenses of gases was carried out specially graduated rotameters RS - 5. On a parity of mass fuel consumption and an oxidizer the chemical composition of products of combustion are determined by thermodynamic calculation (Alemasov et al.., 1972). To absolute calibration of spectra of radiation of a flame are applied spectrameasured lamps SIRSh 8,5-200-1 and globar KIM, preliminary graduated on metrology provided standards.

Measurement of spectra of radiation and spectra of absorption of radiation by a flame allow to define spectral factors of nonequilibrium functions of a source of radiation in flames. Such measurements have revealed considerable nonequilibrium source functions in an ultra-violet part of a spectrum of a flame (the factor of nonequilibrium reaches values 20 – 100). At the same time vibrational-rotary spectra of radiation of water vapor in flames remain equilibrium. Nonequilibrium radiations OH in flames is strongly shown in an ultra-violet part of a spectrum and considerably influences radiative transfer over in flames and in vibrational-rotary bands v_1, $2v_1$, $3v_1$, where v_1 - frequency of normal fluctuation OH. The error of measurements of function of a source makes 30 % for an ultra-violet part of a spectrum and 7-10 % in infra-red bands of radiation of a flame. It is found out also nonequilibrium radiations in electronic bands of oxides of nitrogen.

At measurement in a mode of absorption of radiation the flame modulates radiation of the lighter 1. Nonmodulated radiation of a flame doesn't give constant illumination and isn't registered by receiving-registering system. Modulation of radiation of a flame is created by the modulator 7 ′. Registration of spectra of radiation of flames in vibrational-rotary bands is carried out by Fourier spectrometer FS - 01 which reception module is finished for the purpose of use of more sensitive cooled receivers of radiation. The major advantage of the Fourier spectrometer in comparison with other spectrometers - digital registration of spectra with application of repeated scanning of spectra and a method of accumulation for increase in the relation a signal/noise. Prominent feature of Fourier spectrometer is discrete representation of the measured spectrum of radiation of a flame with the step equal to the spectral permission. The last has demanded working out of the software for processing of the measured spectra, restoration of true monochromatic spectral factors of absorption and parameters of spectral lines of absorption (radiation), their semiwidth and intensitys. With that end in view measured spectra are exposed to smaller splitting with step $\delta = \Delta/5$, where Δ - the spectral permission of the Fourier spectrometer. Value in splitting points is defined by interpolation.

Reduction of casual noise is reached by smoothing procedure on five or to seven points to splines in the form of a polynom of 5th degree. The spectrum of radiation received in a digital form is exposed to decomposition on individual components of lines.

From the restored contours of spectral lines it is easy to receive intensity and semiwidth of lines. Thus intensity such Lawrence's lines

$$S_m = \int_{-\infty}^{\infty} K_m(\nu) d\nu = K_{\nu m} \pi \alpha_m, \tag{2}$$

where K_m - absorption factor in the center of a contour of a line, α_m - its semiwidth, $K_{\nu m}$ - the restored contour of a spectral line. Thus the condition should be met

$$1 - \int_{-\infty}^{\infty} d\nu \exp\left[-k_m(\nu) w\right] = \int_{-\infty}^{\infty} A_{Im}(\nu) d\nu, \tag{3}$$

where w - the substance maintenance on an optical way, A_{Im} - the measured function of spectral absorption of such line. Parameters of spectral lines of water vapor can be used for temperature control in a flame (Moskalenko & Loktev, 2008, 2009).

On fig. 4 the example of the measured spectrum of the high spectral permission of radiation of a flame for spectral area 3020÷3040 cm^{-1} is resulted. On fig. 5, 6 spectra of radiating ability of a flame in vibrational-rotary bands of water vapor are illustrated at the average spectral permission $\Delta\nu$.

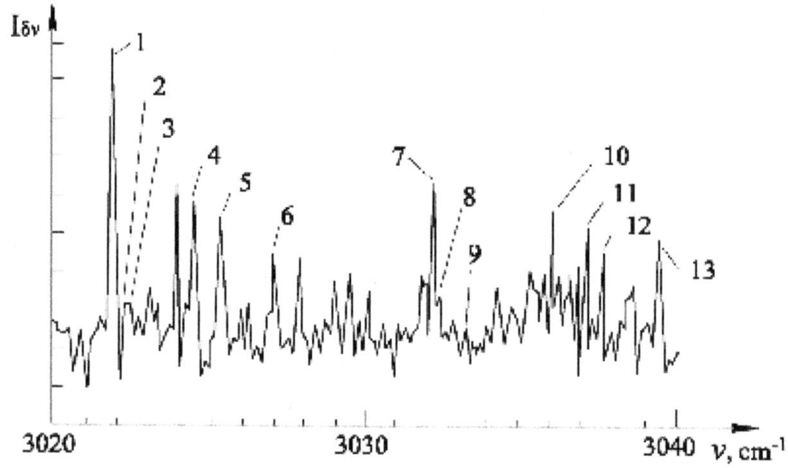

Fig. 4. The record of a high resolution radiative spectrum of the hydrogen – oxygen flame in the range 3020-3040 cm^{-1}. Centers of spectra lines: 1 – 3021,806 (ν_1), 2 – 3022,365 ($2\nu_3$), 3 - 3022,665 ($2\nu_2$), 4 - 3024,369 (ν_1), 5 – 3025,419 ($3\nu_2$ - ν_2), 6 - 3027,0146 (ν_1), 7 - 3032,141 (ν_3), 8 - 3032,498 ($3\nu_2$ - ν_2), 9 - 3033,538 ($3\nu_2$ - ν_2), 10 - 3036,069 ($3\nu_2$ - ν_2), 11 - 3037,099 ($3\nu_2$ - ν_2), 12 - 3037,580 ($3\nu_2$ - ν_2), 13 - 3039,396 (ν_1) cm^{-1}.

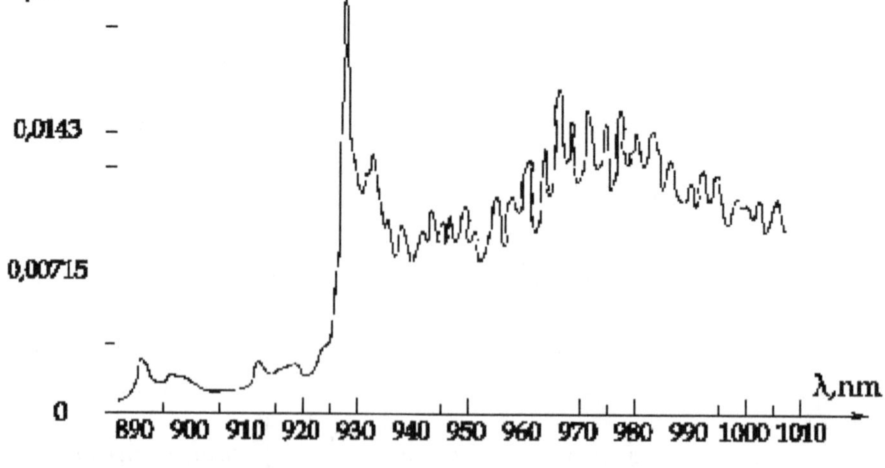

Fig. 5. Spectral emissivity of water vapor at T = 2400K in the band 0,96 μm. ω_{H2O} = 1,59 atm·cm STP, spectral resolution Δv = 10,6 cm^{-1}.

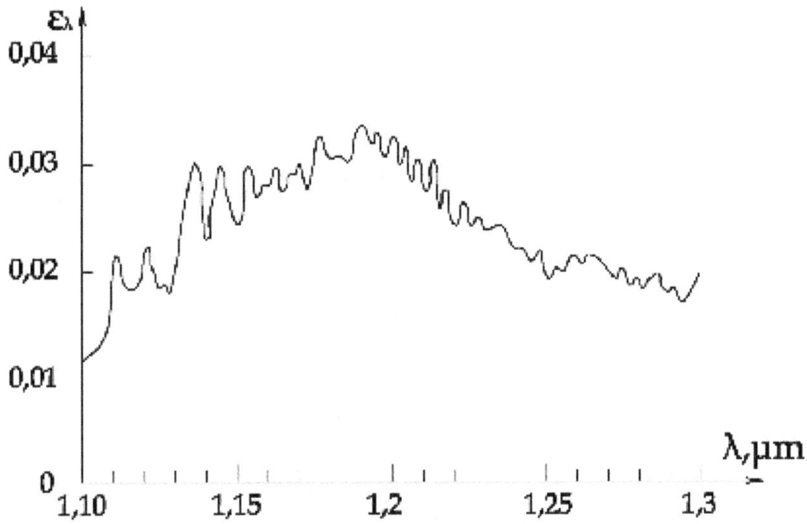

Fig. 6. Spectral emissivity of water vapor in the band 1,14 μm. T = 2400K, ω_{H2O} = 1,59 atm·cm STP, spectral resolution Δv= 15,5 cm^{-1}.

The spectra of radiation of the high spectral permission received in a digital form aren't calibrated on absolute size. Transition from values of relative spectral brightness to absolute radiating ability is carried out on parity

$$A_{\delta v} = \frac{I_{\delta v}(1-\bar{\tau})}{\frac{1}{\Delta v}\int I_{\delta v} dv}, \tag{4}$$

where $\bar{\tau}$ - average value of function spectral transmission for the processed site of a spectrum Δv. Data on $\bar{\tau}$ have been received by us earlier for various products of combustion of flames. Further difficult function $A_{\delta v}$ it is decomposed to separate components, using a method of the differentiated moments, according to which

$$A(v) = \sum_{m=1}^{M} A_m \left[\sum_{n=0}^{N} A_{mn} \left(v - v_m^0\right)^n \right]^{-1}, \tag{5}$$

where A_m - a maximum of intensity of such line, A_{mn} - factors of the generalized contour.

$$q_m = \frac{1}{\sum_n A_{mn}\left(v - v_m^0\right)^n}, \tag{6}$$

Characteristics A_m give the full information on separate contours and are defined as decomposition factors abreast Taylor of some function f_m (v), describing such contour:

$$f_m(v) = \frac{1}{m} \sum_{n=0}^{N} A_{mn}\left(v - v_m^0\right)^n . \tag{7}$$

Value A_m is a maximum of amplitude of a contour. The center ν_m^0 is defined from a condition of equality to zero of factor A_{m1}. Value of semiwidth of a line turns out from a parity

$$\alpha_m = \sqrt{\frac{\sqrt{A_{m2}^2 - 4A_{m4}} - A_{m2}}{2A_{m4}}}, \tag{8}$$

Further the profiles received thus are restored on influence of hardware function of a spectrometer. So, we have separate contours of function of absorption $A_m(\nu)$ from which it is easy to pass to contours of factors of absorption $K_m(\nu)$:

$$K_m(\nu) = \sum_{m=1}^{M} K_m(\nu - \nu_m^0), \tag{9}$$

where M - number of lines in a spectrum, m - line number. On fig. 7 the example of decomposition of function $A_{\delta\nu}$ on individual contours for oxygen-hydrogen of a flame for a spectrum site 3064÷3072 cm^{-1}, and also comparison (a curve 2) and calculated (a curve 3) on the restored contours of spectral lines of function $A_{\delta\nu}$ is presented. Integrated intensity of lines were defined from a parity (2). Detailed processing of spectra of radiation of water vapor in flames which has revealed many lines which were not measured earlier has been executed.

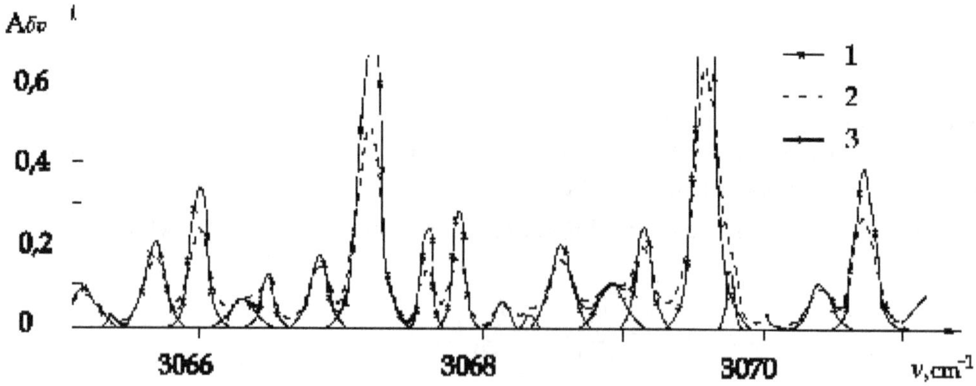

Fig. 7. The expansion of measuring function $A_{\delta\nu}$ on individual contours. 1 – separate components of expansion, 2.3 – function $A_{\delta\nu}$ measuring and calculative by reconstituting parameters of spectral lines accordingly.

In table 1 as an example parameters of spectral lines of water vapor are resulted at temperature T = 2100K for spectral ranges 3271÷3274 and 3127÷3130 cm^{-1}. Recalculation of parameters of lines on other temperatures can be executed under the formula

$$S(T) = S(T_0)\left(\frac{T_0}{T}\right)^{1.5} \frac{Q(T_0)}{Q(T)} \exp\left[-1.439E'\left(\frac{1}{T} - \frac{T}{T_0}\right)\right]. \tag{10}$$

Statistical sum $Q(T)$ in the ratio (10) is calculated in harmonious approach. That circumstance pays attention that the centers of spectral lines measured at temperatures of spectral lines T = 2100 K and temperature T_0 = 1000 K don't coincide that is possible, is caused by the displacement of spectral lines caused by pressure, and also temperature displacement of lines. These distinctions in position of the centers of spectral lines surpass often an error of measurements of the centers of lines which in our experiments makes ±0,02 cm^{-1}. The measured semiwidth of spectral lines of water vapor basically will be coordinated with results of calculations under the theory of the Anderson, executed by us at temperatures 300÷3000 K.

ν, cm^{-1}	S, $atm^{-1}cm^{-1}$	α, cm^{-1}	ν, cm^{-1}	S, $atm^{-1}cm^{-1}$	α, cm^{-1}
3271,731	0,0131	0,075	3127,8714	0,0123	0,129
3271,944	0,00642	0,084	3128,115	0,0015	0,075
3272,101	0,01272	0,080	3128,395	0,0042	0,081
3272,395	0,00408	0,066	3128,600	0,00216	0,076
3272,654	0,00876	0,168	3128,806	0,00277	0,083
3272,811	0,0114	0,080	3129,109	0,00498	0,092
3273,041	0,0236	0,111	3129,273	0,00387	0,091
3273,436	0,033	0,099	3129,589	0,0154	0,105
3273,735	0,0261	0,092	3129,941	0,0130	0,104

Table 1. Parameters of lines of water vapor at T = 2100 K in the hydrogen-oxygen flame for sites of a spectrum 3271 - 3274 and 3127 - 3130 cm^{-1}. STP.

2.2 Device for modeling of the transfer over of selective radiation in structurally non-uniform mediums

The problem of a transfer over of selective radiation of torches and streams of aerocarriers is put in the sixtieth year of XX th century. The transfer over of selective radiation is influenced by following factors: the temperature self-reference of spectral lines of radiation, displacement of spectral lines with pressure, displacement of spectral lines as a result of high speed of aerocarriers (Dopler's effect), the temperature displacement of the spectral lines which have been found out for easy molecules (vapor H_2O, CH_4, NH_3, OH) (Moskalenko et el., 1992). The executed calculations have shown that displacement of spectral lines with pressure in a flame, making thousand shares of cm^{-1}, and doplers displacement of spectral lines in conditions turbulized high-temperature mediums can't render appreciable influence on function spectral transmission. Temperature displacement of spectral lines in a flame make the 100-th shares of cm^{-1} and at high temperatures reach semiwidth of spectral lines and more. It leads to that radiation of a high-temperature kernel of a torch is to a lesser degree weakened by its peripheral layers that strengthens radiating cooling torch kernels. At registration of radiation of a torch of the aerocarrier the effect of an enlightenment of atmosphere is observed more considerably in comparison with the account only the temperature self-reference of spectral lines. If for the temperature self-reference the spectral effect of an enlightenment is observed more intensively for optically thick mediums the effect of an enlightenment of atmosphere at the expense of temperature displacement of spectral lines is shown and for optically thin selective radiators and observed by us earlier at registration of radiation of system «a selective radiator - atmosphere» with the high spectral permission in bands of water vapor.

Earlier the problem of the transfer over of selective radiation was put in interests of the decision of problems of the transfer over of radiation of torches and streams of aerocarriers in atmosphere of the Earth (Moskalenko et el., 1984). It has been found out by numerical modeling that law of the transfer over of radiation in low-temperature and high-temperature mediums considerably differ. Burning and movement of products of combustion in a stream is accompanied by wave processes at which there is high-frequency making (turbulence) and low-frequency (whirls). Thus low-frequency wave processes can make the greatest impact on the transfer over of selective thermal radiation while influence of turbulence on the transfer over of selective radiation can be neglected. At high pressures of the non-uniform medium the thin structure of a spectrum of gas components is greased also with influence of sharp selectivity of spectra of radiation on radiating heat exchange it is possible to neglect.

At low pressure and high temperatures of medium effects of temperature displacement of spectral lines in structurally non-uniform mediums can render the greatest influence on the transfer over of selective radiation, not which account for easy molecules (H_2O, CH_4, NH_3, OH) in settlement schemes can essentially underestimate radiating cooling high-temperature zones of a torch (Moskalenko & Loktev, 2009). On the other hand, sharp selectivity of radiation of the gas medium promotes preservation of heterogeneity a temperature field at movement of products of combustion in a fire chamber owing to decrease in absorption of high-temperature zones of its torch by peripheral low-temperature layers.

Creation non-uniform on temperature of the gas medium in top internal space is promoted also by specificity of radiating heat exchange in top internal space, when speed radiating cooling peripheral zones optically a thick torch above, than in its central part. Even if the burning device forms front of products of combustion homogeneous for temperature in process of movement of gases in a plane, normal to a direction of movement of a stream, there is heterogeneity so heterogeneity of a field of temperature becomes three-dimensional.

Modeling of structurally non-uniform gas mediums is carried out by means of the optics-mechanical device in which the amplitude modeled heterogeneities can varies in a range of temperatures 400÷2500 K. On fig.8 the structure of an optics-mechanical part in section and the top view is shown. Installation contains the lighter with a source of modulated radiation, mirror optical system of repeated passage of a bunch of radiation under White's scheme between which mirrors the block of gas torches mounted on a rack with possibility of change of position and an inclination of a cut of a plane of capillaries of torches concerning a plane of the main sections of mirrors of optical system.

The block of gas torches includes the radiator basis in which branch pipes with capillaries accordingly for combustible and oxidizing gases are built in serially. Cooling of branch pipes with capillaries is carried out by means of radiators of water cooling. Behind a target mirror of optical system are consistently established the mechanical modulator - the breaker of radiation and a spectrometer. The gas torch having possibility of moving on height and a turn in horizontal and vertical planes, together with an optical part make an ardent multiple-pass cell which from above is covered with a metal cap cooled by water.

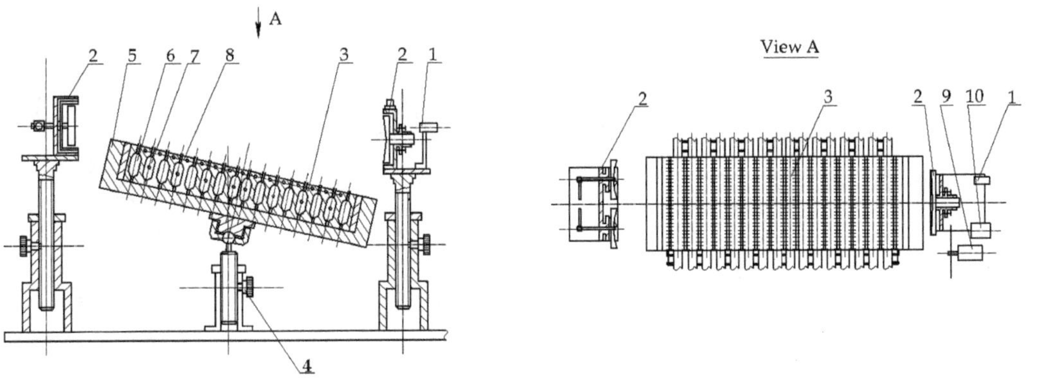

Fig. 8. An optics-mechanical part of a non-uniform radiator: 1 – the lighter; 2 - mirror optical system under White's scheme; 3 - the block of gas torches; 4 – a rack; 5 - the radiator basis; 6 and 7 - capillaries accordingly for combustible and oxidizing gases; 8 – radiators; 9 - the mechanical modulator – the radiation breaker; 10 – a spectrometer.

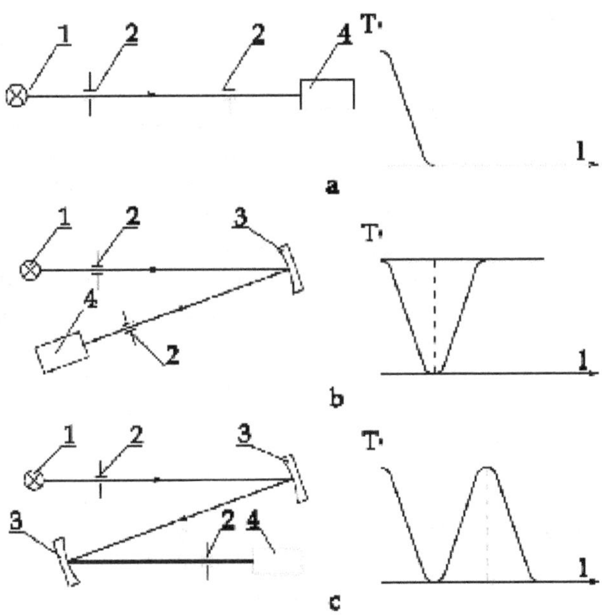

Fig. 9. Formation of profiles of temperature for cases unitary (a), double (b) and triple (c) passages of a beam of radiation through a flame stream. 1 – the lighter; 2 – entrance and target cracks; 3 – spherical mirrors; 4 – the radiation receiver.

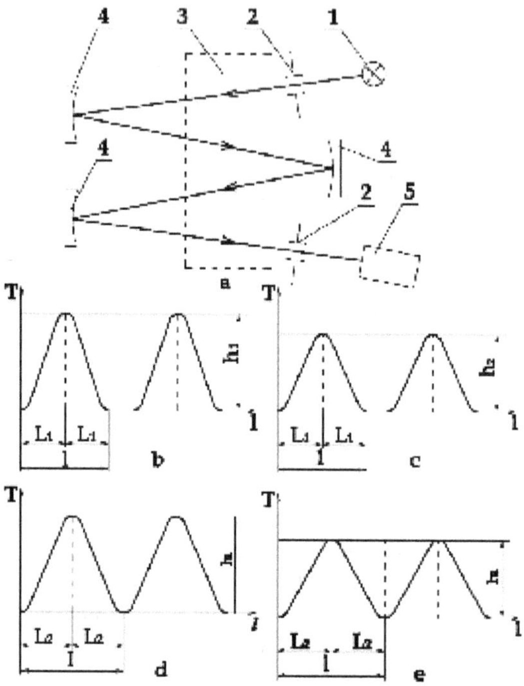

Fig. 10. Formation of profiles of temperature for cases of quadruple passage of radiation through a flame and temperature profiles *T* corresponding to them on an optical way of radiation *l*: 1 - a radiation source; 2 - entrance and target cracks; 3 - a flame zone; 4 - mirrors; 5 - a spectrometer.

For preservation of vertical development of a flame at inclined position burning devices on capillaries it is desirable to establish nozzles with a turn corner (π - α), where α - a corner of a plane of a cut of capillaries burning devices concerning a horizontal plane. On fig. 9 the kind of temperature profiles for various cases of passage of radiation along an optical way through ardent multiple-pass cell is shown. When the plane of a cut of capillaries of a torch is parallel to a plane of the main sections of mirrors of optical system, the bunch of radiation of the lighter passes through the gas medium homogeneous for temperature. Changing height of position of a gas torch, in this case probably to define distribution of temperature depending on height over a plane of cuts of capillaries. Further this information can be used for definition of a profile of temperature non-uniform on temperature of the gas mediums modeled in installation «a non-uniform gas radiator». Optical schemes are presented in the left part of drawing, and temperature profiles *T* on an optical way *l* - in the right part of drawing.

Mirror reflection of these profiles (return temperature profiles) can be received by return turn of a plane of a gas torch concerning a horizontal plane. On fig. 10 the explanatory to formation of profiles of temperature for cases of quadruple passage of radiation through a flame and temperature profiles *T* corresponding to them on an optical way of radiation *l* is presented. Radiation from a radiation source through an entrance crack passes a flame zone, is reflected consistently by mirrors after quadruple passage through a flame zone projected on a target crack of receiving-registering system of a spectrometer. Depending on constructive length *L* zones of a flame along an optical way and height *h* arrangements gas

burning devices over the basis change (to look fig. 10) amplitude of temperature heterogeneity and its half-cycle Δl. On fig. 10 cases are presented, when $h_1 \neq h_2$ and $L_1 \neq L_2$.
For homogeneous system the law of Kirhgof is carried out. In non-uniform medium on structure it is broken, and function spectral transmission in a spectral interval of final width becomes dependent as from thin structure of a spectrum of the radiating volume, and from thin structure of a spectrum of the absorbing medium. Effects of display of sharp selection of spectra of radiating and absorbing mediums on function spectral transmission lead to certain features of radiating heat exchange in a torch and transfer function of distribution of radiation of a torch in medium. So radiation of a kernel of a torch is to a lesser degree weakened by its peripheral layers. In chambers of combustion it leads to increase heat-receptivity by surfaces of heating at the expense of radiating heat exchange, and at distribution of radiation of a torch of the aerocarrier to atmosphere the effect of an enlightenment of atmosphere when atmosphere becomes more transparent for non-uniform high-temperature selective radiators, in comparison with not selective radiators is observed.
Consideration of process of the transfer over of selective radiation in atmosphere allows constructing the following scheme of its account through the factors of selectivity defining the relation of function spectral transmission for selective radiation τ_c to function spectral transmission for not selective radiation. If to enter factor of selectivity for a component i:

$$\eta_{i\lambda c} = \frac{\tau_{i\lambda c}}{\tau_{i\lambda n}}, \tag{11}$$

Then full transmission of mediums for selective radiation $\tau_{\lambda c}$ it is represented in a kind:

$$\tau_{\lambda c} = \prod_i \eta_{i\lambda} \cdot \tau_{i\lambda n} \tag{12}$$

As functions $\tau_{i\lambda n}$ are studied, researches $\tau_{\lambda c}$ is reduced to reception of sizes $\eta_{i\lambda}$ as functions of temperature, an optical thickness of radiating and absorbing mediums, and also pressure which can be defined on the basis of experimental researches or the data of numerical modeling of the transfer over of radiation on thin structure of a spectrum of radiating and absorbing mediums (Moskalenko et el., 1984).

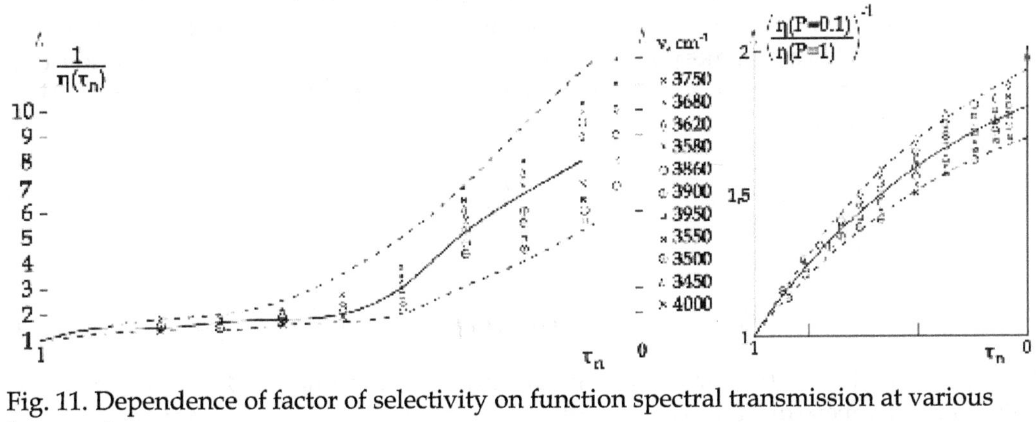

Fig. 11. Dependence of factor of selectivity on function spectral transmission at various frequencies.

The executed experimental researches and results of numerical modeling have shown that sizes $\eta_{i\lambda}$ depend on temperature. At low temperatures of selective radiators, for example streams of turbojets, sizes $\eta_{i\lambda} \leq 1$ and selective radiation is absorbed in atmosphere more intensively than not selective radiation. To calculations of function spectral transmission for not selective radiation it is applied one-parametrical and two-parametrical methods of calculation of the equivalent mass, discussed more low.

Dependence of transfer function on structure of absorbing and radiating mediums is important for considering in problems of remote diagnostics of products of combustion by optical methods and supervision over aerocarriers on their infra-red thermal radiation. The importance of the account of effect of selectivity of radiation on transfer function of atmosphere is illustrated on fig. 11a, on which dependences of spectral factors of selectivity η are presented as function from transmission τ_n for sources of not selective radiation for various sites of a spectrum with the centers v (v - wave number) for optically thin radiator of water vapor. The absorbing medium is atmospheric water vapor. A total pressure P in a selective source and in atmosphere is one atmosphere. The Fig. 11b shows strengthening of display of effect of selectivity with fall of total pressure P to 0,1 atmospheres.

2.3 Functions spectral transmission of vapors H_2O, CO_2 and small components of products of combustion

Let's consider the general empirical technique for calculation of radiating characteristics of a gas phase of products of combustion (Kondratyev & Moskalenko, 1977; Moskalenko et al., 2009), applicable for the decision of problems of radiating heat exchange and the radiation transfer over in torches of aerocarriers, in chambers of combustion of power and power technological units and in the power fire chambers functioning in the conditions of high pressures of a working medium. The developed technique is applicable for function evaluation spectral transmission (the basic radiating characteristic) multicomponent non-uniform on temperature and effective pressure of atmosphere of smoke gases of products of combustion in the chamber of combustion and gas-mains of boilers. A working range of effective pressure $0{,}01 \leq P_e \leq 100$ atm that provides its use at the decision of problems of radiating heat exchange both in modern boilers, and in perspective workings out of power and power technological units.

Generally at function evaluation spectral transmission $\tau_{\Delta v}$ where v - wave number, Δ - the spectral permission, is necessary to allocate contributions to the absorption caused by wings of remote spectral lines of atmospheric gases $\tau^k_{\Delta v}$, by the induced pressure absorption $\tau^n_{\Delta v}$, selective absorption $\tau^c_{\Delta v}$ by the spectral lines entering into the chosen spectral interval. Then for the set component:

$$\tau_{\Delta\nu} = \tau^k_{\Delta\nu} \times \tau^n_{\Delta\nu} \times \tau^c_{\Delta\nu} \,. \tag{13}$$

Function:

$$\tau^k_{\Delta\nu} \cdot \tau^n_{\Delta\nu} = \exp\left[-(\beta_{\nu k}(T) + \beta_{\nu n}(T))\omega P\right], \tag{14}$$

where $\beta_{vk}(T)$ and $\beta_{vn}(T)$ - factors continual and the absorption induced by pressure, depending on temperature T; ω - the component maintenance; P - partial pressure.

For reception of function spectral transmission $\tau^c_{\Delta v}$ it is offered to use the general parity:

$$\left(\frac{1}{\ln\tau_{\Delta\nu}}\right)^2=\left(\frac{1}{\ln\tau'^{C}_{\Delta\nu}}\right)^2+\left(\frac{1}{\ln\tau''^{C}_{\Delta\nu}}\right)^2+\frac{M}{(\ln\tau'^{C}_{\Delta\nu})(\ln\tau''^{C}_{\Delta\nu})},\tag{15}$$

where

$$\tau'^{C}_{\Delta\nu}=\exp[-k_{\nu}(T)\omega]\tag{16}$$

defines function spectral transmission in the conditions of weak absorption and at elevated pressures ($P\geq 10$ atm) in the conditions of the greased rotary structure of a spectrum of absorption,

$$\tau''^{C}_{\Delta\nu}=\exp[-\beta_{\nu c}(T)\omega^{m_{\nu}}P_e^{n_{\nu}}]\tag{17}$$

- function spectral transmission at small P_e <1 atm in the conditions of strong absorption. The M parameter characterizes change of growth rate of function transmission at transition from area of weak absorption in area of strong absorption. Parameters k_v, m_v, n_v, β_{vc}, are defined from the experimental data received by means of described above measuring complexes. In conformity with the theory of modeling representation of spectra of absorption $k_{\nu}=\overline{S}/d$ defines the relation of average intensity to distance between lines, and the size $k_{\nu}\Delta\nu$ - characterizes intensity of group of the spectral lines located in the chosen spectral interval Δv.

It has been shown that the parity (15) describes any modeling structure of a spectrum, including the law of Buger for a continual spectrum of strongly blocked spectral lines. Really, in this case m=1, n=0, $\beta_{vc}=k_v$, M = − 1. The overshoot of spectral lines is stronger, the it is more parameter m and the less parameter n and the closer parameter $|M|$ to unit. For real spectra parameter $M\in\{0,-1\}$. Continual absorption by wings of lines and the absorption induced by pressure is described by a following set of parameters: m=1, n=1, $k_v=\beta_v$, M = − 1.

Let's notice that spectra of the absorption induced by pressure submit to other rules of selection in comparison with vibrational-rotary spectra and the bands of absorption forbidden by rules of selection in vibrational-rotary spectra, become resolved in spectra of the absorption induced by pressure. In this connection the account of the absorption induced by pressure can become necessary in radiating heat exchange in power fire chambers. In power fire chambers the account and continual absorption by wings of strong lines and absorption bands is more important.

With temperature growth the density of spectral lines increases and, hence, parameters m_v, n_v, M_v change. In this connection at calculations $\tau^c_{\Delta v}$ in the conditions of non-uniform on temperature and pressure of medium average values of these parameters in a certain range of temperatures are used.

For calculation $\tau^c_{\Delta v}$ in the conditions of non-uniform on temperature and pressure of medium it is convenient to enter temperature functions:

$$F_{1c}(T)=\frac{K_{\nu}(T)}{K_{\nu}(T_0)},\ F_{2c}(T)=\frac{\beta_{\nu c}(T)}{\beta_{\nu c}(T_0)}.\tag{18}$$

Then

$$-\ln \tau^{c}_{\Delta\nu} = K_{\nu c}(T_0)W_1, \; -\ln \tau''^{c}_{\Delta\nu} = \beta_\nu(T)W_2^{m_\nu}, \tag{19}$$

where

$$W_1 = \int_e \rho(e)F_{1c}[l(T)]dl, \tag{20}$$

$$W_2 = \int_e \rho(e)\left(\frac{P_e(e)}{P_0}\right)^{\frac{n_\nu}{m_\nu}} F_{2c}^{\frac{1}{m\nu}}[l(T)]dl\,. \tag{21}$$

Here effective pressure:

$$P_e = P_{N_2} + B_{O_2}\cdot P_{O_2} + \sum_{i=1}^{N-1} B_{ik}P_{ik}, \tag{22}$$

where P_{N_2} - pressure N_2, P_{O_2} - pressure O_2, B_{ik} – the widening factor (the relation of average semiwidth of lines in the chosen interval of a spectrum for collisions of molecules *i-k* to average semiwidth of spectral lines in case of impact of molecules of type *i* with molecules of nitrogen N_2).
Similarly for induced and continual absorption:

$$\beta_{\nu u}(T) = \beta_{\nu u}(T_0)F_u(T), \qquad \beta_{\nu\kappa}(T) = \beta_{\nu\kappa}(T_0)F_\kappa(T). \tag{23}$$

Temperature functions used for calculations $F_u(T), F_\kappa(T), F_{1c}(T), F_{2c}(T)$ can be presented in the tabular form or in the form of simple analytical approximations, for example, in the exponential-sedate form.
It is experimentally shown that for multicomponent atmosphere full function spectral transmission is defined by the law of product of functions on all gas components:

$$\tau_{\Delta\nu} = \prod_i \tau_{i\Delta\nu}, \tag{24}$$

where *i* - component number.
The parity (24) directly follows from static model of spectra and reflects that fact that the thin structure of spectra of each molecule doesn't depend on other molecules. For induced and continual absorption it is a parity it is carried out owing to absence of rotary structure. Numerical modeling of functions spectral transmission on parameters of thin structure of spectra have shown that the parity (24) is carried out with a margin error no more than 1 %.
Parameters K_ν, β_ν, m_ν, n_ν, M are defined from the measured spectra of radiation and radiation absorption by high-temperature gas mediums, modeling with the help heating cells and fiery measuring complexes.
For definition of parameters of functions spectral transmission the data of experimental researches has been added by results of numerical modeling under high-temperature atlases of parameters of the spectral lines prepared with use of the base data, received by means of measuring complexes of the high spectral permission. For an example on fig. 12 spectral

factors of absorption of water vapor K_v, and on fig. 13 – spectral dependences β_v water vapor in bands 1,37, 1,87 and 2,7 μm on experimental data are led. On fig. 14 spectral dependences of factors of absorption CO_2 in band 2,7 μm is given. On fig. 15 spectral factors of absorption K_v in the basic bands CO and NO according to numerical modeling of thin structure of spectra of absorption are illustrated. For vapor H_2O parameters m_v, n_v, M_v poorly depend on length of a wave a range of temperatures 600-2500K and probably to use average values n=0,45, m=0,65, M = −0,2. Strong temperature dependence of spectral factors of absorption in a range of spectrum 10−20 μm pays attention. At growth of temperature from 300 to 2500K increase intensitys the spectral lines entering into the specified interval of a spectrum, in 6600 times is observed.

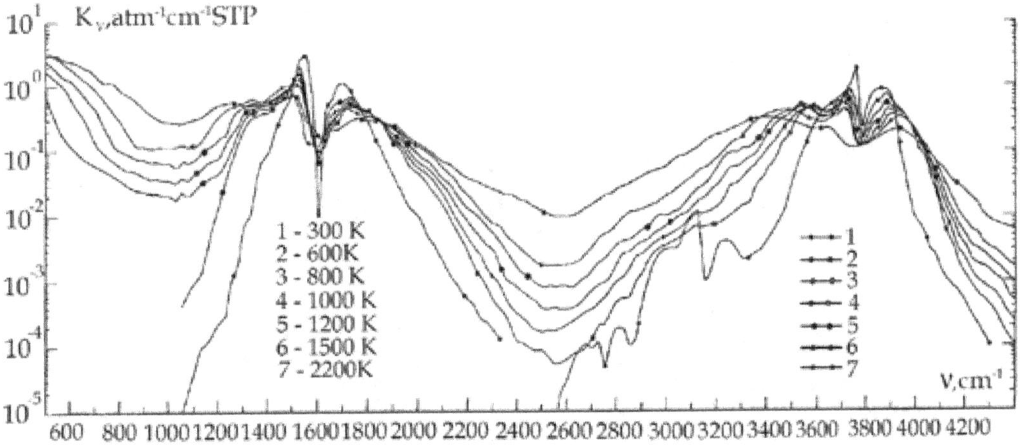

Fig. 12. Spectral factors of absorption K_v water vapor in bands 6,3 and 2,7 μm on experimental data.

Applicability of the received parameterization of functions spectral transmission for the decision of problems of the transfer over of radiation in high-temperature mediums and radiating heat exchange in chambers of combustion with application described above parities for calculations spectral intensitys thermal radiation and nonequilibrium radiation of electronic spectra in non-uniform working mediums under structural characteristics taking into account absorption and scaterring of radiation by a disperse phase has been considered. Main principle of correctness of spent calculations is calculation of equivalent mass on indissoluble trajectories from the radiating volume to a supervision point, including at reflection of radiation from walls and at scattering of radiation by a disperse phase. Streams of thermal radiation on walls of the working chamber are defined by integration spectral intensitys on a spectrum of lengths of waves and a space angle within a hemisphere.

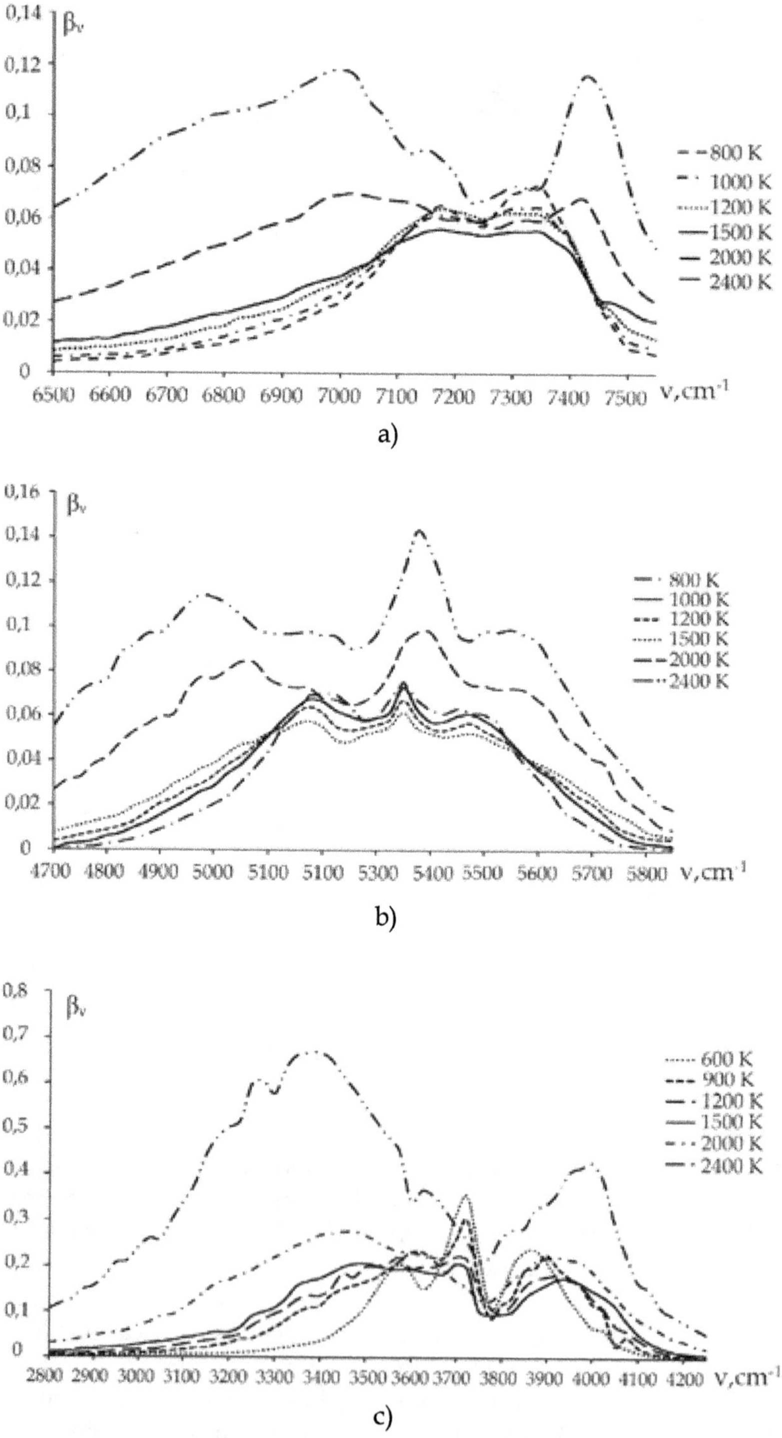

Fig. 13. Spectral dependences of parameter $\beta\nu$ in bands 1,37 (a), 1,87 (b) and 2,7 μm (c) water vapor.

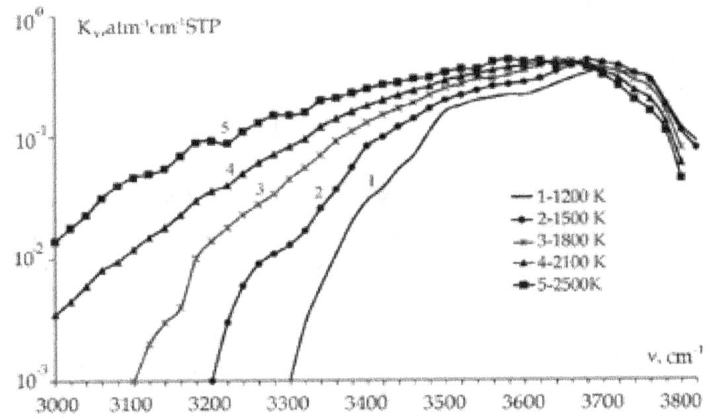

Fig. 14. Spectral dependences of factors of absorption K_ν in band 2,7 μm CO_2 on experimental data.

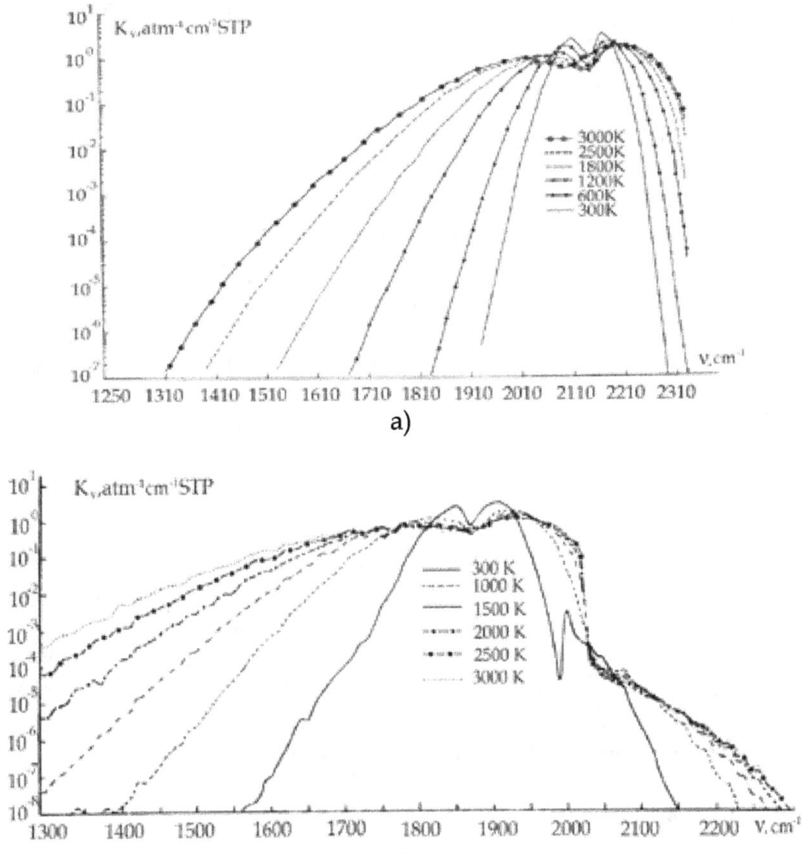

Fig. 15. Spectral dependences of factors of absorption K_ν in the basic bands CO (a) and NO (b) by results of numerical modeling of thin structure of a spectrum.

3. Nonequilibrium processes of radiation in flames

Executed spectraradiometry measurements have revealed presence of nonequilibrium radiation in ultra-violet, visible parts of a spectrum as result of display of effect of chemiluminescence in processes of burning of fuel. The contribution of nonequilibrium radiation in radiating cooling a torch has appeared essential. The basic components defining nonequilibrium radiation, are: OH, CH, NO_2, NO, SO_2, CN. Probably also influence of splinters of difficult hydrocarbonic connections which are formed in dissociated process the difficult hydrocarbonic connections which are in wild spirits.

Nonequilibrium radiation is formed by a kernel of a torch and then extends on all volume of the chamber of combustion, being transformed and participating in process of heating of particles of fuel and heatsusceptibility surfaces. Radiating cooling molecules occurs during their relaxation, making ≤10^{-4} sec, commensurate in due course courses of chemical reaction, and reduces adiabatic temperature of products of combustion in peaks of chemical reactions. Out of zones of chemical reactions radiation is equilibrium.

3.1 Definition nonequilibrium radiating cooling a flame from experimental data

Nonequilibrium radiation is generated by mainly electronic bands of radiation of the raised molecules of products of the combustion, lying in ultra-violet and visible parts of a spectrum and in vibrational-rotary bands. Temperature T of a zone of burning was measured by optical methods with a margin error no more than 2 %. A known chemical composition of gas fuel allows to calculate adiabatic temperature of zones of chemical reactions and to define size $\Delta T = T_a - T$, characterizing radiating cooling zones of active burning. Radiating cooling can be equilibrium and nonequilibrium. Equilibrium radiating cooling ΔT_e it is possible to calculate on absolute spectra of radiation of a flame and on the measured temperature and a chemical composition of products of combustion, speed of the expiration of a stream that allows defining radiating cooling ΔT_n, caused by nonequilibrium radiation.

Nonequilibrium radiating cooling $\Delta T_n = \Delta T - \Delta T_e$ is convenient for characterizing in size $\xi = \Delta T_n / T_a$ that which according to our measurements varies in a range of values (0,02-0,13), and increases with growth of temperature T_a.

Features of registration of average temperature of a flame an optical method have demanded working out of a method of definition ΔT_n in conformity with absolute spectra of the radiation registered by the spectral device, allocated optical systems in volumes of the radiating medium.

For definition nonequilibrium radiating cooling of flames results of measurements by optical methods of temperature hydrogen-oxygen, hydrogen-air, the propane-butane-oxygen, the propane-butane-air, acetylene-oxygen flames, formed by burning of gas fuel of a controllable chemical composition in air or oxygen have been used. Specially developed burning devices provided formation homogeneous for temperature flames in optical measuring channels. The temperature of a homogeneous flame was measured by a method of the self-reference of spectral lines of water vapor and on spectral brightness of radiation of a flame in "black" lines of water vapor or CO_2. Really, for a flame homogeneous for temperature and the gas medium which are in thermodynamic balance, spectral brightness of radiation $B_{\Delta\lambda}$ will be defined by a parity:

$$B_{\Delta\lambda} = B_{\lambda}^{abb}\left(1 - \tau_{\Delta\lambda}\right), \tag{25}$$

where $\tau_{\Delta\nu}$ – function of spectral transmission a flame for a small range of length of a wave, B_λ^{abb} – spectral brightness of radiation of absolutely black body on length of a wave λ, Δ - semiwidth of hardware function of a spectrometer. If $\tau_{\Delta\nu}$ =0 (a black line of radiation), $B_{\Delta\lambda} = B_\lambda^{abb}$ and the size $B_{\Delta\lambda}$ unequivocally defines temperature of a homogeneous flame.
The adiabatic temperature of a flame is calculated on a known chemical composition of products of combustion of burned gas and factors of surplus of oxygen and air α. Comparison of calculated values T_a with the measured optical methods in temperatures of flame T has shown that observable temperature T of a flame always more low T_a. The only thing the reason leading to lower value of size T in comparison with in adiabatic temperature T_a, can be effect radiating cooling a flame.
For the homogeneous radiating medium speed radiating cooling will be defined by the formula:

$$\frac{\partial T}{\partial t} = \frac{\oint\limits_S \left(F^{\uparrow}(S) - F^{\downarrow}(S) \right) dS}{\bar{C}p(T)\rho(T)V}, \tag{26}$$

where integration is made on the closed surface S covering all radiating volume V. The size $F^{\downarrow}(S)$ defines the integrated flux of equilibrium radiation entering into radiating volume V in point S; $F^{\uparrow}(S)$ represents an integrated flux of the equilibrium radiation leaving radiating volume V in point S; $\bar{C}p(T)$ - a specific thermal capacity of medium at constant pressure:

$$\bar{C}p(T) = \frac{\sum\limits_i C_{pi}(T) P_i(T)}{\sum\limits_i P_i(T)}, \tag{27}$$

where P_i - partial pressure of i-th component. Summation in the ratio (27) is made on all gas components which are a part of products of combustion; ρ (T) - the density of the gas medium, which dimension is defined by dimension $\bar{C}p(T)$,

$$F^{\uparrow}(S) = \int\limits_0^{\pi/2} \int\limits_0^{2\pi} \int\limits_0^{\infty} J_\lambda^{\uparrow}(S,\theta,\phi) \sin\theta \cos\theta \, d\theta \, d\phi \, d\lambda, \tag{28}$$

$$F^{\downarrow}(S) = \int\limits_0^{\pi/2} \int\limits_0^{2\pi} \int\limits_0^{\infty} J_\lambda^{\downarrow}(S,\theta,\phi) \sin\theta \cos\theta \, d\theta \, d\phi \, d\lambda. \tag{29}$$

In parities (28), (29) antiaircraft corner θ is counted from a normal to the closed surface in a point of supervision S. Boundary conditions at the decision of the equation of the transfer over were set in conformity with the constructional decision of measuring complexes (Moskalenko et al., 1992). Functions spectral transmission were calculated with use of a two-parametrical method of equivalent mass on indissoluble optical ways from the radiating volume to a supervision point (Kondratyev & Moskalenko, 2006).
At data processing of measurements were considered flames horizontal development of two counter streams of the flame surrounded with the quartz heat exchanger, reducing radiating

cooling a flame and promoting preservation of uniformity of temperature within an optical way of thermal radiation to the optoelectronic device, and also excluding formation sooty ashes as a result of process of pyrolysis of combustion products at burning hydrocarbonic fuels. The geometrical way of each of counter streams of a flame made 20 cm at speed of a current of a stream ω =8-10 m/s. The geometrical way to devices with vertical development of a flame made 6 cm at speeds of a current ω =10-15 m/s. Researchers have shown that in devices with water cooling of the case of the chamber of combustion (Moskalenko et al., 1992) at burning of hydrocarbonic fuel it is formed sooty ashes. Therefore in these devices as fuel the pure hydrogen excluding possibility of formation sooty ashes in products of combustion was used.

On fig. 16 and in table 2 results of definition of size $\xi=\Delta T_n/T_a$ for hydrogen-oxygen, hydrogen-air, the propane-butane-oxygen, the propane-butane-air, acetylene-oxygen flames are shown. With increase adiabatic temperatures T_a the size ξ increases and in a range of temperatures T_a from 1800 to 3200K varies from 2 % to 13 %.

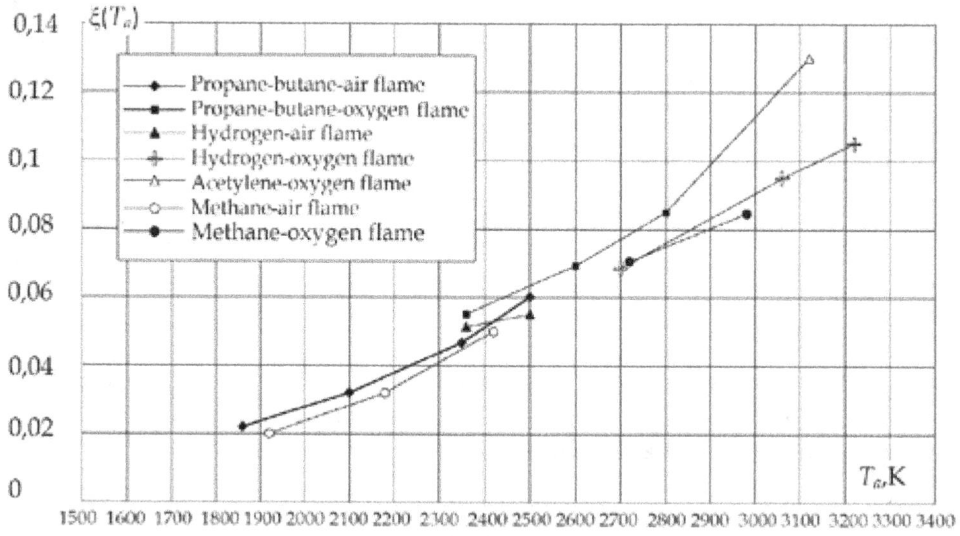

Fig. 16. Dependence of parameter ξ=ΔTn/Ta from adiabatic temperatures Ta.

As have shown results of experimental definition of function of source B_λ and factors of nonequilibrium $\eta_\lambda = B_\lambda(T)/B_\lambda^{abb}(T)$, where $B_\lambda^{abb}(T)$ is spectral brightness of absolutely black body (Planck's function), nonequilibrium radiation is formed mainly in electronic spectra of radiation of the molecules located in ultra-violet and visible parts of a spectrum. The effect nonequilibrium radiations in the vibrational-rotary bands lying in infra-red area of a spectrum, is shown considerably only at adiabatic temperature T_a> 2500 K.

The obtained data of experimental researches concerns optically thin torch when influence nonequilibrium is shown in a greater degree. Therefore the data presented on fig. 16 and in tab. 2, can be used for an estimation of the maximum size of energy which is transferred on heatsusceptibility to a surface in case of burning of gaseous fuel. It can be estimated on radiation of gas products of combustion of a kernel of a torch and on change enthalpy of combustion products. In the first case we will use the law of Stefan-Boltzman's defining an integrated hemispherical stream of radiation of absolutely black body,

$$F(T) = \sigma T^4 . \tag{30}$$

T_a, K	$\xi = \Delta T_n / T_a$	Type of flame
1860	0,021	propane-butane-air
2100	0,032	propane-butane-air
2350	0,048	propane-butane-air
2500	0,06	propane-butane-air
2360	0,055	propane-butane-oxygen
2600	0,069	propane-butane-oxygen
2800	0,085	propane-butane-oxygen
3120	0,13	acetylene-oxygen
2360	0,051	hydrogen-air
2500	0,056	hydrogen-air
2700	0,068	hydrogen-oxygen
3060	0,095	hydrogen-oxygen
3220	0,105	hydrogen-oxygen
1920	0,020	methane-air
2180	0,033	methane-air
2420	0,050	methane-air
2720	0,071	methane-oxygen
2980	0,094	methane-oxygen

Table 2. Influence of nonequilibrium processes of radiation on radiating cooling a flame. T_a is adiabatic settlement temperature; ΔT_n is radiating cooling the homogeneous flame, caused by nonequilibrium radiation. Factor of surplus of air and oxygen $\alpha > 1$.

Let's enter integrated function transmission:

$$\tau(T) = \int_0^\infty B_\lambda(T)\tau_\lambda(T)d\lambda \Big/ \int_0^\infty B_\lambda(T)d\lambda . \tag{31}$$

As the range of changes of temperature (T_a-T) is insignificant, in the specified range of temperatures of a flame $\in \{T_a - T\}$ we will accept τ_n=const. Then a parity:

$$\frac{\Delta F}{F(T_a)} = \frac{F(T_a) - F(T)}{F(T_a)} = 1 - \frac{F(T)}{F(T_a)} = 1 - \frac{F(T_a - \Delta T_n)}{F(T_a)} =$$

$$= 1 - \frac{\sigma(T_a - \Delta T_n)^4}{\sigma T_a^4} = 1 - \left[1 - \left(\frac{\Delta T_n}{T_a}\right)^4\right] \approx \frac{4\Delta T_n}{T_a}. \tag{32}$$

Thus, in a case optically a thin torch on heatsusceptibility surfaces without easing can get from 8 % to 30 % of full radiation of a torch.

Change enthalpy of combustion products will be defined by the formula:

$$\Delta H = \sum_i \left(\int_T^{T_a} C_{pi}(T)dT \right) V_i , \tag{33}$$

where $C_{pi}(T)$ - a specific thermal capacity at constant pressure for a component i combustion products, V_i - led to standard temperature and pressure volumes of components of products of combustion (P_{0i}=1 atm, T_{0i}=273K). Summation is carried out on all components i combustion products.

Volumes V_i are calculated on a known chemical composition of gas fuel and value of factors of surplus of air or oxygen. Size estimations $\Delta H/H_a$ yields the results close to sizes $\Delta F/F$ (T_a).

In case of mediums with the big optical density effects of influence nonequilibrium radiations are extinguished owing to suppression processes of chemiluminescence, caused by absorption of nonequilibrium radiation and display radiationless transitions in molecules. Because nonequilibrium radiation is formed in ultra-violet and visible parts of a spectrum, and also in infra-red area of a spectrum, in the transfer over of nonequilibrium radiation it is important to consider radiation absorption thing-dispersion ashy fraction and the hydrocarbonic connections absorbing radiation in spectral area of generation of nonequilibrium radiation. At absorption of nonequilibrium radiation there is a heating of particles of fuel and their ignition, and absorption of nonequilibrium radiation by heatsusceptibility surfaces weakens.

3.2 Definition of spectral factors of nonequilibrium vibrational temperature of components in flames

Quantum-mechanical consideration of a problem of formation of spectra of radiation of flames shows that presence nonequilibrium in radiation in electronic spectra (Moskalenko et al., 2009) it is accompanied by the nonequilibrium mechanism of radiation and in vibrational-rotary spectra as cores, overtone and composition bands, as have confirmed results of our measurements. The mathematical model of the transfer over of radiation is developed for the analysis of spectra of radiation flames in nonequilibrium radiating biphasic mediums in approach of unitary scattering of radiation (Moskalenko et al., 2009; Moskalenko & Loktev, 2009). The Chemical composition of products of combustion was calculated (Alemasov et al., 1973). Measurements were carried out on the spectral installations described in section 2.1 in spectral area 0,2÷25 μm in a range of temperatures 1500-2900 K. Hydrogen, methane, the propane-butane and acetylene in oxygen and in air were burned. The spectral resolution made Δ=3÷5 cm^{-1}. Besides, in the field of a spectrum 2,5÷3,5 μm were used results of the measurements executed with the high spectral resolution Δ=0,05 cm^{-1} by means of the Fourier spectrometer that has allowed to reveal influence nonequilibrium a flame in the basic band OH 2,8 μm. Results of the executed researches have shown that radiation of water vapor in vibrational-rotary bands is equilibrium. Nonequilibrium character of radiation CO_2 starts to be shown at thermodynamic temperatures of flame T_t>2100 K. Nonequilibrium character of radiation in electronic spectra of oxides of nitrogen in a spectral range 0,3÷0,7 μm was detected. Diffused radiation NO_2 has poorly expressed maximum of radiation in a spectrum range 0,55÷0,58 μm. At burning CH_4, propane-butane, C_2H_2 there is a formation of particles of soot to optical density on length of a wave λ=0,55µm accordingly equal 0,098; 0,2; 0,4 m^{-1} in a flame. Spectral dependences of factors of absorption and easing sooty ashes are used by us for restoration of its microstructure with the subsequent calculation of matrixes light scattering on a radiation spectrum (Moskalenko et al., 2010). We will more low stop on the analysis of spectra of nonequilibrium radiation of hydroxyl OH.

For research of radiating characteristics of a hydroxyl it is expedient to use hydrogen-oxygen a flame in order to avoid overlapping of bands of radiation OH other radiating components. Also it is expedient to choose modes of burning with factors of surplus of an oxidizer α_{ox}<1. Spectra of radiation H_2-O_2-flame are measured on installation SUVGI (Moskalenko et al., 1980; Moskalenko et al., 1992) and laboratory breadboard models of this installation at temperatures 1500-2500K. Hydrogen-oxygen flames, used in the present researches, have is bluish-white coloring with less contrast external zone of orange color. The flame height strongly depends on a mode of burning and on the average makes 30 cm. At heights more than 25 mm over hydrogen and oxygen capillaries ardent streams see height more merge, forming a continuous radiating torch. The maximum concentration of the basic radiating components of the hydrogen-oxygen flame: water vapor and a hydroxyl makes accordingly 58,46 % (at α_{ox}=1) and 12,21 % (at α_{ox}=1,3÷1,5). Thermodynamic parameters of products of combustion of the hydrogen-oxygen flame, calculated at pressure P=1atm, are presented in table 3. In spectra of radiation of a flame the most intensive bands are observed in a range 261÷360 nm. In flames with α_{ox} <0,5 in combustion products molecular oxygen is absent and in spectra of radiation of a band of Schuman-Runge of hot molecular oxygen weren't observed.

Unlike a propanee-butane-oxygen flame, a hydrogen-oxygen flame is transparent, since its emissivity in the visible is $\varepsilon_\lambda \leq 0.003$. Weak emission bands belonging to water vapor are observed in the neighborhoods of λ = 577.5, 579.7, 693, 718, 766, and 813 nm. The Na doublet was also detected from the slight concentration of sodium in the hydrogen. The emission peak of nitrogen dioxide (NO_2) is observed in the range of 550–580 nm. Members of the Delandres scheme of the ground state of the OH bands make a small contribution to the visible emission from the hydrogen-oxygen flames.

Table 3 lists the main thermodynamic parameters of the combustion products of the hydrogen-oxygen flame, including the thermodynamic temperature T of the flame and the chemical composition of the combustion products in the form of the volume concentrations of the components, as functions of the oxygen excess coefficient α_{ox}.

The OH radical is formed in all combustion processes for hydrogen-containing fuels and plays a significant role in the chemical chain reaction mechanism (Kondratyev, 1958; Kondratyev & Nikitin, 1981). It is a component of the combustion products in the exhaust plumes of ballistic rockets. Intense UV emission of the OH group is an important signifier of the launching of rockets and the firing of artillery systems. Large amounts of the OH groups are produced during nuclear explosions in the atmosphere.

The emission and absorption spectra of OH in hydrogen-oxygen flames has been studied for several decades. Small sized premixed flames have been used for this. The results indicate a disequilibrium over the internal degrees of freedom in the reaction zone of the flame. The emission from hydrogen flames at low pressures is essentially pure chemiluminescence. The most important criterion for chemiluminescent radiation is that the absolute intensity of the OH group radiation must exceed its equilibrium value because of a nonequilibrium vibrational and rotational energy distribution which leads to anomalously high vibrational and rotational temperatures. Increasing the pressure and introducing inert gases into the combustion zone cause quenching of the chemiluminescence on the part of hydroxyl groups. At low pressures the emission intensity of the OH groups in hydrogen low-temperature flames is many orders of magnitude greater than the equilibrium emission intensity. Here we measure the UV emission and absorption of OH in a hydrogen-oxygen flame. This flame can be used to model the temperatures in the exhaust plumes of rocket engines.

α_{ox}	T, K	K_1	H_2	H_2O	H	OH	O	O_2
0,1	1089	0,79	0,9	0,1	-	-	-	-
0,15	1442	1,19	0,85	0,15	-	-	-	-
0,2	1763	1,59	0,7998	0,2	0,002	-	-	-
0,25	2047	1,98	0,7483	0,2497	0,0019	-	-	-
0,3	2272	2,38	0,6917	0,2977	0,0099	0,0006	-	-
0,35	2475	2,78	0,6354	0,3444	0,0181	0,0019	-	-
0,4	2620	3,18	0,5756	0,3871	0,0379	0,0051	0,0001	-
0,45	2733	3,57	0,5166	0,4256	0,0468	0,0102	0,0005	0,0001
0,5	2822	3,97	0,4604	0,4595	0,0606	0,0174	0,0014	0,0005
0,6	2945	4,76	0,3617	0,5129	0,0809	0,0368	0,0049	0,0027
0,7	3017	5,56	0,2842	0,5483	0,0896	0,0584	0,011	0,0083
0,8	3055	6,35	0,2256	0,5694	0,0895	0,0784	0,0186	0,0183
0,9	3071	7,14	0,1818	0,5804	0,0843	0,0946	0,0261	0,0326
1,0	3074	7,94	0,1486	0,5846	0,0768	0,1065	0,0328	0,0504
1,1	3068	8,73	0,1231	0,5843	0,0688	0,1146	0,0382	0,0708
1,2	3057	9,52	0,1031	0,5809	0,0609	0,1196	0,0422	0,0931
1,3	3043	10,32	0,0871	0,5755	0,0536	0,1221	0,045	0,1164
1,5	3008	11,91	0,0636	0,5609	0,0413	0,1221	0,0476	0,1643
2	2907	15,87	0,0317	0,5171	0,0212	0,107	0,0435	0,2792
3	2699	23,81	0,0093	0,4357	0,0055	0,0664	0,0245	0,4582
5	2291	39,69	0,0007	0,3215	0,0002	0,0172	0,0038	0,6563
10	1546	79,37	-	0,1817	-	0,0002	-	0,8181

Table 3. Thermodynamic parameters of the combustion products of hydrogen-oxygen flames at a pressure of P=1atm.

The UV spectrum of the OH is formed by the electronic transition $A^2\Sigma - X^2\Pi$ consisting of the four most intense sequences Δv = 0, ±1, ±2 lying in the region 261–360 nm. The 0–0 band is the most intense. The nonequilibrium OH emission of the flame is characterized by the spectral nonequilibrium coefficient of the radiation defined η_λ as the ratio of the intensity of the nonequilibrium emission to the equilibrium intensity at wavelength λ. The spectral brightness of the equilibrium OH emission $B_{\lambda e}^{OH}$, for a uniform medium is given by:

$$B_{\lambda e}^{OH} = B_{\lambda}^{0}\left(1-\tau_{\lambda}^{OH}\right), \tag{34}$$

where τ_λ^{OH} is the spectral transmission function of the OH groups in the flame; $B_{\lambda e}^{OH}$ is the absolute spectral brightness of the OH emission expressed in energy units, and is the spectral brightness of an absolute black body at the thermodynamic temperature of the flame.

In accordance with the definition, the spectral nonequilibrium coefficient is written in the form:

$$\eta_\lambda = B_{\lambda n}^{OH} / B_{\lambda e}^{OH} = B_{\lambda n}^{OH} \Big/ B_{\lambda e}^{0}\left(1-\tau_{\lambda}^{OH}\right), \tag{35}$$

Where $B_{\lambda n}^{OH}$ is the spectral brightness of the nonequilibrium emission from the OH groups.

An examination of the theory of nonequilibrium radiation transfer (Moskalenko et al., 2006) shows that, under the given conditions for formation of the flare, η_λ depends on the optical thickness of the emitting medium and reaches a maximum for small optical thicknesses of OH, when the spectral transmission function $\tau_\lambda^{OH} \rightarrow 1$. If η_λ is obtained at τ_λ^{OH}, then $\eta_{\lambda\max}^{OH} = \eta_\lambda^{OH} / \tau_\lambda^{OH}$.

The quantity $\eta_{\lambda\max}$ shows up in the computational integro-differential equations for modelling radiative transfer in nonequilibrium radiating media (Moskalenko & Zaripov, 2008). Studies of the integral effects of nonequilibrium radiative processes on radiative transfer have shown that the effect of the nonequilibrium in the radiative processes increases with the adiabatic temperature T_a of the flame; this is also confirmed by the present spectral measurements of the nonequilibrium OH emission. The values of $\eta_{\lambda\max}$ can be used for determining the spectra (wavelength dependence) of the vibrational temperatures (Moskalenko et al., 2010).

In the faint nonequilibrium emission bands, the spectral transmission $\tau_\lambda \rightarrow 1$ and small values of $(1 - \tau_\lambda)$ cannot be observed experimentally. In that case, the vibrational temperatures obtained for the fainter bands can be used to find the spectral absorption coefficients and integral intensities of the faint emission bands whose absolute emission spectra have been recorded by measurement devices.

In order to determine the degree of nonequilibrium of the OH group emission, their transmission spectra were measured under the same conditions as those for measuring the emission spectra. The structure of the transmission spectra correlates with the structure of the emission spectra. The resulting spectral variations in η_λ and $\eta_{\lambda\max}$ (fig. 17) are rapidly varying. The degree of variation in the 281–296 nm range is slightly higher than in the 306–328 nm range. The maximum value of $\eta_{\lambda\max}$ for T=2380 K is 100. In the 306-328 nm band, $\eta_{\lambda\max}$ reaches 50 (fig. 17).

It should be noted that the spectral dependences of η_λ for different temperatures are essentially different. From our standpoint, the reason for this is that it is impossible to make the conditions for obtaining the emission and transmission spectra completely identical, since they have were not recorded simultaneously. The existence of a spectral variation in $\eta_{\lambda\max}$ indicates that the vibrational temperature may depend on the vibrational and rotational quantum numbers. In addition, the shape of the spectrum lines may have an effect on the value of η_λ. Turbulent inhomogeneity during burning of the flame may also have some effect on η_λ. For practical calculations it is appropriate to introduce the average degree of nonequilibricity $\bar{\eta}_{\lambda max}$ which for, say, the sequence is $\bar{\eta}_{\lambda max}(\Delta\upsilon = 0) = 18$, and for the sequence Δυ = +1, is $\bar{\eta}_{\lambda max}(\Delta\upsilon = +1) = 28$ at T=2380 K. According to the theory of formation of nonequilibrium emission process $\bar{\eta}_{\lambda max}(\Delta\upsilon = -1) = 28$ for the sequence Δυ = −1. This value of $\bar{\eta}_{\lambda max}$ and the absolute spectral intensities of the nonequilibrium emission from the hydroxyl group have been used to determine the spectral absorption coefficients and the integrated absorption intensity of the OH for the Δυ = −1 sequence (Fig.18).

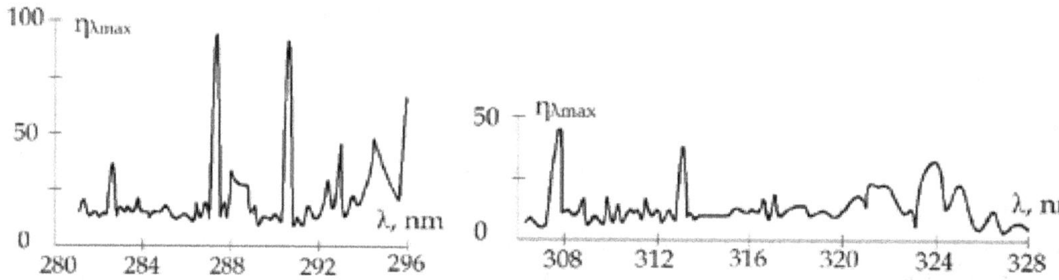

Fig. 17. Spectral dependence of the maximum degree of nonequilibrium in the emission from OH groups at a temperature of 2380 K.

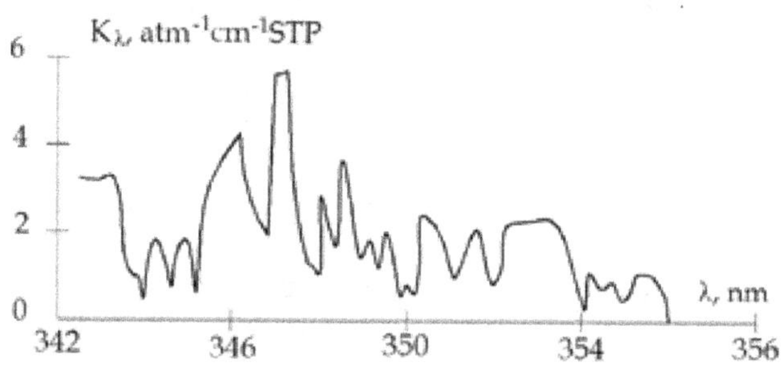

Fig. 18. Spectral dependence of the absorption coefficient for the emission from OH groups in the 342–356 nm range at 2380 K obtained from spectra of the nonequilibrium flame emission.

Our measurements of the absolute intensity spectrum can be used to determine the OH vibrational temperature by the method of (Broida and Shuler, 1952) using the formula:

$$0{,}624 \cdot \frac{E_1 - E_2}{T} = 4 \cdot \lg\frac{\lambda_2}{\lambda_1} + \lg\frac{P_1}{P_2} + \lg\frac{J_2}{J_1}, \quad (37)$$

where E_1 and E_2 are the energies of the upper vibrational levels of the chosen emission sequences; λ_1 and λ_2 are the
locations of the centers of the bands or of their edges; P_1 and P_2 are the vibrational transition probabilities; J_2 and J_1
are the emission intensities of the individual vibrational bands of the chosen sequences.
The values of J_2 and J_1 are determined experimentally after energy calibration of the emission spectra. The position of the edges of the bands can be determined fairly accurately after spectral calibration of the measured emission spectra. The vibrational transition

probabilities P_1 and P_2 differ considerably according to the data of various authors. Thus, it has been proposed (Moskalenko and Loktev, 2009) that the vibrational temperatures be measured from $\eta_{\lambda\max}$, and the average vibrational temperature, from $\bar{\eta}_{\lambda\max}$. It should be noted that the integrated intensities of the electronic absorption spectra for the $\Delta\upsilon$=0,±1 sequences can be used in an experimental determination of the transition probabilities P for these sequences. When Eq.(37) is used legitimately, the vibrational temperatures found by the two different methods (Broida & Shuler, 1952; Moskalenko & Loktev, 2009) should coincide.

The results of this determination of the vibrational temperatures indicate a nonequilibrium distribution of the energy released during the combustion reaction over the vibrational levels of the electronically excited state $A^2\,\Sigma^+$. The integrated intensities of the OH group bands for the sequences Δv = 0, +1, –1 are equal, respectively, to 30592, 14472, and 1980 atm^{-1}·cm^{-2} at STP. According to the present measurements, the total intensity of all the absorption bands of the OH groups in the 281–333 nm range is 47044 atm^{-1}·cm^{-2} at STP. The intensities are in the proportion $S(\Delta v = 0) : S(\Delta v = +1) : S(\Delta v = -1)$ = 1:0.49:0.065.

Figure 19 shows the measured spectral dependence of the absorption coefficients of the hydroxyl groups and Fig. 20 the spectral dependence of the vibrational temperature $T_{\lambda k}$ for the Δv = 0, +1 sequences of the hydroxyl bands according to the formula (Moskalenko & Loktev, 2009):

$$T_{\lambda k} = \frac{C_2}{\lambda}\left\{\ln\left[\frac{\exp(C_2/\lambda T)-1}{\eta_{\lambda\max}}+1\right]\right\}^{-1}, \tag{38}$$

where λ is the wavelength, T is the flame temperature, and C_2 is the second constant in the spectral brightness of a black body (Planck function).

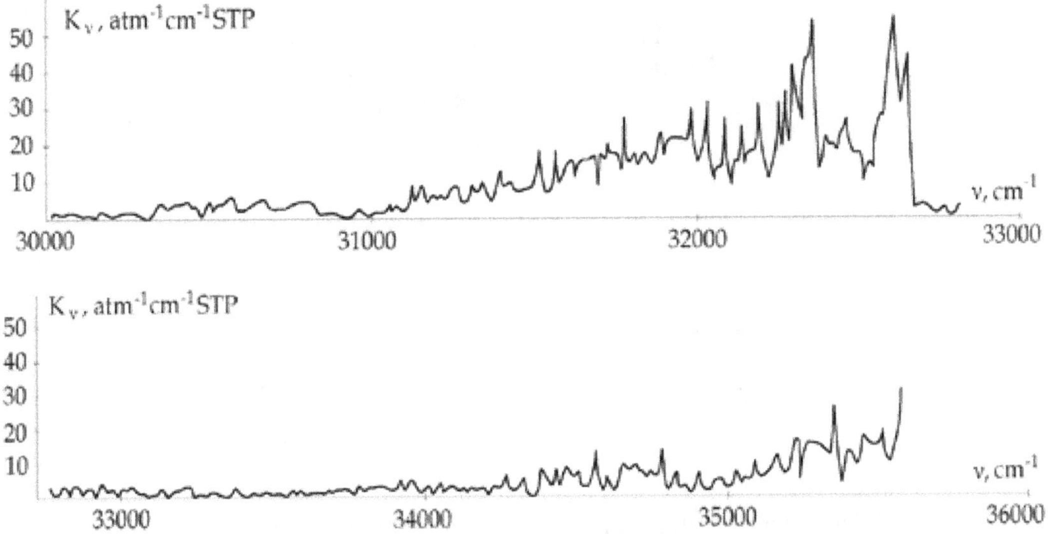

Fig. 19. Spectral dependence of the absorption coefficient $K\nu$ for OH groups at 2450 K in the 281–333 nm range (30000–36000 cm^{-1}).

The data obtained here can be used to study the mechanisms responsible for the nonequilibrium character of the emission from hydrogen-oxygen flames, calculating the spectral brightness of the UV emission from rocket exhaust plumes, solving radiative heat transfer problems in power generation furnaces, and in atmospheric nuclear explosions. The present study of the effect of nonequilibrium processes on the formation of the emission spectra of hydrogen-oxygen flames applies to a total pressure of $P = 1{\cdot}10^5$ Pa (1 atm) in the flames. The radiative nonequilibrium coefficients are higher at lower flame pressures, as confirmed by experiments on the effect of nonequilibrium radiative processes on radiative cooling of flames (Moskalenko and Zaripov, 2008). These data show that nonequilibrium radiative processes must also have an effect on the vibrational-rotational emission bands of OH, and the degree of this influence will depend on the relaxation time of hydroxyl from an excited electronic state into an equilibrium state and will show up in the form of additional bands that are shifted toward the IR relative to the main transition. Because of overlap of the main vibrational-rotational band of OH with the intense emission spectrum of water vapor, it will be necessary to measure the emission spectra of flames with high spectral resolution in order to establish the effect of nonequilibrium processes on the formation of the IR emission spectrum of OH. However, the present analysis of the emissivity spectra of hydrogen-oxygen flames has revealed the existence of continuum absorption in the 4.1–4.4 μm range owing to vibrational-rotational transitions of OH groups in excited electronic states (See Fig. 21).

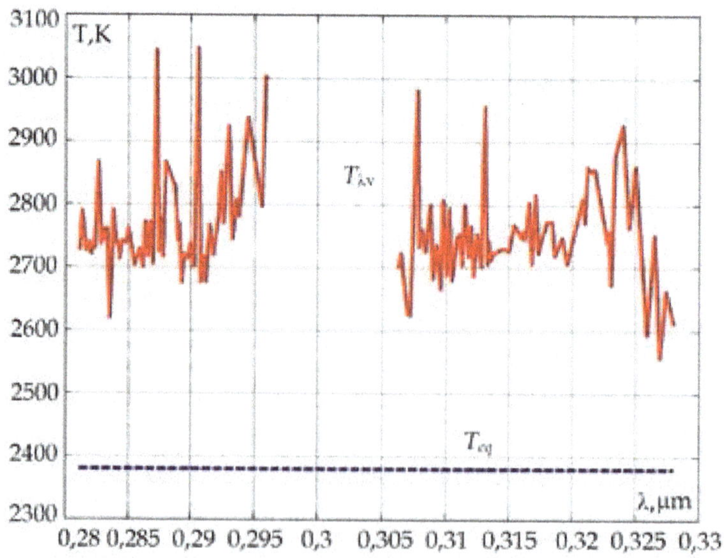

Fig. 20. Spectral dependence of the equilibrium thermodynamic temperature T_{eq} and the vibrational temperature $T_{\lambda v}$ for the $\Delta v = 0, +1$ sequences of the electronic bands of the hydroxyl group.

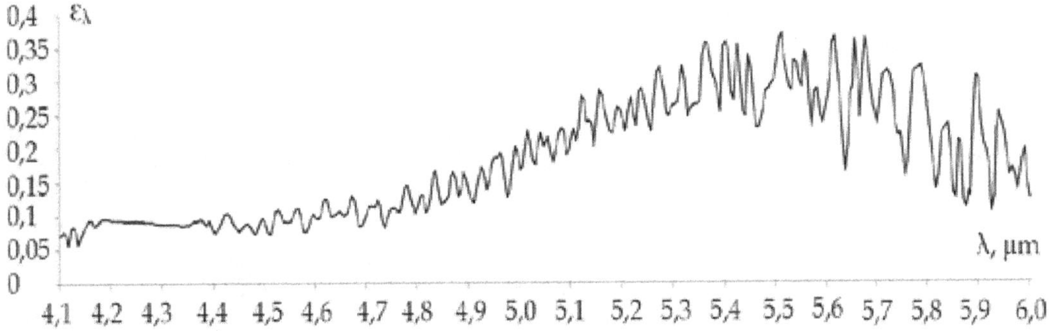

Fig. 21. Spectral emissivity of a hydrogen-oxygen flame with continuum emission of the hydroxyl group in an excited electronic state in the 4.1–4.4 μm range.

The analysis of spectra of radiation of the high spectral permission the hydrogen-oxygen flame in a spectral range of 3000-3500 cm^{-1} received at the spectral permission Δ = 0,1 cm^{-1} (Moskalenko et al., 1992), has revealed presence of nonequilibriumnes in the basic band v_1 hydroxyls OH. In a considered range of a spectrum radiation is formed by vibrational-rotary bands of vapor H_2O and hydroxyl OH. The spectrum of radiation of a flame was energetically not normalized. Besides, because of presence in experimental installation not blown and anonvacuum a site of an optical way the true spectrum of radiation of a flame is transformed by lines cold (T=291 K) water vapor and carbonic gas. For definition of vibrational temperature of hydroxyl OH the true spectrum of radiation corrected from influence of hardware distortion and atmospheric absorption of radiation by vapor H_2O and CO_2 at first is restored. Are most informative for definition of vibrational temperature of a Q-branch of band OH which vibrational temperature surpasses equilibrium temperature of a flame in a spectrum of wave number (vibrational temperature of vapor H_2O in a flame) on size ΔT_v=350÷850 K at average size ΔT_v=650K. This value ΔT_v is recommended to be used at definition of the contribution of radiation OH in radiating heat exchange in high-temperature mediums, including chambers of combustion of boiler installations. We will notice that nonequilibrium spectra of radiation OH in the basic vibrational-rotary band 0 – 1 it is caused by formation as a result of chemical reactions of molecules OH in the basic electronic state with superfluous population of the raised vibrational levels in comparison with equilibrium and transitions of molecules OH with the nonequilibrium populated raised electronic state to selection rules on vibrational and rotary quantum numbers. Definition of vibrational temperature from spectra of radiation is more preferable, as nonequilibrium radiation amplifies not only at the expense of increase in intensity of lines with growth of vibrational temperature ΔT_k, but also growth of function of source B_λ (T_v) with increase T_v.

In spectra of radiation of a flame influence nonequilibrium radiations OH in vibrational-rotary bands of the first and second overtones of the basic and raised electronic states is revealed. Their influence on radiating heat exchange high-temperature of flames can be more essential in connection with growth integrated intensitys bands $2v_1$, $3v_1$ OH with increase in temperature and weaker overshoot of spectra OH lines of water vapor. Bands $2v_1$ and $3v_1$ OH the raised electronic state settle down in a vicinity of lengths of waves 2,1 and 1,4 μm, and for the basic electronic state they are in a vicinity of lengths of waves 1,43 and 1 μm.

4. Results of definition of a microstructure of sooty ashes and its optical characteristics

Sooty ashes it is generated at burning of any hydrocarbonic fuel, including wood and an industrial production waste. The nobility suffices for the decision of the scalar equations of the transfer over of radiation spectral factors of absorption, scattering and indicatryss of scattering ashes and their spatial distribution on the radiating volume. The last depends on a chemical composition and a microstructure sooty ashes which are defined by a kind of fuel and a mode of its burning. The least concentration sooty ashes are observed at fuel burning in oxygen with surplus factor $\alpha \geq 1$ and at most high temperatures of a flame (Moskalenko et al., 2010). Also more thin microstructure ashes in this case takes place. At burning firm power fuel in fire chambers of boiler installations and power technological units the microstructure ashes in top internal volume changes in process of combustion of particles and is multimodal in connection with simultaneous generation sooty ashes in the course of fuel combustion. Thus a roughly-dispersion the fraction ashes is defined by characteristics of grinding devices. For gaseous fuel optical characteristics strongly depend on fuel structure. At hydrogen burning sooty ashes this kind of fuel isn't formed also is non-polluting at its burning in oxygen atmosphere. At burning of gas hydrocarbonic fuel a microstructure sooty ashes and its mass concentration in a flame depends on structure of gas fuel and can change largely (Moskalenko et al., 2010).

4.1 Results of definition of a microstructure of sooty ashes at burning of gas fuel

Variety used fuels and technologies of their burning studying of influence of a chemical composition of fuel, a microstructure of its scattering, a way of giving of fuel and an oxidizer (air), a fire chamber design, burning cause of devices on radiating heat exchange. Last is defined by a chemical composition, a microstructure and a phase condition of hot particles of fuel, a chemical composition of a gas phase of products of combustion and radiating characteristics ashes which is generated in the course of burning of fuel or as a result of ionic nucleation from a gas phase of products of combustion. Process of burning of fuel is accompanied by simultaneous radiating heat exchange and mass exchange combustion products on volume of a fire chamber and a heat transfer to evaporating surfaces. In the present work possibilities of definition of microstructural characteristics of ashes from data on easing of radiation by products of combustion and from radiation optical characteristics of flames which are included into the decision of the scalar equation of the transfer over of radiation are considered and are necessary for modeling of radiating heat exchange in high-temperature diphasic mediums, for example, in chamber fire chambers of power and power technological units. Thereupon optical characteristics industrial ashes are analysed, including sooty particles of the most probable not spherical forms. In case of spherical and cylindrical particles calculations were carried out under the theory of Mi. In case of not spherical particles optical characteristics were calculated in approach of geometrical optics and effective sections of absorption of scattering and scattering indicatryss are received by their averaging on space within a space angle 4π in the assumption that spatial orientation depending on a radiation direction is equiprobable. This assumption is proved enough in case of the decision of problems of radiating heat exchange in combustion chambers where strong turbulence of products of combustion and presence of whirlwinds is observed. For the analysis results of measurements of spectra of radiation of radiation and radiation easing flames, received on measuring complexes (Moskalenko et al., 1980, 1992) and their breadboard models.

Microstructure restoration ashes from the given spectral measurements of radiation of radiation of a flame and easing of radiation by a flame is a return task of remote sounding and demands for the decision of data on spectral dependences of optical characteristics ashes from distribution of particles in the sizes. Ashe of combustion products is polydisperse and on the modeling chosen by us it is represented superposition of separate fractions, microstructure N (r) which is defined by parity (Kondratyev & Moskalenko, 1983):

$$N(r)=\sum_{i=1}^{N}N_i(r)=\sum_{i=1}^{N}A_i\cdot r_a^{a_i}\cdot\exp\left[-b\cdot r^{c_i}\right], \tag{39}$$

where A_i, a_i, b_i, c_i - parameters of i-th fraction, N - number of fractions. As well as in (Kondratyev & Moskalenko, 1983) decisions on the restored spectral factors of easing and radiation absorption ashes we will represent from them normalized for length of a wave λ =0.55 μm, and the mass maintenance ashes on an optical way we will express in an optical thickness $|-\ln\tau|$, where τ - function spectral transmission on length of a wave λ =0.55 μm. The essence of a problem of restoration of a microstructure ashes consists in definition of weight functions $N_i(r)$ which in the best way will describe spectral dependence of the measured factors of absorption and radiation easing ashes (see fig. 22).

It is necessary to notice that flames form strongly absorbing gas medium so lengths of waves for the decision of a return task should be chosen so that radiation of gases in them was minimum. Inclusion in processing of lengths of waves with λ> 4.2 μm only increases errors of restoration of a microstructure of ashes.

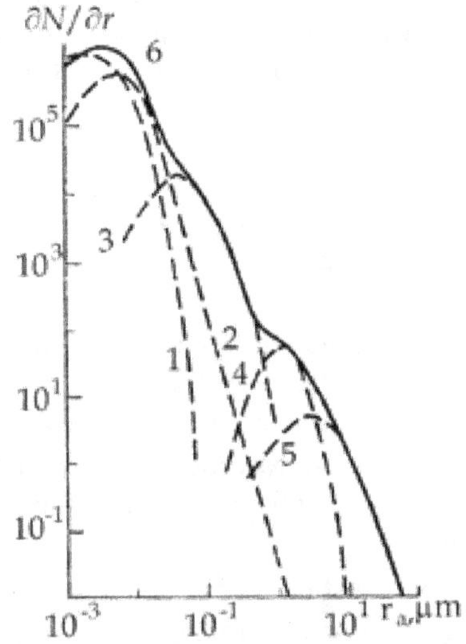

Fig. 22. An example of construction of model of microstructural characteristics (a curve 6) sooty ashes in the form of superposition of definitely weighed fashions curves 1, 2, 3, 4, 5.

Experience of restoration of a microstructure sooty ashes in flames has shown that in flames is present thin-dispersion sooty primary ashes which is generated in flames from a gas phase of products of combustion (Moskalenko et al., 2009) as a result of ionic nucleation, the microstructure and which optical density strongly depend on a chemical composition of gas fuel. The most probable modal radius of particles of this fraction ashes r_m=0.003 μm is received at methane burning in air and in oxygen. A population mean of optical density $(\partial\bar{\tau}/\partial L)=0.099$ m^{-1}. At propane-butane burning in air the increase in optical density, so $(\partial\bar{\tau}/\partial L)=0.2$ m^{-1} is observed. The highest values $(\partial\bar{\tau}/\partial L)=0.4$ m^{-1} is observed in flames at burning of acetylene C_2H_2 when the microstructure sooty ashes is observed most roughly-dispersion. At burning C_2H_2 higher value unburning fuel is observed also.

Microstructure restoration sooty ashes on spectral dependence of factors of radiation and easing of radiation by a flame demands inclusion in consideration of lengths of waves in an ultra-violet part of a spectrum, visible and near infra-red ranges. For definition of a microstructure of huge particles of mineral fraction ashes the additional information on scattering indicatryss of radiation is required. Mineral ash in flames is present at a liquid phase, the complex which indicator of refraction remains to unknown persons that complicates reception of matrixes of scattering of radiation for the purpose of their further use at definition of a microstructure mineral ash in flames. It is important to remember that mineral ash in flames and gas-mains boiler installations it is enriched by carbon (soot) and its radiating properties strongly depend on a subtlety of crushing of firm fuel. Than is thinner highling crushing, the disperse phase of products of combustion has especially high volume factors of absorption (radiation). In process of burning out of particles of firm fuel their size decreases and the specific density of particles that leads to reduction of their radiating ability increases. However, as a result breaking the thaw in the course of them degassing, is generated more a thin dispersion fashion mineral ash with modal radius $r_m \approx 0.3$-1 μm, strengthening radiating ability of products of combustion in visible and near infra-red spectrum ranges. Optical characteristics of some equivalent ensemble of spherical particles are defined by a complex indicator of refraction $n=m-ix$, where m and x – its valid and imaginary parts which for a multicomponent chemical composition of particles are described by a parity:

$$m=\frac{\sum_i m_i\cdot\rho_i}{\sum_i \rho_i};\ x=\frac{\sum_i x_i\cdot\rho_i}{\sum_i \rho_i}, \tag{40}$$

where ρ_i – mass concentration of connection i in a particle. Summation extends on all connections which are a part of a particle, including air for friable particles, and also a moisture at the influence account condensation processes on optical characteristics of a disperse phase in atmosphere. In table 4 parameters of models of a microstructure of the particles used as the aprioristic information at restoration of a microstructure of a disperse phase of products of combustion are resulted.

For microstructure definition sooty ashes in flames it is necessary to restore at first its spectral optical density. Function spectral transmission gas components of products of combustion for the chosen optical channels of sounding (optimum for sounding of lengths of waves) and their contribution to spectral intensity of radiation of a flame is for this purpose calculated. Further function of spectral easing of radiation ash and function spectral transmission ashes $\tau^a_{\lambda a}$, caused only is defined by radiation absorption. On the measured spectral functions of easing of radiation τ^a_λ and spectral radiating ability ashes (1-$\tau^a_{\lambda a}$) it is defined then spectral factors of easing σ^a_λ and absorption (radiation) $\sigma^a_{\lambda a}$ ashes. The restored spectra σ^a_λ and $\sigma^a_{\lambda a}$ serve for microstructure definition ash by optimum adjustment under spectral dependences σ_{λ}^{a} and $\sigma_{\lambda a}^{a}$. In table 5 the measured spectra σ_{λ}^{a} and $\sigma_{\lambda a}^{a}$ for methane-oxygen, the propane-butane - oxygen and acetylene-air flames are resulted. The restored distributions of a microstructure sooty layer in considered flames are presented in table 6.

№	a	b	c	r_m, µm
1	1	50	0.5	$1.6 \cdot 10^{-3}$
2	0.2	6	0.5	$4.4 \cdot 10^{-3}$
3	1	9	0.5	$4.9 \cdot 10^{-2}$
4	1	7.5	0.5	$9.0 \cdot 10^{-2}$
5	2	12	0.5	$8.9 \cdot 10^{-2}$
6	1.5	6	1	$2.5 \cdot 10^{-1}$
7	0.1	0.5	2	0.3
8	1	1	2	0.3
9	3	6	0.5	1.0
10	3	4	0.5	2.25

Table 4. Parameters of models of a microstructure of particles modified scale - distributions.

п/п	λ, µm	0.287	0.304	0.338	0.55	0.82	1.08	2.62	4.15	9.1
A	σ_{λ}^{a}, m^{-1}	0.161	0.154	0.144	0.099	0.072	0.058	0.029	0.019	0.008
	$\sigma_{\lambda a}^{a}$, m^{-1}	0.105	0.099	0.092	0.065	0.049	0.041	0.022	0.015	0.007
B	σ_{λ}^{a}, m^{-1}	0.234	0.230	0.225	0.204	0.174	0.164	0.103	0.082	0.044
	$\sigma_{\lambda a}^{a}$, m^{-1}	0.126	0.126	0.120	0.115	0.098	0.091	0.065	0.052	0.031
C	σ_{λ}^{a}, m^{-1}	0.385	0.398	0.405	0.400	0.380	0.372	0.271	0.225	0.156
	$\sigma_{\lambda a}^{a}$, m^{-1}	0.190	0.197	0.208	0.210	0.200	0.196	0.158	0.135	0.096

Table 5. The restored spectral optical density sooty ashes σ_{λ}^{a} and $\sigma_{\lambda a}^{a}$ for methane-oxygen (A), the propane-butane - oxygen (B) and acetylene-air (C) flames.

Results of the present measurements of a microstructure sooty ashes are used in calculations of optical characteristics, including spectral factors of easing, scattering, absorption and scattering indicatryss (Moskalenko & Loktev, 2009), in radiating heat exchange and problems of remote sounding.

Sooty ashes in combustion products define unburning fuel and reduce thermal efficiency.

On a microstructure sooty ash probably to define volume of particles ashe:

$$V = A \cdot \int_{r} \frac{4}{3} \cdot \pi \cdot r^{3} \cdot \frac{\partial N}{\partial r} dr, \tag{41}$$

where A – normalizing multiplier which is defined on optical density on length of a wave λ =0.55 μm. Knowing density sooty ashes ρ, it is possible to define chemical unburning of fuel $\rho_C \cdot V$.

r, μm	∂N/∂r			r, μm	∂N/∂r	
	A	B	C		B	C
0.001	$7 \cdot 10^5$	$1 \cdot 10^5$		0.15	$5 \cdot 10^1$	$2 \cdot 10^3$
0.002	$1.1 \cdot 10^6$	$3 \cdot 10^5$		0.2	$1 \cdot 10^1$	$7 \cdot 10^2$
0.003	$1 \cdot 10^6$	$5 \cdot 10^5$		0.3	$2 \cdot 10^0$	$1.6 \cdot 10^2$
0.004	$9 \cdot 10^5$	$6 \cdot 10^5$	$1 \cdot 10^3$	0.4	$8 \cdot 10^{-1}$	$4.0 \cdot 10^1$
0.006	$8 \cdot 10^5$	$6.5 \cdot 10^5$	$2 \cdot 10^3$	0.6	$3 \cdot 10^{-1}$	$1.5 \cdot 10^1$
0.008	$6 \cdot 10^5$	$6 \cdot 10^5$	$3 \cdot 10^3$	0.8	$1 \cdot 10^{-1}$	6
0.01	$4 \cdot 10^6$	$5 \cdot 10^5$	$5 \cdot 10^3$	1	$3 \cdot 10^{-2}$	2
0.015	$8 \cdot 10^5$	$3 \cdot 10^5$	$1 \cdot 10^4$	1.5		0.6
0.02	$1 \cdot 10^4$	$1 \cdot 10^5$	$1.3 \cdot 10^4$	2		0.2
0.03	$1 \cdot 10^3$	$1.6 \cdot 10^4$	$2 \cdot 10^4$	3		0.06
0.04	$1 \cdot 10^2$	$3 \cdot 10^3$	$1.3 \cdot 10^4$	4		0.02
0.06	$1 \cdot 10^1$	$1 \cdot 10^3$	$8 \cdot 10^3$	6		0.01
0.08	$1 \cdot 10^0$	$6 \cdot 10^2$	$5 \cdot 10^3$	8		0.003
0.1	0.1	$1.5 \cdot 10^2$	$3 \cdot 10^3$	10		0.001
				15		0.0001

Table 6. The restored distributions of a microstructure sooty ashes in methane-oxygen (A), the propane-butane - oxygen (B) and acetylene-air (C) flames.

Concentration of sooty ashes in combustion products strongly depends on structure and a mode of burning of gas fuel and changes in a range of values $4\times10^{-7}\div2.6\times10^{-4}/m^3$. The minimum concentration of soot is observed in flame methane-oxygen at factor of surplus of oxygen α =1,03. The maximum concentration is measured for flame acetylene-air.
The received data on microstructure ashes for various flames are used for calculations of matrixes light-scatterig and spectral dependences of factors of easing, absorption and scattering, scattering indicatryss on a spectrum of lengths of waves with their subsequent use in calculations of radiating heat exchange in chambers of combustion of boilers and power technological units. Normalizing optical characteristics sooty ashes it is desirable to carry out with a binding on optical density on length of a wave λ =0.55 μm that allows to use the measured values received in the present work, at calculations of radiating heat exchange in projected objects. Thus it is possible to believe that burning of various components of fuel occurs independently. The weight of each microstructure in optical density sooty ashes is defined by a share of each component in weight of burned fuel.
The microstructure of sooty ashes at movement of products of combustion changes in connection with its burning out thin-dispersion fraction and capture of small particles of soot by large particles in the course of coagulation. These effects can be considered by change of weight of optical density of various fractions sooty ashes on fire chamber volume (Moskalenko et al., 2009).
Trustworthy information on concentration and a microstructure of ash can be received also by the analysis of the selected tests of products of combustion at lower temperatures with

use warmed up multiple-pass cell as the working chamber (Moskalenko et al., 2010). Spectral dependence of easing of radiation bears the information on a microstructure ashes and, hence, about mass concentration ashes.

The knowledge of distribution of mass concentration and fields of temperatures of a disperse phase is of great importance by optimization and heat exchange intensification in top internal devices at coal and black oil burning.

5. Spectral and spatial distribution of thermal radiation in multichamber fire chambers

In the present section possibility of creation of boiler installations with the multichamber fire chambers is considered, allowing essentially to intensify radiating heat exchange in fire chambers and to raise vapor-productivity boiler installations. Now there are two possibilities of increase of efficiency of radiating heat exchange in fire chambers: increase in number of torches in a fire chamber with creation of strong radiating contrasts separate flames on the general thermal background of products of combustion in a fire chamber; creation of multichamber fire chambers where the increase in efficiency radiating cooling is reached at the expense of preservation of dimensions of the top internal chamber by a multichamber design.

The second variant is more effective and supposes wide variety of constructional decisions and allows to design vapor boilers with natural and compulsory circulation, and also to apply the mixed forms of circulation. Multichamber fire chambers can be used for burning firm hydrocarbonic fuels. In this case application in boiler installations of burning of fuel in the cyclonic prefire chambers is perspective, allowing to clear products of combustion of a cindery component. The top internal chamber in this case is actually radiating heat exchanger which for increase of efficiency radiating cooling combustion products is carried out by the multichamber.

The design of a multichamber fire chamber with ascending movement of products of combustion in a fire chamber and vertical development of a flame is most effective (Moskalenko & Zaripov, 2008; Moskalenko et al., 2008) with matrix burning device, the general gas collector for oxidizer giving (air or oxygen) and a collector for giving of gas fuel to multirow to torches. Burning devices are expedient for the transfer out with a radiator for its cooling by water on an independent circulating contour. At use the hearth burning devices providing maximum time of stay of products of combustion in a fire chamber. For the given height of a fire chamber and its sizes increase vapor-productivity the boiler can be reached by increase in speed radiating cooling $\partial T/\partial t$ which is reached by decrease in cross-section section of separate chambers in a multichamber fire chamber.

On fig. 23 dependence of the relation of speed radiating cooling $\partial T(L)/\partial t$ by the speed $(\partial T(L)/\partial t)_{max}$ from distance L between walls of a multichamber fire chamber is presented.

From the presented drawing it is visible that speed $\partial T(L)/\partial t$ reaches the maximum value at small distances L between fire chamber walls when function spectral transmission aspires to unit. With increase in distance L between walls of a fire chamber speed radiating cooling $\partial T(L)/\partial t$ decreases and, hence, for achievement of temperature T_{uh}=const on an exit from a fire chamber time $\Delta t(L)$ decreases with increase vapor-productivity a boiler. The less size L in a multichamber fire chamber, the at smaller height H from hearth fire chambers is reached value T_{uh}. The full value radiating cooling ΔT=T_a-T_{uh} depends on an arrangement of the torches forming the ascending stream of a flame.

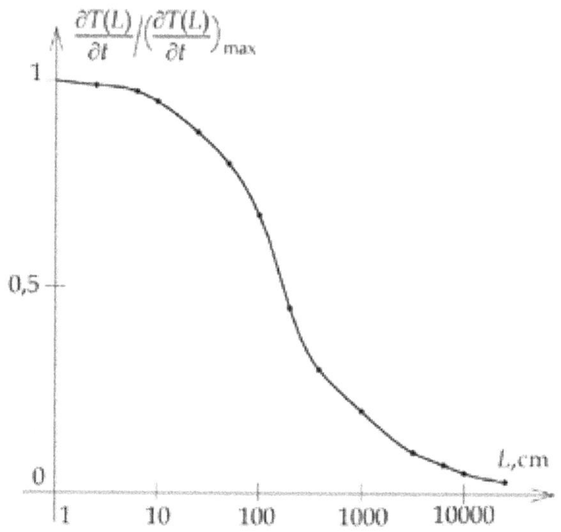

Fig. 23. Relative dependence of average speed radiating cooling $\frac{\partial T(L)}{\partial t} \bigg/ \left(\frac{\partial T(L)}{\partial t}\right)_{max}$ from distance L between walls heatsusceptibility surfaces. Fuel – natural gas, the hearth burning device of matrix type with multirow torches, factor of surplus of air α =1,03.

For formation short flames it is expedient to use multirow burning devices the height of an ardent zone in which depends on diameter of tubes. For definition of height of an ardent zone in the presented calculations the theory of similarity and really observable heights of a flame was used. At diameter of tubes D=8 mm through which fuel (diameter of an individual nozzle) height of an ardent zone of products of combustion arrives makes 70 cm. With increase in diameter D the height of an ardent zone increases and, hence, its speed nonequilibrium radiating cooling decreases.

At modernization of existing steam-boilers with their transfer into an operating mode of multichamber fire chambers it is possible to increase in 2-3 times them vapor-productivity. Application of multichamber fire chambers in case of use of gas fuel allows to reduce steam-boiler dimensions (at the same vapor-productivity).

Designing of multichamber fire chambers demands consideration of technological aspects of their manufacturing about economic efficiency of manufacture. From our point of view manufacturing of a boiler of the cylindrical form of tower type with multichamber fire chambers and rectangular multichamber fire chambers should be rather effective. High thermal loadings of the bottom part heatsusceptibility surfaces of a steam-boiler demand optimization of loading of circulating contours of the heat-carrier and designs heatsusceptibility the surfaces providing their steady hydrodynamic characteristics.

It is necessary to notice that in a homogeneous medium of the infinite sizes of radiating heat exchange doesn't exist.

5.1 Mathematical model of the transfer over of nonequilibrium radiation in flames and high-temperature diphasic mediums

Errors of calculations at the decision of problems of radiating heat exchange are defined by function spectral transmission in structurally non-uniform multicomponent mediums.

Generally at function evaluation spectral transmission $\tau_{\Delta v}$ it is necessary to allocate contributions to the absorption, caused by wings of the remote spectral lines of the various atmospheric gases $\tau_{\Delta v}^{k}$, the absorption induced by pressure $\tau_{\Delta v}^{n}$, selective absorption of the spectral lines entering into the chosen spectral interval (owing to distinctions in these cases of function spectral transmission $\tau_{\Delta v}$ from the maintenance absorbing (radiating) gas, effective pressure *P*, and temperature *T*). Then for the set component function spectral transmission is defined as product of three considered above functions. Similar division allows providing universality of the description $\tau_{\Delta v}$ for any almost realized atmospheres of top internal devices of the present and the future workings out.

For multicomponent atmosphere $\tau_{i\Delta v}$ it will be defined as product $\prod_i \tau_{i\Delta v}$, where *i* - component number. Legitimacy of this law is checked up experimentally and follows from independence of thin structure of spectra of the various absorbing (radiating) components which are a part of torches and oven atmosphere.

Let's believe that structural characteristics of the top internal chamber are known. For the account of nonequilibrium processes of radiation in a torch we will express function of a source for nonequilibrium radiation of a component *i* a torch in a kind

$$B_{i\lambda}(T) = B_{\lambda}^{abb}(T) \cdot \eta_{i\lambda}(T), \tag{42}$$

Where $\eta_{i\lambda}(T)$ – factor of nonequilibrium radiations for a component *i*.

Let's consider at first the elementary case of the absorbing medium: radiation scattering is absent or radiation scattering is neglected. We will assume that the temperatures of walls T_g is known, distribution of temperature *T* on volume of the top internal chamber and a field of concentration of gas and disperse components are set. Let *O* - a supervision point in the top internal chamber, *K* - a point of intersection of a vector of supervision ***l*** with a surface of the top internal chamber. A vector of scanning of volume of space from point *K* we will designate ***L***. We will assume also that a wall surface is Lambert's. Then spectral intensity of thermal radiation in a direction l will be defined by a parity:

$$\begin{aligned} J_{\Delta\lambda}(l) = & \int_0^{L_k} \sum_i B_\lambda\left[T(l)\right]\eta_{i\lambda}\left[T(l)\right]\frac{d\tau_{i\Delta\lambda}}{dl}\prod_{k\neq i}\tau_{k\Delta\lambda}(l)\,dl + \left(1-\delta_\lambda^k\right)B_\lambda\left(T^k\right)\tau_{\Delta\lambda}\left(l_0^k\right) + \\ & + \frac{\delta_\lambda^k}{\pi}\int_0^{2\pi}\int_0^{L_k} B_\lambda\left[T(L)\right]\sum_i \eta_{i\lambda}\left[T(L)\right]\frac{d\tau_{i\Delta\lambda}}{dl}\left[L(\theta,\varphi)+l_0^k\right]\cdot\prod_{k\neq i}\tau_{k\Delta\lambda}\left[L(\theta,\varphi)+l_0^k\right]dL\,d\Omega + \\ & + \frac{\delta_\lambda^k}{\pi}\int_0^{2\pi}\left[1-\delta_\lambda^g\right]B_\lambda\left[T\left(L_g\right)\right]\tau_{\Delta\lambda}\left(L_0^k + l_k\right)d\Omega, \end{aligned} \tag{43}$$

where *T*(*l*) – temperature of medium along an optical way *l*; $B_\lambda\left[T(l)\right]$ – spectral brightness of radiation of absolutely black body at temperature *T* in a point *l*; δ_λ – spectral factor of reflection of a wall; l_0^k – an optical way between points *O* and *K*; T^k – temperature in point *K*; $\tau_{\Delta\lambda}(l)$ – function spectral transmission for an optical way *l* in a spectral interval in width $\Delta\lambda$; λ – length of a wave; T (L_g) – temperature in a point of intersection of vector ***L*** with a wall surface; $d\Omega$ – a space angle element; θ, φ – antiaircraft and azimuthally corners,

accordingly; $\tau_{\Delta\lambda}(L+l)$ means function spectral transmission along an optical way (*L*+l); the index «g» means wall border.
In the ratio (43) product undertakes on all components $k \neq i$, including ashes,

$$\tau_{\Delta\lambda} = \prod_i \tau_{\Delta\lambda i}, \tag{44}$$

where $\tau_{\Delta\lambda i}$ - function of spectral transmission for *i*-th component as gas, so disperse phases of top internal atmosphere. For gas components function $\tau_{\Delta\lambda i}$ are calculated on a two-parametrical method of equivalent mass, considered in section 2.2.
For the account of absent-minded radiation, we will choose the beginning of coordinates at the bottom of a fire chamber. An axis of coordinates *z* we will choose in conformity with symmetry of an ascending stream of products of combustion. We will enter polar system of coordinates. We will designate a supervision point z_n with antiaircraft θ_0 and azimuthal φ_0 supervision corners; θ, φ - flowing antiaircraft and azimuthally corners of integration on space. Then any point in fire chamber space will be characterized by height *z* concerning a bottom of a fire chamber and corners θ, φ, and a surface limiting space of a fire chamber - coordinates z_g, θ, φ. The radiation going to the top hemisphere from a point of supervision z_n , we will name ascending with intensity $J\uparrow$. The radiation going to the bottom hemisphere with intensity $J\downarrow$ we will name descending. The corner of scattering of radiation $\Psi(\theta_0, \varphi_0, \theta, \varphi)$ depends as on a supervision direction θ_0, φ_0, and current corners of integration θ, φ of absent-minded radiation. We will assume further that the fire chamber surface has temperature $T(z_g, \theta, \varphi)$ and spectral factor of reflection $\delta_\lambda(z_g, \theta, \varphi)$, λ - length of a wave of radiation.
Let's enter further scattering indicatryss $f(z,\Psi)$ in such a manner that

$$\int d\phi \int f(z,\Psi)\sin\Psi d\Psi = 1\,. \tag{45}$$

Let $\tau_{\Delta\lambda a}(l)$ - function spectral transmission at the expense of absorption of radiation of a gas phase of top internal atmosphere and its disperse phase, $\tau^a_{\lambda s}(l)$ - function spectral transmission (easing) only at the expense of scattering of radiation of a disperse phase of top internal atmosphere, $\tau^a_{\lambda a}(l)$ - function spectral transmission at the expense of absorption of radiation by aerosols for which following parities are fair:

$$\tau^a_{\lambda a}(l) = \exp\left[-\sigma^a_{\lambda a}\int_l \sigma^a_0(l)dl\right], \tag{46}$$

$$\tau^a_{\lambda s}(l) = \exp\left[-\sigma^a_{\lambda s}\int_l \sigma^a_0(l)dl\right], \tag{47}$$

$$\tau^a_{\Delta\lambda a}(l) = \tau^a_{\lambda a}(l)\prod_i \tau^a_{\Delta\lambda i}(l), \tag{48}$$

where l - an optical way which runs radiation beam, $\sigma^a_{\lambda a}, \sigma^a_{\lambda s}$ - spectral normalizing volume factors of absorption and aerosol scattering, $\sigma^a_{\lambda 0}(l)$ - volume factor of easing of

radiation by an aerosol on length of a wave λ =0,55 μm, $\tau_{\Delta\lambda i}(l)$ - function spectral transmission for *i*-th component of a gas phase of atmosphere for spectral intervals in width Δ with the center λ which are calculated on a two-parametrical method of equivalent mass.
Let's choose a supervision direction *l* which will cross borders of a surface of a fire chamber for the top hemisphere $z_g^{+}(\theta_0,\phi_0)$ and for the bottom hemisphere $z_g^{-}(\theta_0,\phi_0)$ concerning a horizontal plane $z = z_n$.

Then for intensity of ascending radiation $J_{\Delta\lambda}^{\uparrow}$ in approach of unitary scattering in a direction θ_0, φ_0, in a point z_n

$$J_{\Delta\lambda}^{\uparrow} = J_{1\Delta\lambda}^{\uparrow} + J_{2\Delta\lambda}^{\uparrow} + J_{3\Delta\lambda}^{\uparrow} + J_{4\Delta\lambda}^{\uparrow} + J_{5\Delta\lambda}^{\uparrow}, \tag{49}$$

and for intensity of descending radiation

$$J_{\Delta\lambda}^{\downarrow} = J_{1\Delta\lambda}^{\downarrow} + J_{2\Delta\lambda}^{\downarrow} + J_{3\Delta\lambda}^{\downarrow} + J_{4\Delta\lambda}^{\downarrow} + J_{5\Delta\lambda}^{\downarrow}, \tag{50}$$

Where $J_{1\Delta\lambda}^{\downarrow}$ - own descending radiation of the medium of the top internal chamber in a supervision point; $J_{2\Delta\lambda}^{\downarrow}$ - radiation of a wall of the top internal chamber in the supervision direction, weakened by top internal atmosphere; $J_{3\Delta\lambda}^{\downarrow}$ - disseminated in a direction of supervision the radiation which is starting with volume of top internal atmosphere (from point volume); $J_{4\Delta\lambda}^{\downarrow}$ - absent-minded radiation of all walls of the fire chamber, reflected from a point l^g on a wall in a supervision direction; $J_{5\Delta\lambda}^{\downarrow}$ - own radiation of all walls of the chamber, weakened by oven atmosphere and reflected from a point on a wall in a supervision direction. The physical sense of components intensitys in the ratio (49) for ascending radiation is similar.
For nonequilibrium radiation source function is various for various radiating components and can change within top internal volume and on a spectrum of lengths of waves of electron-vibrational transitions of molecules. If sizes $\eta_{i\lambda}$ for components *i* are known, in the intensity equations it is necessary to enter summation of radiations on components *i* under the badge of integrals, having replaced size $B_\lambda(T)$ in size $B_\lambda(T)\cdot\eta_{i\lambda}$. Then for intensity of ascending radiation:

$$J_{1\Delta\lambda}^{\uparrow} = \int_{z_g^-}^{z_n} \sum_i \frac{B_\lambda\left[T(z),\theta_0,\phi_0\right]\cdot\eta_{i\lambda}\left[T(z),\theta_0,\phi_0\right]\tau_{\lambda s}^{a}(z,\theta_0,\phi_0)\dfrac{\partial\tau_{i\Delta\lambda a}(z,z_n,\theta_0,\phi_0)}{\partial z}}{\prod_{k\neq i}\tau_{i\Delta\lambda a}^{a}(z,z_n,\theta_0,\phi_0)}\,dz, \tag{51}$$

$$J_{2\Delta\lambda}^{\uparrow} = B_\lambda\left[T_g\left(z_g^-,\theta_0,\phi_0\right)\right]\tau_{\lambda s}\left(z_g^-,\theta_0,\phi_0\right)\tau_{\Delta\lambda a}\left(z_g^-,\theta_0,\phi_0\right), \tag{52}$$

$$J_{3\Delta\lambda}^{\uparrow} = \int_0^{2\pi}\partial\phi\int_{-\pi/2}^{\pi/2}\sin\theta\,\partial\theta\int_{z_g^-}^{z_n} f_\lambda\left[z,\psi\left(\theta,\phi,\theta_0,\phi_0\right)\right]\frac{\partial}{\partial z}\tau_{\lambda s}\left[z,z_g,\left(\theta,\phi\right),\theta,\phi;z,z_n,\theta_0,\phi_0\right]\cdot$$

$$\cdot \int_{z_g^+}^{z} \left\{ \sum_i B_\lambda[T(z,\theta,\phi)] \eta_{i\lambda}[T(z,\theta,\phi)] \frac{\partial}{\partial z} \tau_{i\Delta\lambda a}[z,z',\theta,\phi;z_n,z,\theta_0,\phi_0] \prod_{k\neq i} \tau_{k\Delta\lambda a}(z,z',\theta,\phi;z_n,z,\theta_0,\phi_0) \right\} dz' dz, \quad (53)$$

$$J_{4\Delta\lambda}^{\uparrow} = \frac{\delta_{\lambda g}(z_g^-,\theta_0,\varphi_0)}{\pi} \int_0^{2\pi} d\varphi \int_{-\pi/2}^{\pi/2} \sin\theta d\theta \int_{z_g^-}^{z_n} f_\lambda[z,\Psi(\theta,\varphi,\theta_0,\varphi_0)] \frac{\partial}{\partial z} \tau_{\lambda s}[z,z_g,(\theta,\varphi),\theta,\varphi;z,z_n,\theta_0,\varphi_0] \cdot \quad (54)$$
$$\cdot \tau_{\Delta\lambda a}[z,z_g,(\theta,\varphi),\theta,\varphi;z,z_n,\theta_0,\varphi_0] \cdot [1-\delta_{\lambda g}(z,\theta_0,\varphi_0)] B_\lambda^g[T(\theta,\varphi)] dz,$$

$$J_{5\Delta\lambda}^{\uparrow} = \frac{\delta_\lambda(z_g^-,\theta_0,\varphi_0)}{\pi} \int_0^{2\pi} d\varphi \int_{-\pi/2}^{\pi/2} \{1-\delta_\lambda(z_g,(\theta,\varphi))\} \cdot B_\lambda^g[T(\theta,\varphi)] \tau_{\lambda s}[z_g^-,z_g(\theta,\varphi),\theta,\varphi;z_g^-,z_n,\theta_0,\varphi_0] \cdot \quad (55)$$
$$\cdot \tau_{\Delta\lambda a}[z_g^-,z_g(\theta,\varphi),\theta,\varphi;z_g^-,z_n,\theta_0,\varphi_0] d\theta,$$

where summation is carried out on all components i, and product – on all components $k \neq i$; z_g^- means that fire chamber borders are located below supervision height z_n; $\delta_{\lambda g}^+$ means reflection factor on border of the fire chamber located at height $z_g > z_n$.

For intensity of descending radiation $J_{1\Delta\lambda}^{\downarrow}$ it is easy to write parities, similar (50-54),

$$J_{1\Delta\lambda}^{\downarrow} = \int_{z_g^+}^{z_n} \sum_i B_\lambda[T(z),\theta_0,\varphi_0] \cdot \eta_{i\lambda}[T(z),\theta_0,\varphi_0] \tau_{\lambda s}^a(z,\theta_0,\varphi_0) \frac{\partial \tau_{i\Delta\lambda a}(z,z_n,\theta_0,\varphi_0)}{\partial z} \prod_{k\neq i} \tau_{i\Delta\lambda a}^a(z,z_n,\theta_0,\varphi_0) dz, \quad (56)$$

$$J_{2\Delta\lambda}^{\downarrow} = B_\lambda[T_g(z_g^+,\theta_0,\varphi_0)][1-\delta_\lambda(z_g^+,\theta_0,\varphi_0)] \tau_{\lambda s}(z_g^+,\theta_0,\varphi_0) \tau_{\Delta\lambda a}(z_g^+,\theta_0,\varphi_0), \quad (57)$$

$$J_{3\Delta\lambda}^{\downarrow} = \int_0^{2\pi} d\varphi \int_{-\pi/2}^{\pi/2} \sin\theta d\theta \int_{z_g^+}^{z_n} f_\lambda[z,\Psi(\theta,\varphi,\theta_0,\varphi_0)] \frac{\partial}{\partial z} \tau_{\lambda s}[z,z_g,(\theta,\varphi),\theta,\varphi;z,z_n,\theta_0,\varphi_0] \cdot \quad (58)$$

$$\cdot \int_{z_g^+}^{z} \left\{ \sum_i B_\lambda[T(z,\theta,\phi)] \eta_{i\lambda}[T(z,\theta,\phi)] \frac{\partial}{\partial z} \tau_{i\Delta\lambda a}[z,z',\theta,\phi;z_n,z,\theta_0,\phi_0] \prod_{k\neq i} \tau_{k\Delta\lambda a}(z,z',\theta,\phi;z_n,z,\theta_0,\phi_0) \right\} dz' dz, \quad (59)$$

$$J_{4\Delta\lambda}^{\downarrow} = \frac{\delta_{\lambda g}(z_g^+,\theta_0,\varphi_0)}{\pi} \int_0^{2\pi} d\varphi \int_{-\pi/2}^{\pi/2} \sin\theta d\theta \int_{z_g^+}^{z_n} f_\lambda[z,\Psi(\theta,\varphi,\theta_0,\varphi_0)] \frac{\partial}{\partial z} \tau_{\lambda s}[z,z_g,(\theta,\varphi),\theta,\varphi;z,z_n,\theta_0,\varphi_0] \cdot \quad (60)$$
$$\cdot \tau_{\Delta\lambda a}[z,z_g,(\theta,\varphi),\theta,\varphi;z,z_n,\theta_0,\varphi_0] \cdot [1-\delta_{\lambda g}(z,\theta_0,\varphi_0)] B_\lambda^g[T(\theta,\varphi)] dz,$$

$$J_{5\Delta\lambda}^{\downarrow} = \frac{\delta_\lambda(z_g^+,\theta_0,\varphi_0)}{\pi} \int_0^{2\pi} d\varphi \int_{-\pi/2}^{\pi/2} \{1-\delta_\lambda(z_g,(\theta,\varphi))\} B_\lambda^g[T(\theta,\varphi)] \cdot \tau_{\lambda s}[z_g^+,z_g(\theta,\varphi),\theta,\varphi;z_g^+,z_n,\theta_0,\varphi_0] \cdot \quad (61)$$
$$\cdot \tau_{\Delta\lambda a}[z_g^+,z_g(\theta,\varphi),\theta,\varphi;z_g^+,z_n,\theta_0,\varphi_0] d\theta.$$

Processes of nonequilibrium radiation at burning hydrocarbonic fuels practically aren't developed also their influence on radiating cooling a torch of top internal space practically isn't studied. From the most general reasons of formation of electron-vibrational spectra it is possible to draw a conclusion that the greatest influence on process of radiating heat exchange in fire chambers nonequilibrium renders radiations at burning of gaseous hydrocarbonic fuel and black oil which incorporate S-contents and N-contents the components forming nonequilibrium radiation of a high-temperature kernel of a torch.

At the decision of problems of radiating heat exchange in boilers operate integrated intensity thermal radiation which are defined by integration spectral intensitys thermal radiation on a spectrum of lengths of waves λ:

$$J^{\uparrow}(z_n,\theta_n,\phi_n)=\int_0^{\infty} J_{\lambda}^{\uparrow}(z_n,\theta_n,\phi_n)d\lambda\,, \tag{62}$$

$$J^{\downarrow}(z_n,\theta_n,\phi_n)=\int_0^{\infty} J_{\lambda}^{\downarrow}(z_n,\theta_n,\phi_n)d\lambda\,. \tag{63}$$

Knowing sizes $J^{\uparrow\downarrow}(z_n,\theta_n,\phi_n)$, it is possible to define streams of thermal radiation on any direction including on heatsusceptibility surfaces, having executed spatial integration $J^{\uparrow\downarrow}$ within a space angle 2π . In particular, for streams of descending and ascending radiation

$$F^{\uparrow\downarrow}(z)=\int_0^{2\pi} J^{\uparrow\downarrow}(z,\theta,\phi)d\Omega\,, \tag{64}$$

where $d\Omega$ - a space angle element. Radiating change of temperature will be defined from a parity

$$\frac{\partial T(z,\theta,\phi)}{\partial t}=\frac{1}{\rho(z,\theta,\phi)C_p(z,\theta,\phi)}\cdot\frac{\partial F(z,\theta,\phi)}{\partial z}\,, \tag{65}$$

where $\rho(z,\theta,\phi),C_p(z,\theta,\phi)$ - accordingly density and a thermal capacity in a local point with coordinates z, θ, φ,

$$F(z,\theta,\varphi)=F^{\uparrow}(z,\theta,\varphi)-F^{\downarrow}(z,\theta,\varphi).$$

If heat exchange process is stationary, $dT(z,\theta,\phi)/dz=const$ for any local volume with coordinates z, θ, φ. If heat exchange process is not stationary there are time changes of temperature in the local volumes which time trend can be calculated by application of iterative procedure of calculations on each time step *i* so

$$T_{i+1}(z,\theta,\phi)=T_i(z,\theta,\phi)+\Delta t\frac{dT_i(z,\theta,\phi)}{dt}\,. \tag{66}$$

However thus it is necessary to take into consideration and influence of other mechanisms of heat exchange: diffusion, turbulent diffusion, convective heat exchange.

Most intensively radiating cooling it is shown in a torch kernel, in this connection its temperature always below theoretical on 15-20 %. The last means that during combustion of fuel the torch considerably cools down as a result of radiating cooling. Degree radiating cooling a torch is maximum, if the stream expires in free atmosphere. In the closed volume of a fire chamber radiating cooling increases with growth of temperature of a torch, degree of its blackness at the expense of absorption of radiation by gas and disperse phases of products of combustion and decreases at rise in temperature heatsusceptibility surfaces and their factors of reflection. In cold zones of a fire chamber can take place and radiating heating if in them active components contain optically. If there are temperature inversions in temperature distributions in zones of temperature inversions radiating heating or easing radiating cooling also can be observed.

Full radiating cooling combustion products in a fire chamber depends on time of their stay in top internal volume and, hence, from speed of movement of products of combustion $V(z)$ in a fire chamber which can change on fire chamber height. Full radiating cooling combustion products ΔT it is defined by the formula:

$$\Delta T = \int_0^H \frac{1}{V(z)} \cdot \frac{\partial T}{\partial z} dz\,, \tag{67}$$

where H is fire chamber height.

Let's analyze the physical processes proceeding in the top internal chamber under the influence of nonequilibrium short-wave radiation which is generated in ultra-violet and visible parts of a spectrum as a result of a relaxation of the raised molecules formed at burning of fuel. If the difficult molecule is formed in wild spirits and dissociates on unstable short-living splinters also its splinters will be in wild spirits and to generate nonequilibrium short-wave radiation. Owing to small time of life of these connection spectral lines of nonequilibrium radiation will be much wider, than for equilibrium radiation, and can create the diffuse spectra of radiation which are not dependent from widening of pressure. Functions spectral transmission for the nonequilibrium medium submits to the law of Buger:

$$\tau_{\Delta v}(L) = \int_L \exp\left[-k_v(L)\right] dL\,, \tag{68}$$

where $k_v(L)$ - absorption factor, v - the wave number, and integration is carried out along optical way L to a torch kernel. Vibrational and rotary structures of a spectrum of nonequilibrium radiation it will be washed away and poorly expressed. There is a basis to believe, as nuclear spectra of elements also can be nonequilibrium that proves to be true on an example of nuclear spectra of the sodium which lines of radiation have appeared nonequilibrium and at high temperatures can't be used for definition of temperature of a flame. Hence, probably to expect presence of photochemical reactions under the influence of the short-wave radiation, products of combustion essentially influencing a chemical composition in the top internal chamber.

Feature of nonequilibrium processes of radiation is considerable cooling zones of chemical reactions in time $\approx 10^{-4}$ sec, commensurable in due course courses of chemical reactions. In this connection the equilibrium temperature of a flame considerably decreases that leads to much lower concentration of a monoxide of nitrogen NO. Really, it agree (Zel'dovich et al., 1947)

$$[NO]_{\max} = 4.6\sqrt{C_{N_2} C_{O_2}} \exp\left(\frac{-21500}{RT_{\max}}\right), \tag{69}$$

where at fuel burning in air

$$C_{O_2} = 21(\alpha - 1)V_0 / (1 + \alpha V_0)$$
$$C_{N_2} = 79\alpha V_0 / (1 + \alpha V_0) \quad (70)$$

Here T_{max} - the maximum absolute temperature in peaks of volumes of chemical reactions, R - a gas constant, V_0 - theoretically necessary quantity of air for fuel burning, α - factor of surplus of air. At high temperatures real concentration NO in combustion products on an order and lower, than intended under formulas (69, 70) that from our point of view is caused nonequilibrium radiating cooling peaks of chemical reactions. And real concentration NO can depend on depth of turbulence burning and a spectrum of whirls.

By consideration of radiating heat exchange in the top internal chamber with torch burning of firm fuel in the twirled streams it is necessary to take into consideration the phenomenon of separation of particles when the largest particles are taken out in peripheral zones of a fire chamber where, settling, can grasp sooty ashes, formed as a result of pyrolysis in cold zones of a fire chamber, and then to flow down in a cold funnel.

Considering dependence of absorption of nonequilibrium radiation by combustion products, we will pay attention to strengthening of absorption with increase in capacity of the top internal chamber. Hence, with increase in capacity of a fire chamber nonequilibrium radiation in a greater degree passes in thermal energy of particles of fuel and thermal energy of products of combustion. Nonequilibrium radiating cooling decreases also and concentration NO_x increases with increase in capacity of a fire chamber that is really observed by results of statically provided supervision.

Let's pay attention to results of measurements of a chemical composition of products of combustion of wood (Moskalenko et al., 2010) when raised concentration NO_2 have been found out. If at burning of black oil and gases the relation of concentration $C(NO_2)/C(NO) \approx 0.1$, at burning of wood the relation of concentration $C(NO_2)/C(NO) \approx 1/3$. It means that the increase in concentration NO_2 causes increase in intensity of the nonequilibrium radiation reducing temperature of a flame, and, hence, leads to reduction of concentration NO.

Considering optical properties of a disperse phase depending on a microstructure of liquid or firm fuel at chamber burning, it is necessary to notice that concentration NO will increase in smoke gases with increase in a subtlety of scattering of liquid fuel and crushing of firm fuel. From the point of view of ecological influence of atmospheric emissions on flora and fauna expediently chamber burning of fuel of rough crushing and scattering. Besides, from the point of view of minimization of anthropogenous influences on medium it is expedient to burn fuel at lower pressure as nonequilibrium radiating cooling amplifies with pressure decline in a fire chamber (process of suppression of a chemical luminescence with pressure decline it is weakened).

Presence chemical unburning leads to formation of heavy hydrocarbons in combustion products (especially benzologies) that causes suppression of nonequilibrium radiation in a fire chamber. The last can be formed in interfaces of the top internal chamber and weaken heatsusceptibility of screens owing to strong absorption of ultra-violet radiation.

Presence of connections of sulfur in fuel leads to occurrence of nonequilibrium radiation SO_2 in the field of a spectrum $\lambda<0.4$ μm which reduces flame temperature, and, hence, and concentration NO_x.

5.2 Modelling of radiating heat exchange in multichamber fire chambers

Let's consider results of modeling of radiating heat exchange of multichamber fire chambers taking into account nonequilibrium processes of radiation (Moskalenko et al., 2009), section 5.1 executed on algorithms for diphasic structurally non-uniform medium of top internal space of the chamber of combustion.

On fig. 24 for an example results of calculations of vertical profiles of speed radiating cooling $\partial T(z)/\partial t, \partial T(z)/\partial z$ and stationary distribution of temperature $T(z)$ from fire chamber height z over cuts of capillaries matrix burning devices are illustrated. Fuel is natural gas of a gas pipeline of Shebalovka-Brjansk-Moskva, the size of horizontal section of a cell of a multichamber fire chamber 1,25x1,6 m^2.

Speed of giving of products of combustion on an initial site of a fire chamber makes values υ_0=25 m/s and υ_0=20 m/s at pressure in a fire chamber $1 \cdot 10^5$ Pa. Height of an ardent zone Δz = 0,7 m. In calculations are considered equilibrium and nonequilibrium processes of radiation on the algorithms considered above. It is supposed that process of burning of various components of gas fuel occurs independently at optimum value of factor of surplus of air α =1,03.

The microstructure sooty ashes is measured at burning (to look section 4) methane, propane-butane and acetylene (Moskalenko et al., 2010). Optical characteristics sooty ashes are calculated for the measured microstructures of a disperse phase of products of combustion. Volume factors of easing, absorption and scattering normalized on the measured values of optical density ash (Moskalenko et al., 2009).

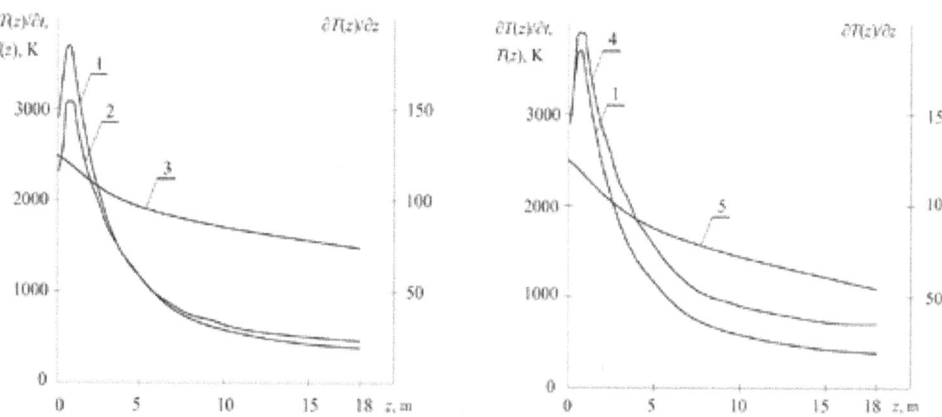

Fig. 24. Results of calculation of radiating heat exchange in a multichamber fire chamber with the size of horizontal section of a cell 1,25x1,6 m^2 for initial average speed of a current of products of combustion of 25 m/s (a) and 20 m/s (b). $\partial T(z)/\partial t, \partial T(z)/\partial z$ - speeds radiating cooling, $T(z)$ – a temperature profile of average on section of temperature depending on height z over cuts of capillaries multirow torches. 1– $\partial T(z)/\partial t$; 2 – $\partial T(z)/\partial z$ for initial average speed of a current of products of combustion of 25 m/s; 3 - $T(z)$ for initial average speed of a current of products of combustion of 25 m/s; 4 – $\partial T(z)/\partial z$ for initial average speed of a current of products of combustion of 20 m/s; 5 - $T(z)$ for initial average speed of a current of products of combustion of 20 m/s.

On fig. 25 examples of spectral and spatial distributions of thermal radiation on heatsusceptibility surfaces of a cell of a multichamber fire chamber by results of the closed modeling of process of radiating heat exchange with calculation of speed radiating cooling products of combustion and their temperature depending on height over cuts of capillaries multirow a torch forming ascending streams of a flame are resulted. Horizontal section of a cell of a multichamber fire chamber - a square with the party of 1,4 m. Fuel - natural gas of a gas pipeline of Shebalovka-Brjansk-Moskva, factor of surplus of air α = 1,03. Average initial speed of products of combustion makes 25 m/s. Pressure in a fire chamber - 10^5 Pa. The executed calculations of heatsusceptibility surfaces show that the greatest thermal loading the bottom part of lateral screens and heatsusceptibility is exposed to a surface hearth of fire chambers. So, on the central axis of the lateral screen at heights 1, 7, 17 meters from a cut of capillaries of a torch falling streams of heat make accordingly 260,313; 99,709; 48,387 kW/m^2. For the center hearth of fire chambers the falling stream of heat answers value of 249,626 kW/m^2, and the ascending stream of heat at height h = 18 m on an axis of a cell of a fire chamber makes 41,115 kW/m^2. A full stream

$$F = \int_S F(S)dS = \sum_i V_i \int_0^h C_{ip}[t(z)]dz\,, \tag{71}$$

where C_{ip}, V_i - accordingly a thermal capacity at the constant pressure, answering to temperature t in a point z and volume for a component i combustion products. This condition at the closed modeling of heat exchange is carried out with a margin error 1 %. In approach of "gray" radiation when calculations are carried out under the law of Buger, overestimate heatsusceptibility on 15 % is observed. The account of effective pressure reduces an error of calculation full heatsusceptibility by 5-6 %. At use of a two-parametrical method of equivalent mass in calculations of function spectral transmission at modeling of a disperse phase of products of combustion in the present calculations it is supposed that burning of each component of fuel occurs independently that allows to use optical density sooty ashes by results of measurements on ardent measuring complexes. For methane, propane-butane, acetylene the optical density on length of a wave 0,55 μm is accepted according to equal 0,1; 0,2; 0,4 m^{-1} in an ardent zone. Above an ardent zone it is observed exponential recession of numerical density thin-dispersion ashes with height in connection with its burning out. More rougly-dispersion fractions 2,3 sooty ashes don't burn out, and their distribution doesn't depend on height. The contribution of each fraction ashes is normalized according to volume concentration CH_4, propane-butane, C_2H_2.
On fig. 26 distribution of an integrated stream of the radiation calculated taking into account absorption (radiation) by basic optically by active components of products of combustion on lateral walls of a cell of a multichamber fire chamber depending on height of a fire chamber in case of weak approximation is illustrated. On fig. 27 distribution of an integrated stream of radiation to lateral walls of a cell of a multichamber fire chamber depending on height the fire chambers calculated with use of function spectral transmission on a two-parametrical method of equivalent mass is presented. On fig. 28 distribution of an integrated stream of the radiation calculated taking into account absorption (radiation) by basic optically active components of products of combustion, but without effective pressure is resulted. For the given design of a multichamber fire chamber the contribution of nonequilibrium radiation to radiating heat exchange makes 7,5 % from a full stream. Absence of the account of

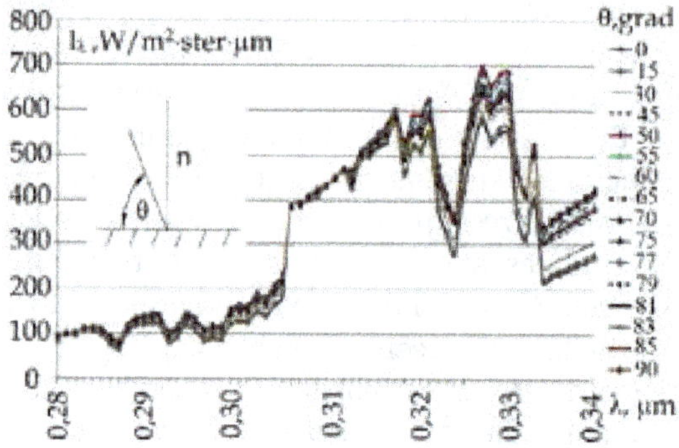

A-a)

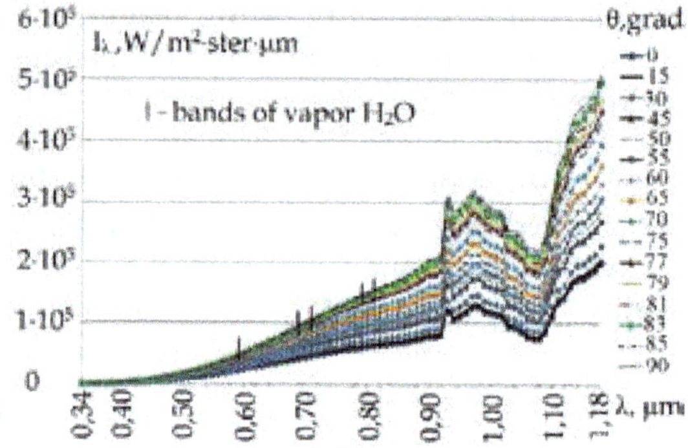

A-b)

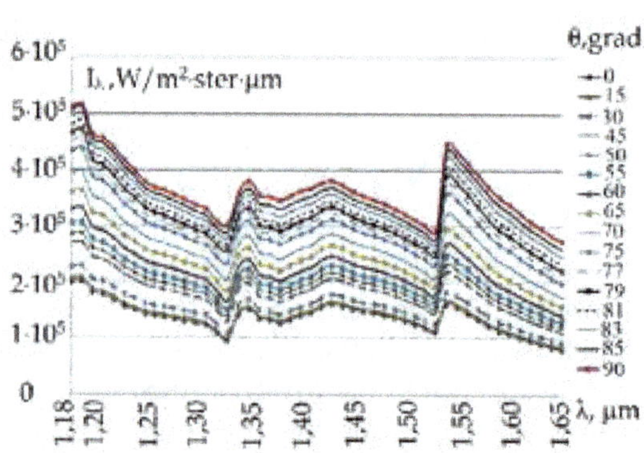

A-c)

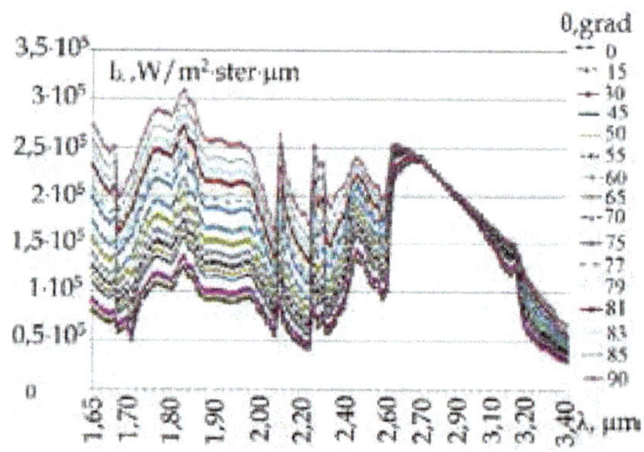

A-d)

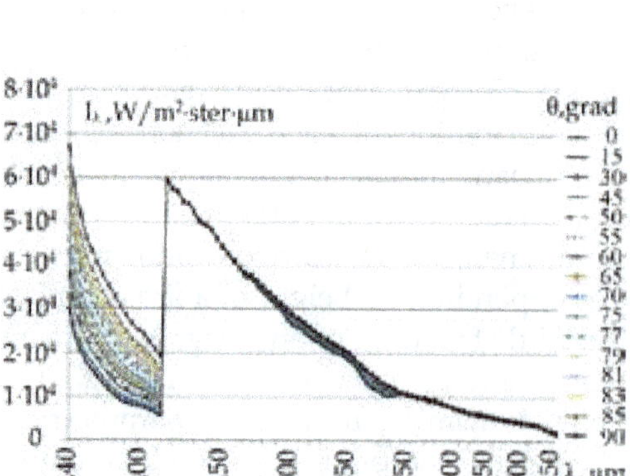

A-e)

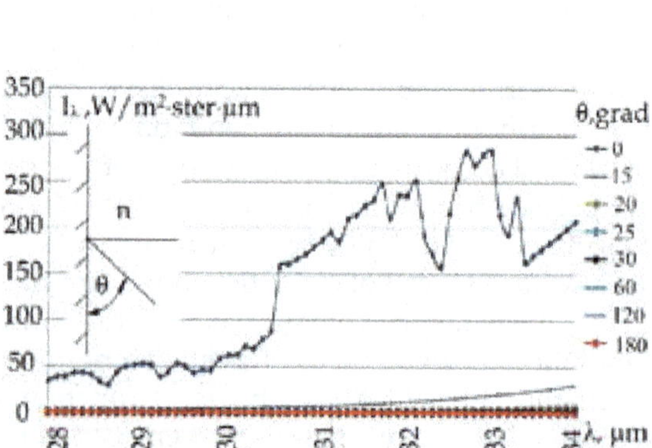

B-a)

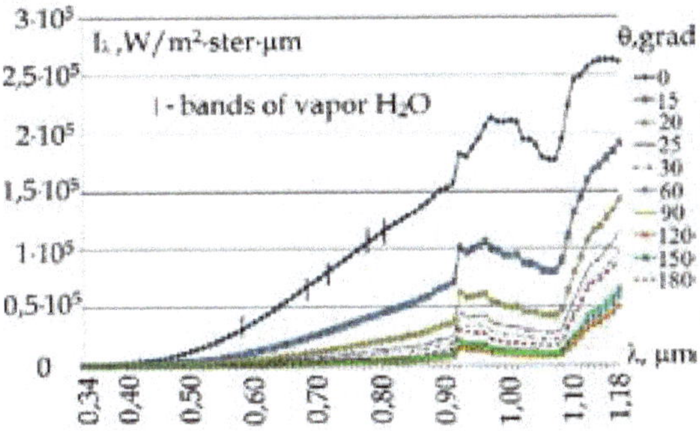

B-b)

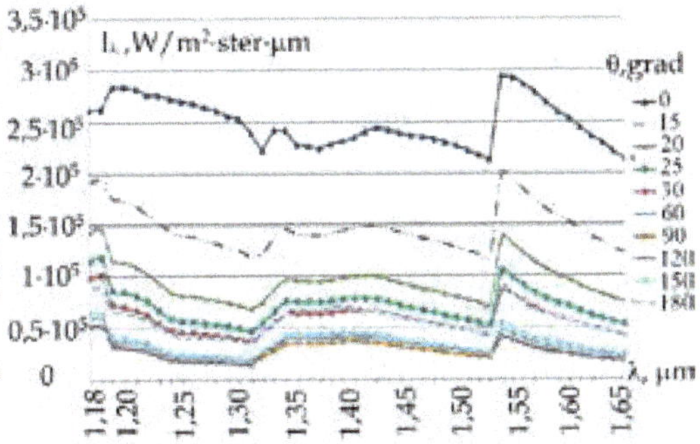

B-c)

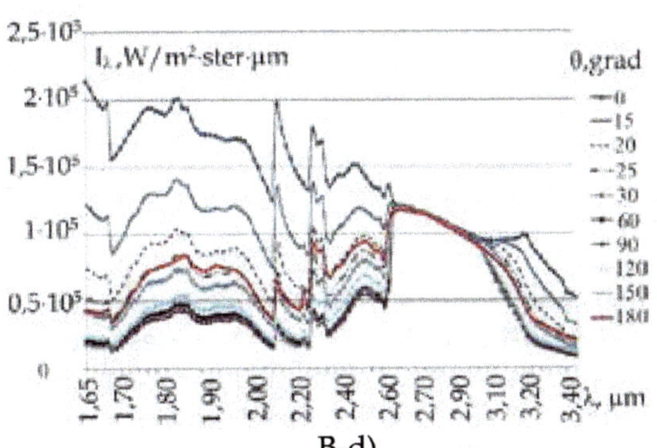

B-d)

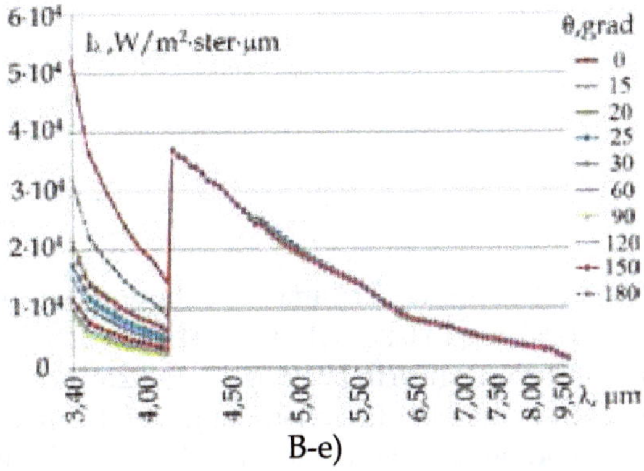

B-e)

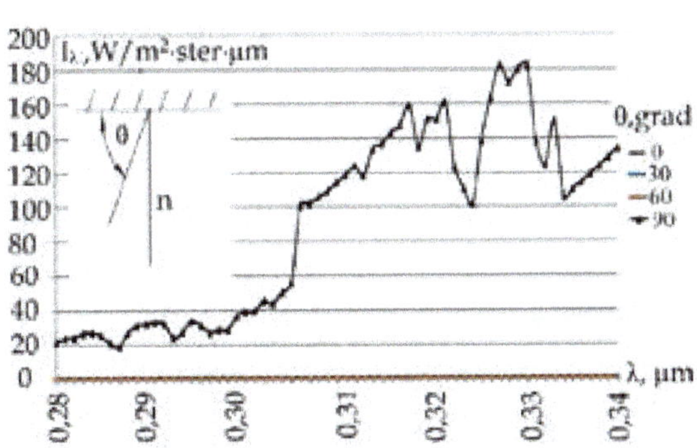

C-a)

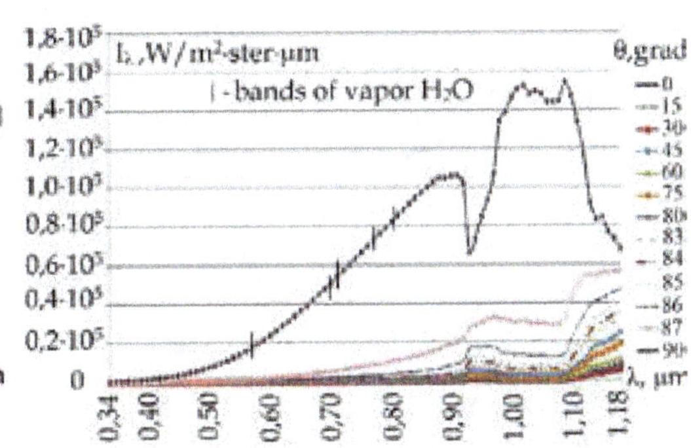

C-b)

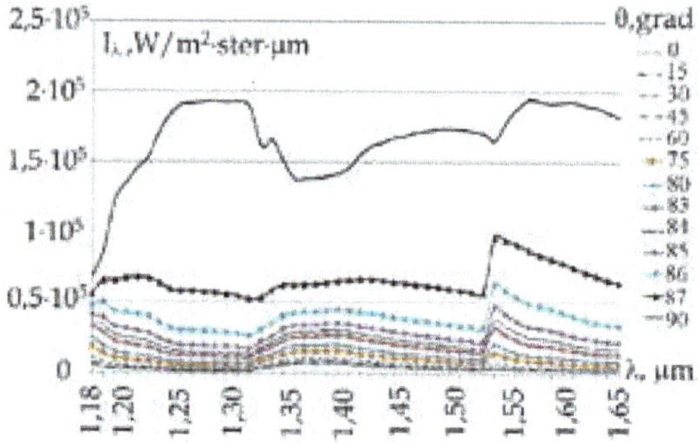

C-c)

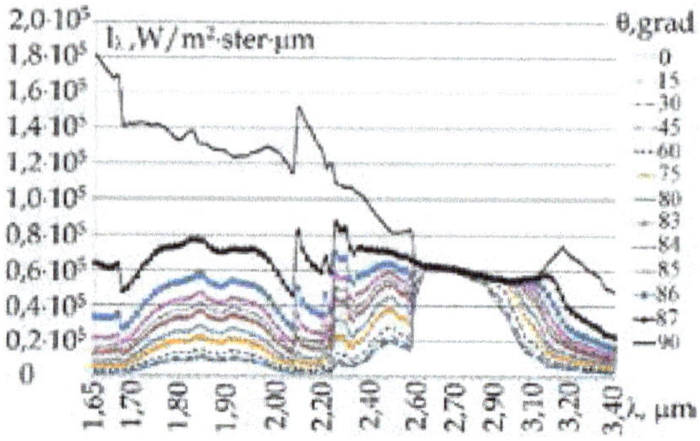

C-d)

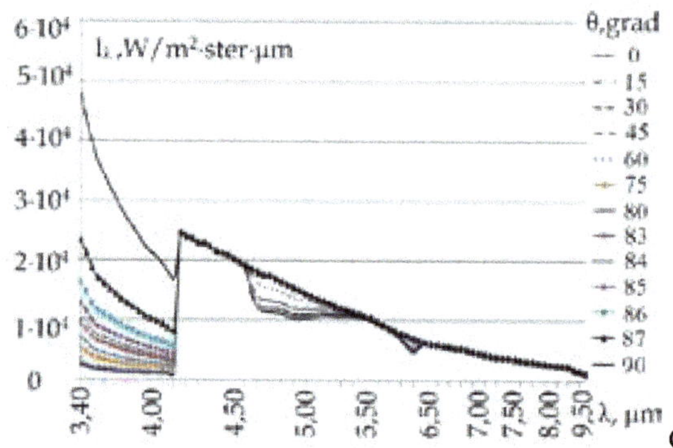

C-e)

Fig. 25. Spectral and spatial distribution of thermal radiation in spectrum ranges: a) 0,28÷0,34 μm; b) 0,34÷1,18 μm; c)1,18÷1,65 μm; d) 1,65÷3,4 μm; e)3,4÷9,5 μm. A – descending radiation on a hearth heatsuscebility surface, B – falling radiation on lateral screens of a cell of a multichamber fire chamber at level 7 meter from a cut of capillaries multirow torches, C – ascending radiation at level of 18 meter from a cut of capillaries multirow torches of a cell of a multichamber fire chamber

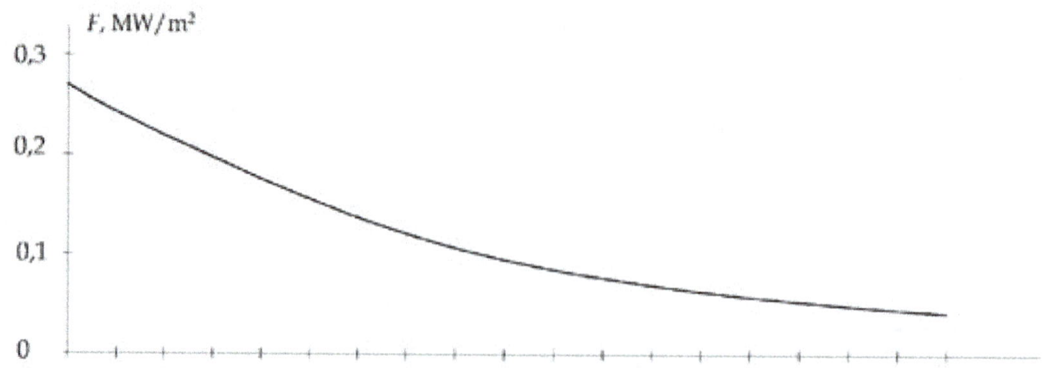

Fig. 26. Distribution of an integrated stream of the radiation depending on height of a fire chamber in case of weak approximation.

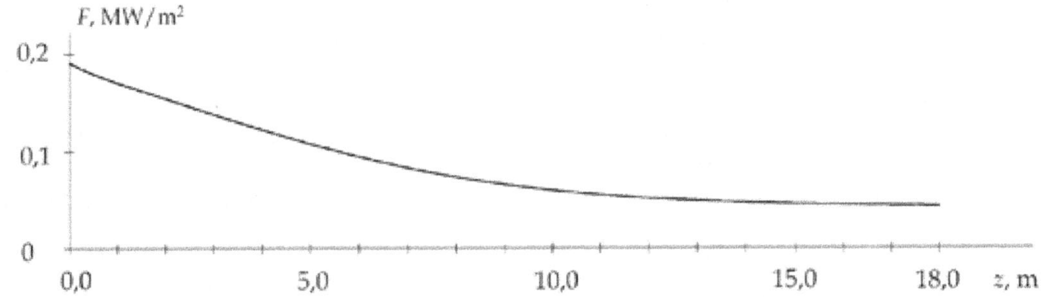

Fig. 27. Distribution of an integrated stream of the radiation depending on height the fire chambers calculated with use of function spectral transmission on a two-parametrical method of equivalent mass.

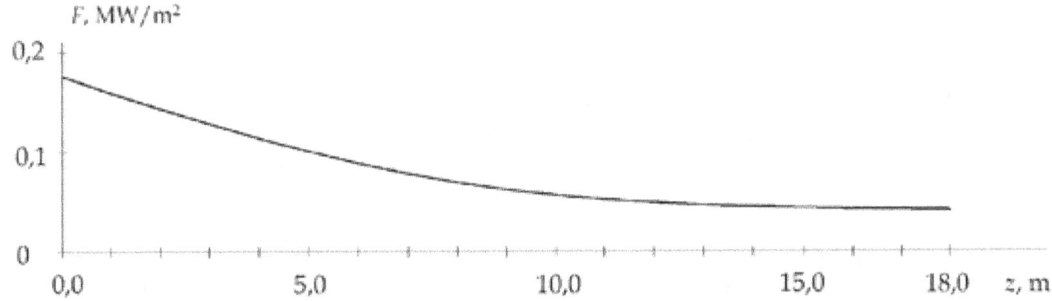

Fig. 28. Distribution of an integrated stream of the radiation calculated without effective pressure.

effective pressure in functions spectral transmission gas components underestimates radiating heat exchange on 5-6 %. The disperse phase of products of combustion influences radiating heat exchange at the expense of radiation ashes. Radiation scattering ashes poorly influences radiating heat exchange in strongly absorbing top internal atmosphere. Reflection of radiation from walls of the top internal chamber leads to reduction of speed radiating cooling in top internal volume.

Generally at heat exchange calculations it is necessary to consider the transfer over of heat at the expense of recirculation of products of combustion in a fire chamber and massexchange owing to diffusion which influence temperature distribution on volume of the chamber of combustion. With the advent of supercomputers there is possible an application of numerical methods of the decision of problems of the transfer over of radiation (Marchuk & Lebedev, 1981; Surgikov, 2004; Moskalenko et al., 1984) which restrain insufficient reliability of data on parameters of spectral lines of gas components of products of combustion.

6. Conclusion

In the conclusion we will stop on the basic results received in the course of the present work.
1. The developed measuring optic-electronic complexes for research of optical characteristics of high-temperature mediums and flames have allowed to spend registration of spectra of absorption and spectra of radiation various flames with the average and high spectral

permission at various lengths of an optical way from 0,2 to 16 m. Uniformity of temperature flames provided possibility of measurement of their temperature by optical methods with a margin error ±2 %. The method of definition nonequilibrium radiating cooling a flame from experimental data on its temperature is developed. Data on a role of nonequilibrium processes on radiating cooling optically the thin torch are received, allowing estimating influence of nonequilibrium processes of radiation in ultra-violet, visible and infra-red parts of a spectrum on radiating heat exchange of torches of aerocarriers and in top internal chambers.

2. The analysis of results of long-term measurements of radiating characteristics of gas and disperse phases of products of combustion is made and radiating characteristics various optically active components of products of combustion, including the cores (vapor H_2O and CO_2) and small components are received. Data on a microstructure sooty ashes and to its optical characteristics are received at burning of various gas hydrocarbonic components in oxygen and in air. Strong dependence of a microstructure sooty ashes from molecular structure of gas fuel and a burning mode is observed. Mass concentration sooty ashes is minimum at burning of methane CH_4 and is maximum at burning of acetylene C_2H_2. The microstructure sooty ashes at black oil burning is close to its microstructure at acetylene burning. Parameterization of gas components of products of combustion is executed on a two-parametrical method of equivalent mass.

3. The method of modeling of the transfer over of thermal radiation in nonequilibrium to radiating multicomponent non-uniform atmosphere under structural characteristics of top internal space is developed. The design of multichamber fire chambers with ascending movement of products of combustion in a fire chamber and vertical development of a flame of the hearth multirow torches forming uniform for all chambers of a multichamber fire chamber burning device of matrix type with the general gas collector for giving of gas fuel and a collector for giving of an oxidizer (air or oxygen) is offered. The burning device is expedient for the transfer out with a radiator for cooling by its water on an independent circulating contour. The design of a multichamber fire chamber at use of gas fuel allows to raise efficiency on 2-3 % and to increase it vapor-productivity in 2-3 times at preservation of parameters of vapor and boiler dimensions.

4. The closed modeling of radiating heat exchange in the chamber of combustion of a multichamber fire chamber with horizontal section of a cell of a boiler 1,25x1,6 m and 1,4x1,4 m is executed at factor of surplus of air α =1,03 and average initial speed of a current of products of combustion of 25 m/s and 20 m/s. Data in the speeds radiating cooling $\partial T(z)/\partial t, \partial T(z)/\partial z$ and to temperature profile T (z) depending on height z over cuts of capillaries matrix burning devices are received. Calculations heat susceptibility on heatsusceptibility to surfaces of the top internal chamber is executed. Full stream F of thermal radiation on a fire chamber surface will be coordinated with change enthalpy on an exit from the top internal chamber with a margin error 0,3%. Nonequilibrium radiating cooling makes 7,5 %. The account of effective pressure in a fire chamber leads to growth of a full stream of radiation F on 5÷6 %.

5. Consideration of optical properties of gas and disperse phases of products of combustion hydrocarbonic fuels and the offered algorithms of numerical modeling allows to draw following conclusions:

- nonequilibrium radiation reduces concentration of harmful component NO_x;
- nonequilibrium radiation leads to heating of particles of fuel and accelerates their ignition by that more intensively, than more small particles and then more their section of absorption of radiation;

- in case of burning of gas fuel nonequilibrium radiation practically isn't transformed by products of combustion and without easing is absorbed by walls of the chamber of combustion (screens);
- in case of presence of a disperse phase nonequilibrium radiation is absorbed by aerosols (soot and fuel particles) and its role in processes of radiating heat exchange is weakened, as energy of nonequilibrium radiation as a result of absorption passes in thermal energy of particles;
- nonequilibrium radiation (especially rigid ultra-violet radiation) can to initiate photochemical reactions in processes of combustion and to influence radiating heat exchange through changes of radiating properties of products of combustion.

6. The basic component defining nonequilibrium radiation in flames is hydroxyl OH. Factors of absorption OH in ultra-violet and infra-red areas of a spectrum are defined. Quantum-mechanical consideration of formation of spectra of nonequilibrium radiation shows that nonequilibrium radiation is shown both in electronic, and in vibrational-rotary spectra of molecules OH which is in raised and basic electronic conditions: bands v_1, $2v_1$, $3v_1$, where v_1 – frequency of normal vibration. Nonequilibrium radiation OH is revealed in a vicinity of lengths of waves 1; 1,43; 2,1; 2,7; 4,1 μm in flame hydrogen-oxygen. The method of definition of vibrational temperature in radiation spectra flames is developed. Presence of spectral structure of vibrational temperature testifies to its dependence on vibrational and rotary quantum numbers.

7. For a homogeneous mediums the law of Kirchhoff is carried out. In non-uniform medium on structure it is broken also function spectral transmission becomes depending as from thin structure of a spectrum of the radiating volume, and thin structure of a spectrum of the absorbing medium, and differs from function spectral transmission for sources of not selective radiation which are measured in laboratory experimental researches. The transfer over of selective radiation is influenced by following factors: the temperature self-reference of spectral lines of radiation, displacement of spectral lines with pressure, the temperature displacement of the spectral lines which have been found out for easy molecules (vapor H_2O, CH_4, NH_3, OH). Till now influence of last two factors wasn't investigated. Quantummechanics calculations of displacement of spectral lines with pressure make thousand shares of cm^{-1} and in conditions turbulized atmosphere can't render essential influence on function spectral transmision. Temperature displacement of spectral lines in a flame make the 100-th shares of cm^{-1} and at high temperatures reach semiwidth of spectral lines and more. It leads to that radiation of a high-temperature kernel of a torch is weakened to a lesser degree by its peripheral layers that increases heatsusceptibility surfaces of heating at the expense of radiating heat exchange. At registration of radiation of a torch of the aerocarrier in atmosphere the effect of an enlightenment of atmosphere in comparison with the account only the temperature self-reference of spectral lines of radiation of a torch is observed more considerably.

8. The analysis of radiating heat exchange between gas and disperse phases of products of combustion gaseous fuels shows that the temperature sooty particles should be below thermodynamic temperature of gases that weakens influence of a disperse phase on radiating heat exchange in the top internal chamber. On the other hand, absorbing properties of sooty ashes define its role in radiating heat exchange, forming a field of thermal radiation in space of the top internal chamber. Scattering of radiation by particles of a disperse phase of products of combustion shows weak influence on distribution of streams of radiation on heatsusceptibility surfaces of the top internal chamber. Mass concentration of

sooty ashes and its microstructure considerably depend on structure of gas fuel and a burning mode. At performance of calculations of radiating heat exchange the disperse phase of products of combustion is supposed multicomponent and is defined by various mechanisms of its formation. Each fraction of an aerosol has the optical characteristics, normalized on easing factor at length of a wave λ =0,55 μm. Spectral factors of easing, absorption, scattering and indicatryss of scattering are calculated for polydisperse ensemble of spherical particles of the set chemical compound. The electronic database includes three fractions of sooty ashes (primary thin-dispersion sooty ash, fraction of average dispersion, and coagulation fraction of soot of smoke gases), flying fraction ashes and roughly-dispersion fraction of products of combustion of firm fuel. As structural characteristics optical density on length of a wave λ =0,55 μm for various fractions of a disperse phase of products of combustion acts. The real spectral optical characteristics entering into settlement formulas are calculated on an electronic database in the assumption that burning of each component of fuel occurs independently that allows using optical density and a microstructure of sooty ashes by results of measurements on ardent measuring complexes.

7. References

Alemasov, V.E. & Dregalin, A.P. et al. (1972). *Thermodynamic and Physical Properties of Combustion Products,* VINITI, Moscow, Russia

Broida, H.P. & Shuler, K.E. (1952). Kinetics of OH Radical from Flame Emission Spectra. IV. A Study of Hydrogen-Oxygen Flame. *Journ. Chem. Phys.,* Vol.20, No.1, pp. 168-174

Ludwig, C.B. & Malkmus, W. et al. (1973). *Handbook of Infrared Radiation from Combustion Gases,* NASA, Washington, USA

Kondratyev, V.N. (1958). *Kinetics of Gaseous Chemical Reaction,* AN SSSR, Moscow, Russia

Kondratyev, K.Ya. & Moskalenko, N.I. (1977). *Thermal Emission of Planets,* Gidrometeoizdat, Leningrad, Russia

Kondratyev, B.N. & Nikitin, E.E. (1981). *Chemical Process in Gases,* Nauka, Moscow, Russia

Kondratyev, K.Ya.; Moskalenko, N.I. & Pozdnyakov B.N. (1983). *Atmospheric Aerosols,* Gidrometeoizdat, Leningrad, Russia

Kondratyev, K.Ya. & Moskalenko, N.I. (1984). *Hoti-house Effect of Atmosphere and Climate (Vol.12),* VINITI, Moscow, Russia

Kondratyev, K.Ya.; Moskalenko, N.I. & Nezmetdinov R.I. (2006). Role Nonequilibrium Process of Radiative Growing Cold of Combustion Products on Content of Nitrous Oxides in Atmospheric Emission, *Dokl. AN,* Vol.14, pp. 815-817

Marchuk, G.I. & Lebedev, W.I. (1981). *Numerical Methods in Theory Neutrons of Transfer,* Atomizdat, Moscow, Russia

Moskalenko, N.I. & Mirumyanth, S.O. et al. (1976). Installation for Complex Research of Characteristics Molecular Absorption of Radiation by Atmospheric Gases, *Journ. Appl. Spectrosc.,* Vol.19, No.4, pp. 752-756

Moskalenko, N.I. & Cementhov, C.A. et al. (1980). Spectral Installation for Research of Molecular Absorption and Radiation by Gases in Hightemperature mediums, *Journ. Appl. Spectrosc.,* Vol.54, No.2, pp. 377-382

Moskalenko, N.I.; Ilyin, Yu.A. & Kayumova G.V. (1992). Measuring Complex of High Spectral Permission for Research of Flame, *Journ. Appl. Spectrosc.,* Vol.56, No.1, pp. 122-127

Moskalenko, N.I. & Filimonov, A.A. (2001). Modeling of Heat Emission Transfer in Hightemperature mediums, *Problems of Energetic,* No.11-12, pp. 27-41

Moskalenko, N.I. & Chesnokov, S.P. (2002). Thin Parameterization of Gaseous Components Radiative Characteristics of Hydrocarbonful Fueles, *Problems of Energetic,* No.1-2, pp. 10-19

Moskalenko, N.I.; Loktev, N.F. & Zaripov, A.V. (2006). Diagnostics of Flames and Combustion Products by Optical Methods, *Proc. IV-th Russian National Conference on Heat Transfer,* pp. 277-280, Moscow, Russia, October 23-27, 2006

Moskalenko, N.I.; Zaripov, A.V.; Loktev N.F. & Nezmetdinov R.I. (2007). Research of Role of Nonequilibrium Process in Radiative Growing Cold, *Problems of Gas Dynamics and Heatmassexchange,* Vol.2, pp. 47-50, Sankt-Petersburg, Russia, May 21-27, 2007

Moskalenko, N.I. & Zaripov, A.V. (2008). Research of Role of Nonequilibrium Process in Radiative Crowing Cold of Combustion Products of Firing Chamber, In: *Current Problems in Modern Science,* S.S. Chernov, (Ed.), No.3, 45-73, SIBPRINT, ISBN 978-5-94301-044-6, Novosibirsk, Russia

Moskalenko, N.I. & Loktev, N.F. (2008). Thing Parameters of Radiative Characteristics of Combustion Products and its Application in Tasks of Remote Diagnostics, *Materials from International Conference "Energy – 2008: Innovation, Solutions, Prospects",* pp. 2224-2230, Kazan, Russia, September 15-19, 2008

Moskalenko, N.I.; Zagidullin, R.A. & Kuzin, A.F. (2008). Manny Firing Chambers and Heatexchangers as Means Increase of Effectiveness in Heat Engineering, *Materials from International Conference "Energy – 2008: Innovation, Solutions, Prospects",* pp. 230-234, Kazan, Russia, September 15-19, 2008

Moskalenko, N.I. & Loktev, N.F. (2009). Numerical Modeling in Tasks of Remote Diagnostics of Combustion Firing Fuels and Technological Mediums, In: *Technics and Technology in XXI-th Century: Modern Conditions and Prospects of Development: Monograph,* S.S. Chernov, (Ed.), Vol.4, 13-47, SIBPRINT, ISBN 978-5-94301-068, Novosibirsk, Russia

Moskalenko, N.I.; Zaripov, A.V. & Zagidullin, R.A. (2009). Emission Spectrums and Radiative Heatexchange Mediums, Flames and Firing Chamber, In: *Technics and Technology in XXI-th Century: Modern Conditions and Prospects of Development: Monograph,* S.S. Chernov, (Ed.), Vol.4, 48-87, SIBPRINT, ISBN 978-5-94301-068, Novosibirsk, Russia

Moskalenko, N.I. & Loktev, N.F. (2009). Methods of Modeling Selective Radiation Transfer in Structure - nonhomogeneous Mediums, *Thermal Process in Technique,* Vol.1, No.10, pp. 432-435

Moskalenko, N.I.; Loktev, N.F.; Safiullina, Ya.S. & Sadykova, M.S. (2010). Ingredients Identification and Determination of Ingredient Composition of Atmospheric Emission and Combustion Products by Means of Fine Structure Spectrometry Method, *International Journal of Alternative Energetic and Ecology,* Vol.8, No.2, pp. 43-54

Moskalenko, N.I.; Rodionov, L.V. & Yakupova F.S. (1984). Modeling of Transfer over of Touch Radiation of Differet Carriers, *Problems of Special Engineering,* Ser.1, No2, pp. 54-58

Moskalenko, N.I.; Zaripov, A.V. & Ilyin, Yu.A. (2010). Investigation of Nonequilibrium Hydroxyl Emission Spectra, *Russ. Phys. Journ.,* Vol.53, No.2, pp. 107-113

Moskalenko, N.I.; Zaripov, A.V.; Loktev, N.F. & Ilyin, Yu.A. (2010). Emission Characteristics of Hydrogen-Oxygen Flames, *Journ. Appl. Spectrosc.*, Vol.77, No3, pp. 378-385

Surgikov, S.T. (2004). *Thermal Radiation of Gases and Plasma*, MGTU, Moscow, Russia

Young, S.J. (1977). Evolution of Nonithothermal Band Models for H_2O, *Journ. Quant. Spectrosc. Radiat. Transfer.*, Vol.18, No.1, pp. 29-45

Zachor, A.S. (1968). General Approximation for Gaseous Absorption, *Journ. Quant. Spectrosc. Radiat. Transfer.*, Vol.8, No.2, pp. 771-784

Zel'dovich, Ya.B.; Sadovnikov, P.Ya. & Frank-Kamentsky, D.A. (1947). Oxidizing of Nitrogen to Firing, *AN SSSR*, Moscow, Russia

6

Low Frequency Noise as a Tool for OCDs Reliability Screening

Qiuzhan Zhou, Jian Gao and Dan'e Wu
Jilin University
China

1. Introduction

With the rapid development of the information and science, more and more newly semiconductor devices are used in the electronic equipments or systems, and so is the Optoelectronic Coupled Devices (OCDs). Because of excellent characteristics of it such as small size, long life, non-contact mode and strong anti-interference, OCDs can replace many kinds of devices e.g. relays, transformers, choppers when used in switching circuits, A/D conversion, remote transmission, over-current protection and so on. The reliability of OCDs is very important in numerous applications. It has draw great attention in switching circuit, isolation circuit, analog-digital converter, logic circuit, etc.

However in some high reliability fields, such as navigation and communication of the satellite, it is necessary to make sure of the reliability of the OCDs. In the past, the reliability screening of the OCDs contained ageing experiments; physical analysis at high and low temperature as well as static testing which are either expensive, time-consuming or cannot separate the good ones from the bad ones. So some researchers proposed that using low frequency noise as a reliability indicator.

From the ninety of the last century, we do the research of using noise as reliability screening of the OCDs and improve it continually. So in this paper, we will introduce how to use low frequency noise as a tool for OCDs reliability screening, and summarize what all we had done as well as the latest research.

2. Analysis of noise types in OCDs

Noise as a diagnostic tool for quality control and reliability estimation of semiconductor devices has been widely accepted and used, and there are many papers published in this area. It is very useful to describe the judging rules, which enable us to predict the individual quality of electronic components, based on measurements of their noise.

It is known that an OCD is made of two parts: LED and Photo detector, both of which are p-n junction devices. So it can be concluded that the noise in OCDs below 1 MHz mainly consists of shot noise, 1/f noise, generation-recombination noise and burst noise. Among them, shot noise and 1/f noise are fundamental. It should be noted that the noise that we are interested in here has strong relation to some typical defects in a device. For this reason, it is necessary that the generation mechanisms of 1/f noise, g-r noise and burst noise in OCDs are all briefly discussed, especially on what kinds of defects can lead to these three kinds of noises. And the relation between them should be discussed as well.

2.1 1/f noise

1/f noise spectrum is inversely proportional to frequency in a very wide range. In homogeneous semiconductors, its spectrum can be characterized by a parameter α according to

$$\frac{S_R}{R^2}=\frac{a}{fN} \tag{1}$$

where S_R is the power spectral density of the noise in the resistance R, N is the total number of free charge carriers, and f is the measurement spot frequency. The parameter α then is the contribution of one electron to the relative noise at 1 Hz, assuming that the N electrons are uncorrelated noise sources.

In addition, it has been found that α is not a constant, whose value is between 10^{-6} and 10^{-3}, but that α depends on the prevailing type of scattering of the electrons and perfection of the crystal lattice. In recent years much progress has been made and found that it is mainly caused by lattice scattering.

Vandamme has shown that the 1/f noise parameter a increases with the concentration of dislocations and its noise spectrum is proportional to a and inversely proportional to the carrier lifetime. Konczakowska research has indicated that there is a strong relation between bipolar device lifetime and 1/f noise.

Usually 1/f noise in a semiconductor device usually can be divided into fundamental 1/f noise and non-fundamental (or excess) 1/f noise. The fundamental 1/f noise is connected with phenomena which are included in the process of the operation of the electronic component. It is believed that this 1/f noise has no relation to the semiconductor surface and the defects in the bulk.

The 1/f noise which is related to device defects is called non-fundamental 1/f noise, which means that this kind of 1/f noise is caused by device surface or bulk defects in most cases. Thus, it is possible for us to evaluate the device quality and reliability according to its magnitude. From this point of view, non-fundamental 1/f noise is of great value to device quality evaluation and reliability prediction. Most of the evidence suggests that in some types of device it is a surface effect, as in the case of a MOSFET where the semiconductor/oxide interface plays an important role, but in other devices, such as a homogeneous resistor, 1/f noise is thought to be a bulk effect associated with a random modulation of the resistance, implying a fluctuation in either the number or the mobility of the charge carriers. For example, M. Mihaila et al have shown that 1/f noise in a specimen with more dislocations is at least one order of magnitude larger than that of the specimen with fewer dislocations.

Different causes for 1/f noise generation have been reported as follows: (1) the fluctuation of surface recombination velocity in the p-n junction, (2) the fluctuation of trapping in the oxide layer in BJTs or in MOSFETs, (3) dislocation 1/f noise, (4) quantum 1/f noise (in dispute). It is obvious that 1/f noise intensity is related to the generation-recombination center (surface defect) numbers in device oxide layer. Thus, the establishment of a relationship between device surface quality and reliability can help us judge and screen devices according to excess noise intensity.

Therefore it is possible for us to evaluate the device quality and reliability according to its magnitude. From this point of view, 1/f noise is of great value to device quality evaluation and reliability prediction. It is verified that crystal defects cause 1/f noise to increase. The

experimental results have proved that 1/f noise in the specimen with more dislocation is at least one order of magnitude larger than that of the specimen with fewer dislocations.
At present, a major cause of 1/f noise in semiconductor devices is traceable to properties of the surface of the material. The generation and recombination of carriers in surface energy states and density of surface states are important factors, but even the interfaces between silicon surfaces and grown oxide passivation are centers of noise generation. It is obvious that 1/f noise intensity has a relation to the generation-recombination center (surface defect) numbers in device oxide layer. Thus, the relation between device surface quality and 1/f noise is closely related and can be used to screen poor quality devices according to their intensity of excess noise.

2.2 g-r noise

Generation-recombination (g-r) noise distribution is Gaussian and its signal spectrum can be expressed as Lorentzian,

$$\frac{S_R}{R^2}=\frac{S_V}{V^2}=\frac{S_I}{I^2}=\frac{A\tau_0}{1+(\omega\tau_0)^2} \tag{2}$$

Here, $\tau_0=1/\omega_0$ is the characteristic time corresponding to a characteristic (or corner) frequency f_0 or ω_0, and $\omega_0=2\pi f$ is the angular frequency of measurement. g-r noise has a Gaussian amplitude distribution function because it is actually made up from the superposition of a very large number of independent random telegraph signal processes with the same characteristic time. The coefficient, A, is a measure of the number of such individual processes. It depends on g-r center density, current and device structures.
It has been found experimentally that g-r noise is often absent in high quality silicon devices, but not yet in heterostructures, where lattice defects are often a problem. In poor quality devices, the g-r noise is generated at the contacts or at the surface. In better samples, the dominant conductance noise source is in the bulk. Hence, g-r noise used as a useful diagnostic tool to study trap centers in compound semiconductors, is indispensable.
By experimentation it has been shown that the defects (dislocation, deep-level impurities) in the emitter junction and surface are the main g-r noise sources of transistor, especially as a p-n junction is in a forward biased state. Jevtic and Lazovic have shown that excess g-r noise in reverse biased p-n junctions can be caused by g-r centers near the metallurgical junction. These centers may be the impurity in metal clusters associated with dislocations.
Thus, it can be seen that g-r noise in a device has a direct relation to semiconductor defects (impurities, damage etc.). Therefore, it has become an effective method of analyzing bulk defects and reliability screening by means of measuring g-r noise in devices.

2.3 Burst noise

A random telegraph signal (RTS) known as burst noise, is often observed in p-n junction devices such as diodes, transistors and detectors operating under forward biased conditions. All authors have attributed the phenomenon to defects located in the neighborhood of the emitter base junction.
Hsu et al. first presented a physical model of explaining burst noise. In this model, it is thought that heavy metal impurities deposited in the charge region of p-n junction is the major cause of this noise.

But Blasquez has found that so-called ``pure'' lattice dislocation can also cause burst noise even when heavy metal impurity deposits have been removed. Therefore it seems that metal impurity precipitates are not indispensable to produce burst noise. Dai et al. has proposed a new burst model, which emphasizes the built-in electric field in the p-n junction and the variation of the potential barrier near the defects. This model not only is consistent with the experimental results given by Blasquez, but also can explain various burst noise waveforms. In addition, Jevtic has also presented a new physical model of noise sources, which is based on the assumptions that a conduction channel (p-inversion layer) exists in degraded p-n junctions and that the current flow through the defects is modulated by traps adjacent to the defects. The model explains the appearance of two polarity and multi-level pulse noise. Although burst noise spectrum is not Gaussian as are the other types of noise, its current noise spectrum has the shape of Lorentzian,

$$S_I(f) = \frac{A_b \tau_b^2}{1+\omega^2 \tau_b^2} \tag{3}$$

Where A_b is a constant depending on the nature of the defects and τ_b is defined as $1/\tau_b = 1/\tau_1 + 1/\tau_2$. According to the random switch mode, during the time interval dt, an open switch has the probability dt/τ_1 of closing, and a closed switch a probability dt/τ_2 of opening. The information on the defects is contained in the parameter A_b and τ_b.

Besides, burst (or RTS) noise is a problem typical for submicron MOST's or bipolar devices with crystallographic damage in sensitive areas and this noise is also temperature and bias dependent. Many experiments have already shown that lattice dislocation, a serious defect, is the major source of burst noise for both bipolar transistor and integrated circuit. Therefore, devices with burst noise often degrade faster and at least show a poor noise behavior. Fig. 1 shows a noise waveform of time-domain in an OCD with excess noise.

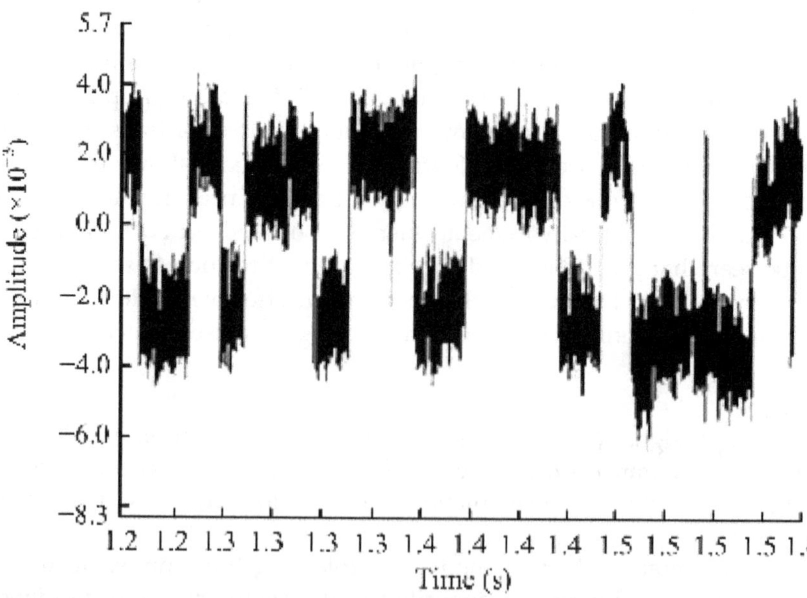

Fig. 1. Noise waveform of time-domain in an OCD with excess noise.

2.4 Brief summary

It can be seen that 1/f noise, g-r noise and burst noise are closely related with some defects such as surface defects, impurities and dislocations, etc. Dislocations and electromigration in metallization affect device reliability, and has been identified as the main source of device failure. Thus, it can be said that 1/f noise is closely related to the surface states of the semiconductor device, g-r noise related to device bulk defects such as impurities, dislocation, etc., and burst noise related to lattice dislocation as well as heavy metal impurity deposits. Besides, emitter region edge dislocation makes both 1/f noise and burst noise increase at the same time in most cases. Strasilla and Struut demonstrated that experimentally observed burst noise consist of a random telegraph signal superimposed on 1/f noise, but the two processes were statistically independent.

Hence in order to exclude these defects and meet high reliability, we can use the three independent noise, 1/f, g-r and burst noise, as reliability indicator for quality estimation of OCDs.

3. The noise measurement and analysis of OCDs

Harder C et al have presented that the noise equivalent circuit of a semiconductor laser diode from the rate equations including Langevin noise sources. This equivalent circuit allows a straightforward calculation of the noise performance of a laser diode combined with extrinsic elements, such as the driving source and the parasitic elements. Recently, using this rate equation, these intrinsic intensity fluctuations in semiconductor laser diode (LD), optoelectronic integrated device (OEID) made by heterojunction phototransistor (HPT) and laser diodes have been analyzed, then the relative intensity noise (RIN) and the correlation between the terminal electrical noise and output optical photocurrent noise have been investigated.

At present, the key to design of low-frequency low noise devices and circuits lies in reducing level of white noise and corner frequency of l/f noise, which has been realized gradually and Whether voltage noise or current noise takes these two parameters as its characteristics. But the present noise measuring apparatus, such as QuanTech2173c/2181 and HP-4470 and so on, only can give out noise of several frequency points or of several fixed frequency, no more give out two pat meters. Thus it can be seen that if one want to understand all-sidelly the low-frequency noise performance of a semiconductor device, to make researches on semiconductor noise mechanism and to apply low-frequency noise to analyzing the inherent defect of a device and its reliability and so forth, one must make study of the measurement of low-frequency noise spectrum and the computation of its parameters. This important work is absolutely necessary to understand noise performance of a device or a circuit and specially to develop low-noise devices.

In this part, the noise equivalent circuit of OCDs have been analyzed; then noise spectrum measurement systems of OCDs based on FFT analyzer (CF-920, made in Japan) and virtual instrument are presented, their measuring range is 0. 25Hz- l00kHz and accuracy is higher than 4%. Moreover, the white noise level and corner frequency are computed by applying weighed least square method.

3.1 The noise equivalent circuit of OCDs

Fig. 2 shows the schematic diagram of the OCDs circuit. The OCDs measured and analyzed in this paper are GD315A, made in China. It is well known that if the input current I_0 of a laser diode is less than the threshold current I_{ph}, the noise equivalent circuit of a semiconductor laser diode can also be used to explain the noise performance of an LED. Fig. 3 shows the noise equivalent circuit of OCDs. It is composed of LED and optotransistor shown in the left portion and the right one, respectively. Where R_s is the source resistance, i_{Rs} is the thermal noise of R_s. The quantities in and v_n are the intrinsic noise sources of LED, the sources in and v_n are partially correlated due to the coupled rates. Their noise spectral densities are shown as follows:

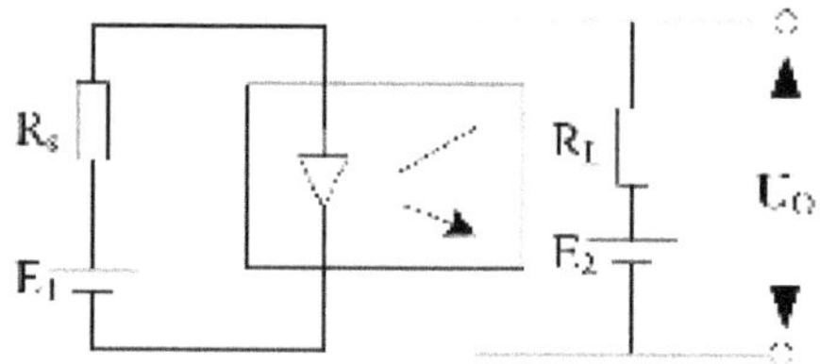

Fig. 2. The schematic diagram of OCDs.

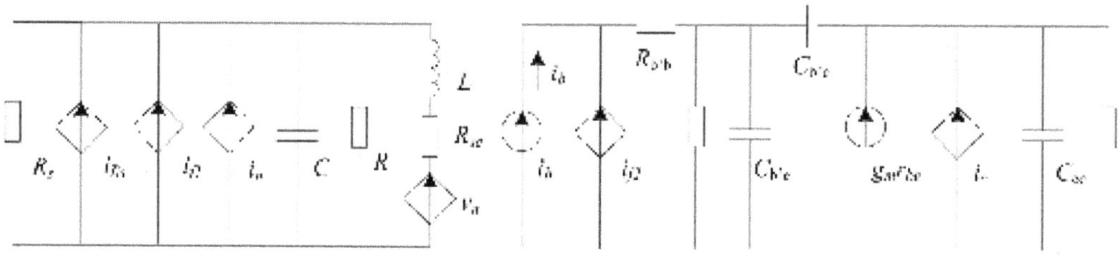

Fig. 3. The noise equivalent circuit of OCDs.

$$S_{i_n}(f) = 2qi_0 + 4q^2 E_{vc} S_0 \tag{4}$$

$$S_{v_n}(f) = \frac{4(mV_T)^2}{\left(n_0\left(AS_0 + (\beta/\tau_s)\right)\right)^2} \frac{1}{\tau_{ph}} + E_{vc} S_0 \tag{5}$$

$$S_{v_n i_n}(f) = 2\frac{mV_T q}{n_0\left(AS_0 + \beta/\tau_s\right)} \frac{1}{\tau_{ph} + 2E_{vc}} S_0 \tag{6}$$

The current i_{f1} denotes the low frequency noise source in LED. The definition of all symbols in Eqs. (4) - (6) and circuit parameters R, C, L, R_{se} in the LED equivalent circuit in Fig. 3 can be found in the references and another publication. In the noise equivalent circuit of the phototransistor, i_b is the base noise current, it is caused by the noise current i_L in LED, hence i_b can be written as γi_L, where γ can be calculated by the current transfer ratio (CTR) of OCDs. Since CTR is defined as I_c/I_0, in the low frequency region all capacitors in the OCDs are omitted; from Fig. 3 we have

$$\mathrm{CTR} = \frac{I_c}{I_0} = \frac{\gamma g_m R_L R_{se}\left(r_{b'b} + r_{be}\right)}{\left(R + R_{se}\right)\left(R_L + R_{ce}\right)} \tag{7}$$

Then γ can be calculated by Eq. (7). In the high frequency region, CTR will decrease due to the influence of circuit capacitance. The CTR of OCDs can be obtained by measurement of OCDs (GD315A) as is shown in Fig. 4, where curve 1 is measured in the condition R_L=1kΩ, and curve 2 is R_L=500Ω.

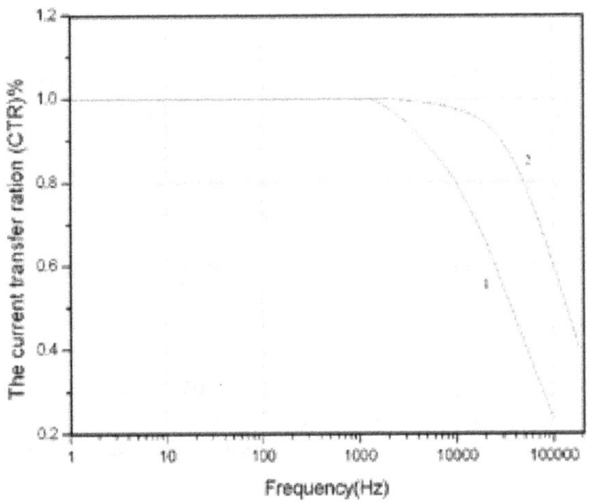

Fig. 4. The CTR of OCDs versus frequency.

In the equivalent circuit of the phototransistor, i_{f2} denotes the low frequency noise source in it. $r_{b'b}$, r_{be}, g_m, $C_{b'e}$, C_{bc}, C_{ce} are the circuit parameters of phototransistor shown in Fig. 3.

3.2 Noise spectrum measurement systems

3.2.1 Noise measurement system based on FFT analyzer

Fig. 5 is the measurement system block scheme of low-frequency noise spectrum testing system we have developed. In our experiments, a dual-channel low frequency amplifier (LNA) chain and the CF-920 cross-spectrum estimator have been used to reduce the background noise of measurement system; hence the noise in the amplifiers will not contribute to the cross-spectrum measured by this system. Therefore, the output voltage noise spectrum of OCDs can be written as

$$S_0(\omega) = \frac{S(\omega)}{K_1 K_2} \tag{8}$$

Where S(ω) is the cross-spectrum measured by CF-920, K_1 and K_2 are the gains of the dual-channel amplifiers LNA I and LNA II, respectively.

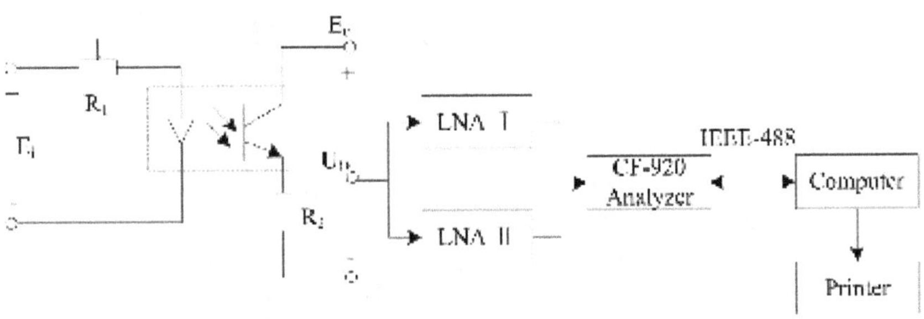

Fig. 5. The noise measurement system of OCDs using CF-920.

In the testing system, the cross-spectrum density estimation method was used to reduce the noise contribution of two preamplifiers. The reason was that two sets of batteries were used as the power supplies for the preamplifiers so that the noises in the two preamplifiers themselves were uncorrelated. So, the measuring system can be used to measure a much smaller signal than usual.

Thus it can be seen that amplifier's self-measurement error has been eliminated basically according to the measuring method in this system. And the measurement accuracy of noise spectrum is mainly decided by CF-920. The measuring range of this system is 0. 25 Hz-100 KHz frequency wide and accuracy is higher than 4%. In the whole measurement process, the measurement and the output of measuring results are controlled automatically by computer.

3.2.2 Noise measurement system based on virtual instrument

Considering the large volume of CF-920 in the above system, which is inconvenient to carry, we design a new measurement system with virtual instrument made by the company of National Instrument and the system block diagram is shown in Fig. 6. It is well known that the virtual instrument platform is widely used in the fields of measurement, auto-control, signal processing and so on, and what the most important is the high precise and small volume.

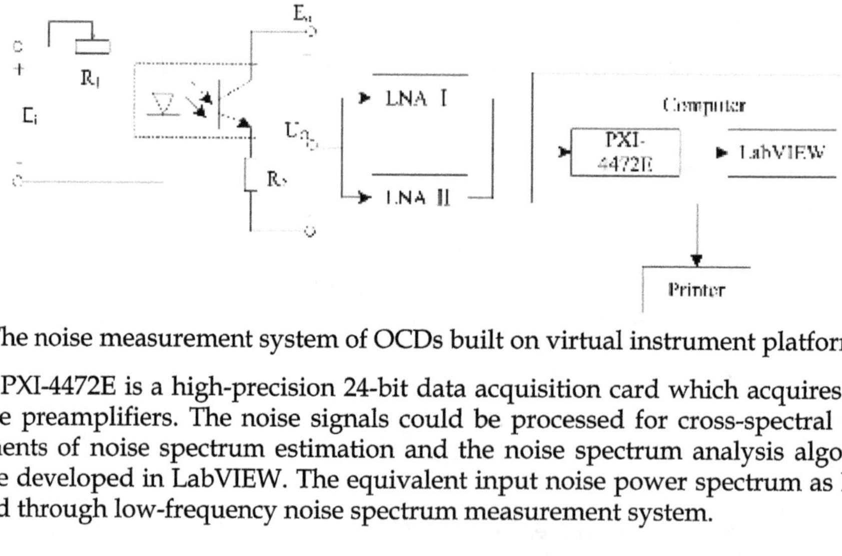

Fig. 6. The noise measurement system of OCDs built on virtual instrument platform.

Where, PXI-4472E is a high-precision 24-bit data acquisition card which acquires the signal from the preamplifiers. The noise signals could be processed for cross-spectral transform, components of noise spectrum estimation and the noise spectrum analysis algorithm in a software developed in LabVIEW. The equivalent input noise power spectrum as Eq. (9) can be tested through low-frequency noise spectrum measurement system.

$$S(f) = A + \frac{B}{f^{\alpha}} + \sum_{i=1}^{N} \frac{\frac{C_i}{f_{oi}}}{1 + (\frac{f}{f_{oi}})^2} \tag{9}$$

Where A, B and C_i are amplitudes of white noise, $1/f$ noise and G-R noise respectively; α is a constant 1; f_{oi} is the corner frequency of excess G-R noise; N is the number of G-R noise sources in devices. As long as parameters A, B and C_i are estimated, the magnitude of each type of noise in the device can be determined. A genetic algorithm which would be discussed below was used to fit the parameters, so the coefficient of white noise, $1/f$ noise and G-R noise were obtained quickly and accurately across the entire spectrum.

3.2.3 Noise analysis of OCDs

According to the noise equivalent circuit given in Fig.3, the noise curves are analyzed in various frequency ranges.

(1) *Low frequency range (1 Hz < f <1 kHz)*: In this frequency range, the 1/f noise and generation-recombination g-r noise are dominant. We measured the noise spectrum of 205 OCDs (GD315A made in China), the noise spectrum for four typical devices are shown in Fig. 7. These devices exhibit various low frequency noises. Using the curve fitting method, the analysis results of noise spectrum for four typical devices shown in Fig. 7 are given as follows:

$$\text{NO.4}\quad S_0(f)=5\times10^{-13}+\frac{0.85\times10^{-10}}{f}+\frac{0.014\times10^{-10}}{1+(f/15)^2}V^2/Hz$$

$$\text{NO.45}\quad S_0(f)=2\times10^{-13}+\frac{4\times10^{-10}}{f}+\frac{0.07\times10^{-10}}{1+(f/15)^2}V^2/Hz$$

$$\text{NO.102}\quad S_0(f)=1.05\times10^{-13}+\frac{1\times10^{-10}}{f}+\frac{2.13\times10^{-10}}{1+(f/15)^2}V^2/Hz$$

$$\text{NO.15}\quad S_0(f)=8\times10^{-13}+\frac{6.28\times10^{-10}}{f}+\frac{0.64\times10^{-7}}{1+(f/2500)^2}V^2/Hz$$

The noise analysis results show that OCD No. 4 exhibits a normal low frequency noise, No. 45 has excess 1/f noise, No. 102 has excess g-r noise (the corner frequency is 15 Hz), and No. 15 has excess 1/f noise and g-r noise with burst noise waveform. It means that OCDs No. 4 is a reliable device, No. 45, 102 and 15 coincide with various defects in OCDs, and i.e. their quality is poor.

(2) *Medium frequency range (1 kHz < f <10 kHz)*: In this frequency range, the low frequency noise current i_{f1} and i_{f2} in the OCDs can be omitted; hence, the OCDs output noise is caused by shot noise $i_c(t)$, thermal noise $i_{Rs}(t)$ and the spontaneous emission noise i_n and v_n. Omitting all capacitors in the OCDs, the output noise spectrum can be obtained:

$$S_0(\omega)=R_L^2(CTR)^2(S_{i_{Rs}}(\omega)+S_{i_n}(\omega)+\frac{S_{v_n}(\omega)}{R_{se}^2}+\frac{2S_{i_nv_n}(\omega)}{R_{se}})+R_L^2S_{ic}(\omega) \tag{10}$$

Where S_{iRs}=$4kT/R_s$, $S_{ic}(\omega)$=$2qI_c$. According to the Reference written by Harder C et al, the noise current source i_n in parallel with the junction represents mainly the fluctuation in electron population. The first term on the right-hand side of Eq. (4) is the usual shot noise term $2qI_0$ of OCDs, the second term results from the fact that the noise is determined by the sum of emission and absorption. The noise voltage source v_n results from the fluctuation of the photon population.

Since the input current I_0 of LED is less than the threshold current I_{ph}, if we omit the second term of in and v_n, Eq. (11) can be written as follows:

$$S_0(\omega)=2qI_0CTR^2R_L^2+2qI_0R_L^2 \tag{11}$$

Let I_0=10mA, CTR=1, R_L=390Ω, R_s=360Ω, I_c=10mA, from Eq. (11) we obtain the output noise voltage spectrum $S_0(\omega)$=9.7×10^{-16}V^2/Hz. The effective value of $S_0(\omega)$ equals $31.2\,nV/\sqrt{Hz}$ which is smaller than the measurement result (287-488 $nV/\sqrt{Hz}$). It means that the second term of i_n and v_n caused by the spontaneous emission in LED cannot be omitted.

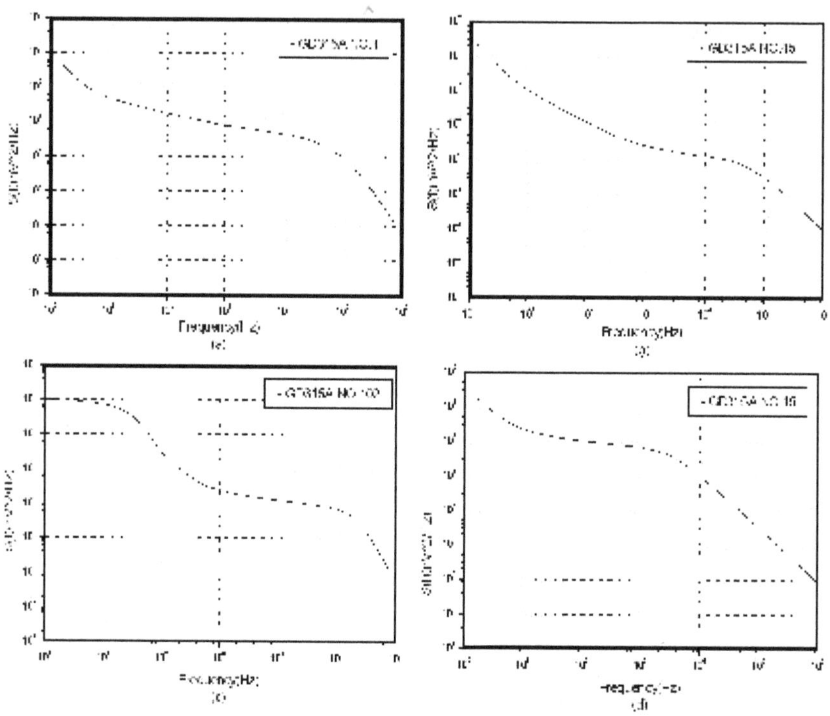

Fig. 7. The noise measurement system of OCDs built on virtual instrument platform.

(3) *High frequency range (10 kHz < f <100 kHz)*: In this frequency range, the white noise is dominant, so that the output noise spectrum of the OCDs is a flat form. However, the noise measurement for 205 OCDs (DG315A) shows that the noise spectrum of all OCDs decreases from 10 to 100 kHz, as shown in Fig. 7. It can be explained as follows: Since $S_0(\omega)$ is directly proportional to CTR2 (Eq. (11)) and CTR decreases from 10 to 100 kHz (Fig. 4), as a consequence, the decrease in noise spectrums of OCDs is caused by the decrease of CTR. Because all capacitors in the OCDs will influence the output noise in this frequency range, the decrease of CTR seems to obey the circuit response.

4. Discussion of the optical screening criterion for OCDs

Generally, it has been already accepted that 1/f noise is closely related to the surface states of the semiconductor device, g-r noise is related to device bulk defects such as impurities, dislocation, and burst noise is related to lattice dislocation as well as heavy metal impurity deposits. From the generation mechanisms of 1/f, g-r and burst noise, it can be seen that the probability to generate these three types of noise by the same defect is quite small although some defects may cause more than one of them simultaneously in some cases.
Hence, in order to exclude these defects and meet high reliability, we can use the three independent noises, 1/f, g-r and burst noise, as reliability indicators for quality estimation of OCDs. In this way even if some defects can cause two or three of them at the same time, such as emitter region edge dislocation which makes both 1/f noise and burst noise increase at the same time in most cases, can also be rejected.
Therefore, because an excess noise is closely associated with some defects in the devices and/or imperfections of technology, noise measurement amplitudes can be used to indicate

the defects. In practice, we found that the device with burst noise can be found from its instantaneous waveform in the time domain. The device with g-r noise can be found through noise component analysis or ratio of noise value at 10 Hz to noise value at 1 Hz (which will be explained and proved in later part), which is used to judge whether there is g-r noise or not. And the device with 1/f noise can be judged by the amplitude of voltage noise value at 1 Hz

Therefore, it is necessary that there be three independent screening conditions to meet the requirement of high reliability to reject the devices with excess 1/f, g-r or burst noise. We take the OCDs of GD315A as an example to present the conditions which are presented as follows.

(1) *The value of the noise spectrum at 1 Hz*: Since 1/f noise depends on the surface and bulk defects in OCDs, and it is dominant in the low frequency region, we select the effective value U_0 of noise spectrum at 1 Hz as the reliability indicator. Fig. 8 shows the histogram of the effective value U_0 of 205 OCDs. We can see that the U_0 of most of OCDs obeys the normal distribution; it can be explained by the discretion of OCDs parameters. However, a few of OCDs exhibit excess 1/f noise, it means that these OCDs may contain surface or bulk defects (such as crystal defects and contamination, emitter edge dislocations, electromigration, imperfection of chip bonding, etc.); hence, they should be recognized as poor quality devices.

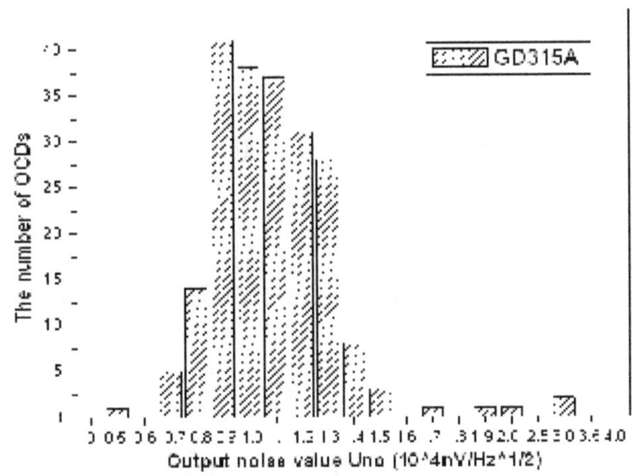

Fig. 8. The histogram of the output noise value $S_0(f)$ of 205 OCDs.

Konczakowska suggested a classification algorithm which has been verified by reliability experiments. On the basis of noise measurement results, we can classify quality groups according to this rule. At first, the mean value and variance of U_0 for 205 OCDs have been calculated, and then the border values for quality groups (i.e., the threshold value) are

$$U_{01} = \bar{U}_0 - \alpha\sigma$$

$$U_{02} = \bar{U}_0 + \alpha\sigma$$

where $\bar{U}_0$ denotes the mean value of U_0, r denotes the variance of U0 and a equals 0.67. In our experiment, the statistical results are $\bar{U}_0 = 11467.5nV/\sqrt{Hz}$, $\sigma = 3194nV/\sqrt{Hz}$, then we obtain $\bar{U}_{01} = 9327.4nV/\sqrt{Hz}$, $\bar{U}_{02} = 13607.7nV/\sqrt{Hz}$. The value of classifying OCDs into three groups according to the 1/f noise is as follows:

- first group $-U_0 \le 9327.4nV/\sqrt{Hz}$, high quality is expected;
- second group $-9327.4nV/\sqrt{Hz} \le U_0 \le 13607.7nV/\sqrt{Hz}$, good quality is expected;
- third group $-U_0 \ge 13607.7nV/\sqrt{Hz}$, poor quality is expected.

(2) *The ratio of noise spectrum at 10 and 1 Hz*: The g-r noise is caused by the defects in the crystal structure of devices and the deep-level impurity in the p-n junction; hence the g-r noise is used as one of the noise reliability indicators. The g-r noise in OCDs can be separated accurately from the output noise spectrum using the curve fitting method; however, it must take a long time and is not suitable for practical application. For the convenience of direct industrial application, we suggest the g-r noise indicator as follows:

$$r = \frac{U_0(f_0)}{U_0(1Hz)} \tag{12}$$

where U_0(1 Hz) is the effective value of noise spectrum at 1 Hz, $U_0(f_0)$ is one at the corner frequency f_0 of g-r noise. From the noise measurement of OCDs, we found that the corner frequency of most OCDs is about 10 Hz, hence we can select f_0=10 Hz.

If the output noise spectrum consists of only white noise and 1/f noise, in the low frequency region, we have

$$S_0(f) = A + \frac{B}{f} \tag{13}$$

If the 1/f noise is dominant, then $S_0(f)=B/f$, the value of r equals $U_0(10Hz)/U_0(1Hz) = 1/\sqrt{10} = 0.33$. If the white noise is dominant, r=0, so that the value of r is in the range of 0-0.33. However, if an OCD exhibits g-r noise, the noise spectrum is shown in a platform below the corner frequency f_0. As a consequence, the value of r is greater than 0.33. In Fig. 9, the histogram of the value r for 205 OCDs is shown. We can see that the value of r for most of the OCDs obeys the normal distribution. The statistical results from 205 GD315A are $\bar{r} = 0.36$ and $\sigma_r = 0.16$. If we select the value of β equal 2, then the threshold value of r is

$$r = \bar{r} + \beta\sigma_r = 0.68$$

The rule of classifying OCDs into two groups according to the g-r noise is as follows:

- $r < 0.68$, first group, high quality is expected;
- $r > 0.68$, second group, poor quality is expected.

(3) *Burst noise in OCDs*: A number of experiments have already shown that heavy metal impurity deposits and lattice dislocations are the major sources of burst noise for a transistor or IC, so that OCDs with burst noise should always be rejected in any case, because it cannot only effect device reliability, but also hinder the normal operation, especially in digital circuit.

The rule of classifying OCDs into two groups according to the burst noise is as follows:

- no burst noise: first group, high quality is expected;
- with burst noise: second group, poor quality is expected.

As a result, the reliability screening conditions of OCDs are

1. $U_0(1Hz) \geq 13607.7nV/\sqrt{Hz}$
2. $r \geq 0.68$
3. with burst noise

The device will be rejected if it meets any condition of the three. The result of the noise measurement for 205 OCDs is that 23 OCDs are rejected by our reliability screening conditions. Among these OCDs, 19 OCDs are rejected by reliability screening condition 1 (here two OCDs exhibit burst noise). Four OCDs are rejected by condition 2 (here one OCD exhibits burst noise).

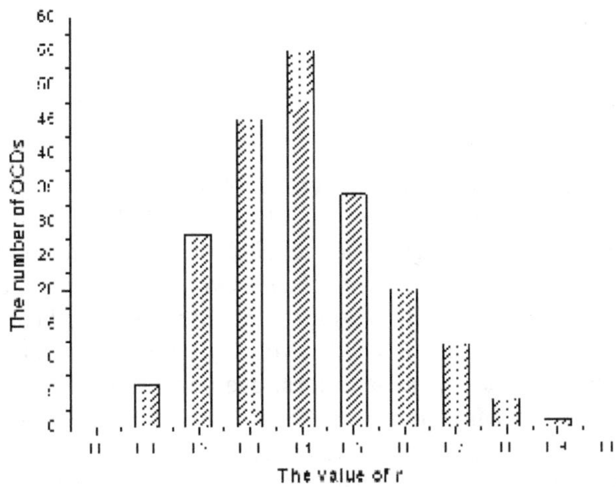

Fig. 9. The histogram of value r of 205 OCDs.

5. Conclusion

In this paper, we analyzed the noise performance of OCDs using the noise equivalent circuit of LED and bipolar transistor.

(1) In the low frequency region (f <1 Hz), the excess 1/f noise and g-r noise can be used as noise reliability indicators; then three reliability indicators have been suggested for quality and reliability screening of OCDs, which have the advantages of general and convenience for direct industrial application.

(2) In the medium frequency region (1 kHz < f <10 kHz), the output noise measurement results show that the fluctuation of the photon population (emission and absorption) in OCDs must be considered. Hence, the value of output noise is larger than usual shot noise in transistor.

(3) In the high frequency region (f > 10 kHz), the output noise will decrease due to the decrease of the current transfer ratio (CTR) of OCDs in this frequency region.

Then the 205 OCDs have been measured by the low noise measurement systems which have high reliability and the satisfactory accuracy. The screening thresholds and experimental results are given. Based on the results, it can be obtained as follows:

(1) It can be found that 1/f noise, g-r noise and burst noise must be used as three independent noise criteria for high reliability estimation.

(2) It is necessary that the estimated error and the maximum value of failure ratio r should be considered together in order to obtain optimal noise thresholds.

(3) In this paper the ratio of Vn(10 Hz) to Vn(1 Hz) was chosen as a screening threshold instead of g–r noise component analysis of noise spectrum, for this reliability indicator is more simple and convenient than the calculation of g–r noise component, especially during the practical screening of a large number of devices.

6. Acknowledgment

Dr. Qiuzhan Zhou would pay the highest respect to his advisor, Prof. Dai Yisong and would like to thank for Prof. Dai's patient guide since 1998. The authors would also thank Prof. Zhang Xinfa for his help on noise parameters measurement. This work is part of Project 60906034 supported by National Natural Science Foundation of China and Project 201115029 supported by Provincial Natural Science Foundation of Jilin Province.

7. References

A. Konczakowska(1995), Quality and 1/f noise of electronic components, *Quality and Reliability Engineering International*, vol. 11, pp. 165-169, ISSN 1099-1638

Bernard O & Philippe S & Jean-Marie P et al(1994), Correlation between electrical laser diodes, *IEEE Transactions on Electron Devices*, vol. 41, pp. 2151-2161, ISSN 0018-9383

Dai.Y& Xu. J(2000). The noise analysis and noise reliability indicators of optoelectron coupled devices, *Solid-State Electronics*, vol. 44, pp. 1495-1500, ISSN 0038-1101

Dai.Y(1991), Optimal low frequency noise criteria used as a reliability test for BJTs and their experimental results, *Microelectronics Reliability*, vol. 31, pp. 75-78, ISSN 0026-2714

Dai.Y(1997), A precision noise measurement and analysis method used to estimate reliability of semiconductor devices, *Microelectronics Reliability*, vol. 37, No. 6, pp. 893-899, ISSN 0026-2714/97

Dai.Y(1991), The performance analysis of cross-spectral density estimator and its applications, *International Journal of Electronics*, vol. 71, pp. 45-53, ISSN 0020-7217

Doru U, Jones BK(1996), Low frequency noise used as a lifetime test of LEDs, *Semiconductor Science and Technology*, vol. 11, pp. 1133-1136, ISSN 0268-1242

Harder C & Katz J & Margalit S et al(1982), Noise equivalent circuit of a semiconductor laser diode, *IEEE J Quant Electron*, vol. 18, pp. 333-337, ISSN 0018-9197

Lei. N,& Du.J & Zhou. Q Z (2009), Fitting Noise Power Spectrum Parameters by Squared Distance Minimization, *Computational Intelligence and Software Engineering, 2009*. pp. 1-4. ISBN: 978-1-4244-4507-3

Levinzon, F. A.(2005), Measurement of low-frequency noise of modern low-noise junction field effect transistors, *Instrumentation and Measurement, IEEE Transactions on*,Vol.54,pp.2427-2432,ISSN 0018-9456

M.M. Jevtic(1995), Noise as a diagnostic and prediction tool in reliability physics, *Microelectronics Reliability*, vol.35, pp. 455-477, ISSN 0026-2714/97

Vandamme L. K. J (1994), Noise as a diagnostic tool for quality and reliability of electronic devices, *Electron Devices, IEEE Transactions on*, vol. 41, pp. 2176-2187, ISSN 0018-9383

Wu,D E &Zhou Q Z &Liu P P(2009) Study of OCDs Reliability Estimation System Based on Bio-Immunology, *Journal of Bionic Engineering*, vol. 6, pp. 306-310, ISSN 1672-6529

Xu.J & Abbott.D & Dai. Y (2000), 1/f, g-r and burst noise used as a screening threshold for reliability estimation of optoelectronic coupled devices, *Microelectronics Reliability*, vol. 40, pp. 171-178, ISSN 0026-2714

7

Anomalous Transient Photocurrent

Laigui Hu[1] and Kunio Awaga[2]

[1]*China*
[2]*Japan*

1. Introduction

The operating principle in conventional optoelectronic devices is based on steady-state photocurrent. In these devices, photogenerated carriers have to travel long distances across the devices. Various dissipation mechanisms such as traps, scattering and recombination dissipate these carriers during transport, and decrease device response speed as well as optoelectronic conversion efficiency, especially in organic devices (Forrest & Thompson, 2007; Pandey et al., 2008; Saragi et al., 2007; Spanggaard & Krebs, 2004; Xue, 2010). Such organic devices have received considerable attention due to their potential for of large-area fabrication, combined with flexibility, low cost (Blanchet et al., 2003), and so on. Efforts to substitute inorganic materials by organic ones in optoelectronics have encountered a serious obstacle, i.e., poor carrier mobility that prevents photogenerated carriers from travelling a long distance across the devices.

Typically, exciton diffusion length in organic materials is approximately 10-20 nm (Gunes et al., 2007). Internal quantum efficiency decreases with the increase in film thickness (Slooff et al., 2007) since recombination will occur prior to exciton dissociation if photogenerated excitons are unable to reach the region near the electrodes. Therefore, though a thicker film can result in an enhanced light harvesting, collecting carriers using electrodes becomes difficult. In addition, the poor mobility of organic materials always triggers the formation of space charges in thin film devices, and the space charges additionally limit the photocurrent (Mihailetchi et al., 2005).

In this chapter, we introduce an anomalous transient photocurrent into optoelectronics based on Maxwell's theory on total current, which consists of conduction and displacement current. In contrast to organic optoelectronic devices based on conduction photocurrent, which sufferrs from poor carrier mobility, the anomalous photocurrent can contribute to optoelectronic conversion and "pass" through an insulator. Though such anomalous photocurrent, or photoinduced displacement current, has received previous attention (Andriesh et al., 1983; Chakraborty & Mallik, 2009; Iwamoto, 1996; Kumar et al., 1987; Sugimura et al., 1989; Tahira & Kao, 1985), its mechanism and characteristics are still largely unresolved. We systematically explained this phenomenon based on our theoretic analyses and experiments on an organic radical 4′4-bis(1,2,3,5-dithiadiazolyl) (BDTDA) (Bryan et al., 1996) thin film device. A double-layer model was introduced, and a new type of device with structure of metal/blocking layer/semiconductor layer/metal was developed to reproduce the anomalous photocurrent (Hu et al., 2010b). The photocurrent transient is observed to involve polarisation in the materials, and stored charges within the phtocells can be released by the time-dependent conduction photocurrent. The formulae derived for this phenomena are promising for the characterisation of carrier transport in organic thin films.

In this chapter, we firstly demonstrate the anomalous photocurrent and steady-state photocurrent in the BDTDA photocells with a structure of ITO/BDTDA (300 nm)/Al (Hu et al., 2010a; Iwasaki et al., 2009). The anomalous photocurrent in the BDTDA films is observed to involve a large polarisation current induced by the formation of space charges near the electrodes. Subsequently, a series of formulae based on the total current equation for a double-layer system have been developed to fit experimental data. The theoretical ideas behind this formula are discussed as well.

Based on the analyses, the metal/blocking layer/semiconductor layer/metal photocell is demonstrated using different organic materials, including insulators and semiconductors, to reproduce the anomalous photocurrent. We introduce the enhancement of anomalous photocurrent by employing a transparent dielectric polymer with a larger dielectric constant (as a blocking layer) since larger polarisation current can be produced. Fast speed can be achieved since the performance is mainly limited by the fast dielectric relaxation (Kao, 2004). These are promising for high-speed operation in optoelectronics. Afterward, the properties of anomalous photocurrent, including light intensity dependencies, are demonstrated.

Finally, we briefly introduce a new method for mobility measurements based on the double-layer model. Unlike the time of flight technique and field effect transistor measurements, this method can be used for an ultra-thin organic semiconductor to check carrier transport along the directions perpendicular to electrodes in photocells. Furthermore, we demonstrate that the technique can be utilised to check the dominant carrier types in a semiconductor. The final section includes the summary and proposals.

2. Anomalous photocurrent in BDTDA photocells

Anomalous transient photocurrent has been independently revealed in organic materials and amorphous inorganic materials. In extant literatures, mechanisms such as trapping/detrapping or electron injection from electrodes were adopted to interpret this behaviour in different materials. A common understanding from previous reports is that the transient photocurrent comes from organic or amorphous materials with poor carrier mobility or large thickness. However, the effects of the dielectric properties on related materials were seldom studied in detail. Moreover, we observed the anomalous transient photocurrent in a radical BDTDA thin film device with a significant imbalance of carrier transports. As a model material, behaviour in the BDTDA devices will be introduced in this section, as well as the physical properties of the pink BDTDA thin films.

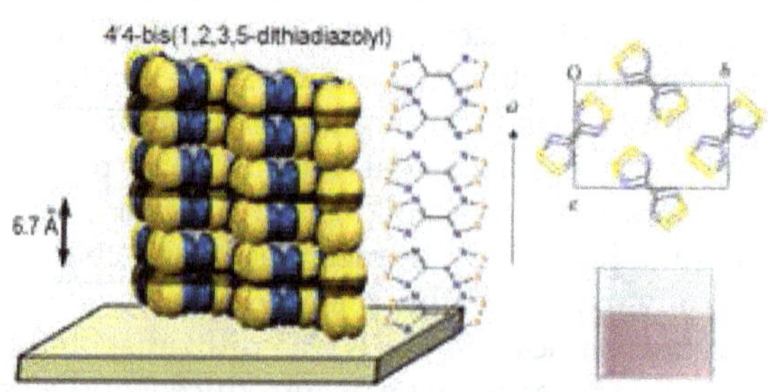

Fig. 1. Molecular π-stacking along the monoclinic a axis of BDTDA, a photograph of a thin film on ITO, and the molecular packing in the bc plane for this material

2.1 Characteristics of BDTDA thin films

2.1.1 Film structures

BDTDA is a disjoint diradical. Molecular orbitals for the two unpaired electrons are localised to separate five-membered rings, and exchange interactions between the two radical centres are very small. Its crystal structure consists of a face-to-face BDTDA dimer, indicating that intermolecular interaction is stronger than intradimer interaction. These dimers show π-stacking along the monoclinic *a* axis. Packing of dimeric stacks produces a herringbone-like motif with electrostatic $S^{\delta+}...N^{\delta-}$ contacts, in which all the molecular planes of BDTDA are parallel to the *bc* plane. It is notable that BDTDA films consist of alternating 1-dimensional π-stacking with molecular planes parallel to the substrates, as shown in Fig. 1 (Iwasaki et al., 2009; Kanai et al., 2009). Therefore, π-stacking can bridge the distance between bottom and top electrodes, which aids photoconduction between the electrodes.

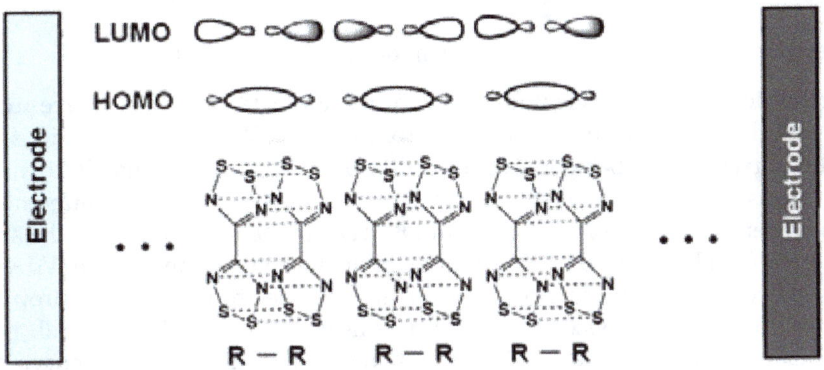

Fig. 2. Bonding and antibonding supramolecular orbitals of π–radical dimer.

2.1.2 Imbalance of carrier transport in BDTDA films

Considering that two π-radical BDTDA molecules exhibit face-to-face overlap, a bonding supramolecular orbital and an antibonding supramolecular orbital are developed (Fig. 2). The population of the bonding supramolecular orbital is concentrated at the centre of the dimer, while that of the antibonding supramolecular orbital spreads outside along the R–R axis. Since these radical dimers create stacking chains with π-π interactions, the antibonding supramolecular orbitals are expected to form a wide band through a large interdimer overlap; population of the lowest unoccupied molecular orbital (LUMO) spreads towards the outside of the dimer. By contrast, the highest occupied molecular orbital (HOMO) forms a narrow band. Therefore, a significant imbalance of carrier transport can be expected, specifically high photoconductivity by the electron migration in the wide LUMO band and poor hole mobility in the narrow HOMO band. In addition, the valence bond image (Iwasaki et al., 2009) suggests that the photoexcited state includes a character of charge transfer, namely, $R{:}R \rightarrow R^{+}R^{-}$, where R is a radical. In other words, electrons will be directly promoted from one molecule to another by photons, which can be regarded as a precursor stage of charge separation. These characteristics are promising for developing photoactivities.

2.1.3 Space charge limited current in BDTDA films

To characterise the diradical film, photocells with a structure of ITO/BDTDA (300 nm) /Al were prepared (Fig. 3) and current-voltage (*J-V*) characteristics were recorded. BDTDA was prepared as described in a previous report (Bryan et al., 1996), and was thermally evaporated onto ITO glasses. As a top electrode, Al was also thermally evaporated onto the thin films. The effective area of this photocell was approximately 0.02 cm^2. The sample was then fixed into a cryostat with a pressure below 1 Pa. During measurement, the Al electrode was grounded, and bias polarity was defined as plus when a positive bias voltage was applied to ITO.

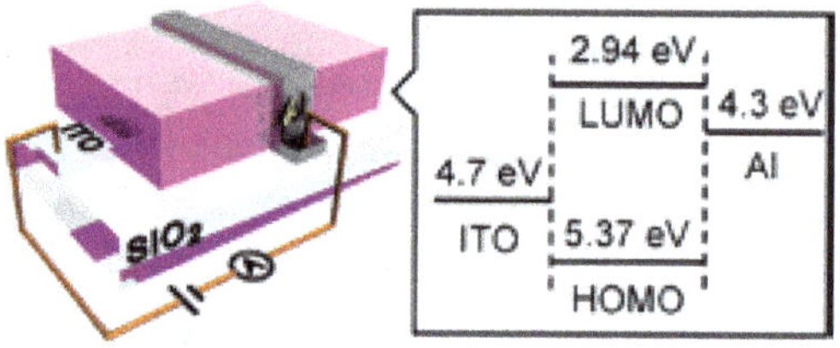

Fig. 3. Schematic views and an energy diagram of BDTDA photocells.

J-V characteristics were investigated using a picoammeter/voltage source under dark conditions and the bias voltage was scanned from -3 V to 3V. As shown in Fig. 4(a), the *J-V* curve exhibits a rectification behaviour, and rectification rate is approximately 10^2 at ±2 V. This behaviour is reasonable, as the work functions of the two electrodes are different, and non-injecting (see energy diagram of electrodes and BDTDA in the inset of Fig. 3). The applied bias V was corrected (van Duren et al., 2003) to compensate for the built-in voltage ($V_{bi} \approx 0.4$ V) that arises from work function difference between the two electrodes. Voltage drop across the series resistance of BDTDA devices was ignored, as its value was negligibly small.

Figure 4(b) exhibits the log (J)-log (V) plots for the data in Fig. 4(a). This curve consists of two regions with a crossover point at ~0.8 V, below which the *J-V* curve demonstrates Shockley behaviour that is ascribed to the injection limited current. At higher voltages (V > 0.8 V), the *J-V* curve shows a linear dependence, and its slope can be estimated as ~4.9. This value indicates that space charge limited current dominates the curve, though the dependence does not satisfy Child's law ($J \propto V^2$) (Coropceanu et al., 2007; Karl, 2003). This is a bulk limited current ascribed to a trap-controlled space charge limited current or a space charge limited current with a field dependence of carrier mobility (Blom et al., 1997; Sharma, 1995). Therefore, space charges are easily generated in this thin film devices, mainly due to significant imbalance of carrier transport and relatively large thickness (300 nm).

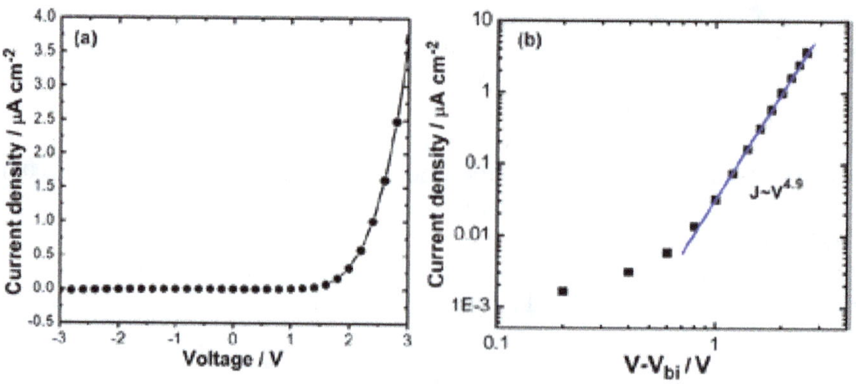

Fig. 4. *J-V* characteristics of a BDTDA photocell under a dark condition; (a) linear plot of J versus V; (b) log (J)-log(V) plot for the data in (a).

2.2 Photoresponses of BDTDA films

To measure the photocurrent of the photocells, a monochromated light, and green laser (532 nm) that can produce a stronger illumination, were employed as light source to irradiate the samples. To match the absorption band of BDTDA thin films, light with a wavelength of 560 nm was chosen for weak illumination to the transparent ITO electrode. We adopted lock-in techniques or an AC method (Ito et al., 2008) for normalised photocurrent-action spectra.

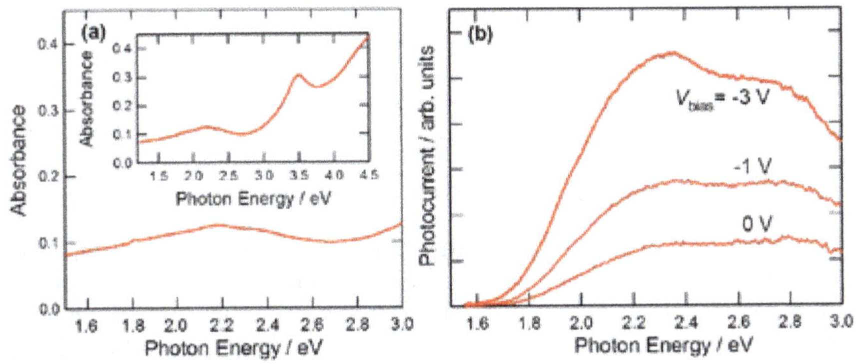

Fig. 5. (a) Absorption spectrum of BDTDA thin film; the inset shows the whole data within the range of 1.4-4.5 eV; (b) photocurrent-action spectra

2.2.1 Steady-state photocurrent of BDTDA films

To determine the optical properties of BDTDA thin films for photocurrent measurements, absorption spectrum of the BDTDA thin film (100 nm) on a quartz substrate within the range of 1.5-3.0 eV was recorded, as shown in Fig. 5(a). The inset shows data in the whole range of 1.2-4.5 eV. It is notable that there is a broad band around 2.1 eV that covers the whole visible range. The molecular orbital calculations indicate that this broad band is a complex of various electronic transitions, including intramolecular-, intradimer-, and interdimer transitions, allowed in the dimeric structure of this disjoint diradical. Subsequently, we examined the photoresponse of ITO/BDTDA (300 nm)/Al sandwich-type photocells.

Figure 5(b) shows the plots of photocurrent versus the photon energy (photocurrent-action spectra) measured by a lock-in technique with bias voltages V_{bias}= -3, -1 and 0 V. Photocurrent is obtained in the whole range of visible light (1.8-3.0 eV), while it shows a quick decrease below 2.2 eV. This decrease is possibly caused by the fact that absorptions below this energy are due to intramolecular excitations. The wide-range response, shown in Fig. 5(b), is advantageous for practical application as photodetectors.

Figure 6 is the photocurrent induced by green laser light illuminating from the ITO side with a small reverse bias voltage V_{bias} of -3V. Upon illumination, conductivity is enhanced with an on/off gain of 1.8×10^2 under an excitation light intensity of 1.59 mW/cm^2. The corresponding photoresponsivity (R_{res}) was calculated to be approximately 3.5 mA/W based on the relation $R_{res} = (I_{ph})/IA$, where A is the effective device area; I_{ph} and I are the photocurrent and the incident light intensity, respectively. The on/off ratio increases with the light intensity, and its maximum value observed in our experiments is approximately 10^3. Meanwhile, the photoresponsivity demonstrates an inverse behaviour, and changes from 10^{-1} to 10^{-4} A/W, which is comparable to that of the most advanced organic polymer photodetectors for visible region (Hamilton & Kanicki, 2004; Narayan & Singh, 1999; O'Brien et al., 2006; Xu et al., 2004).

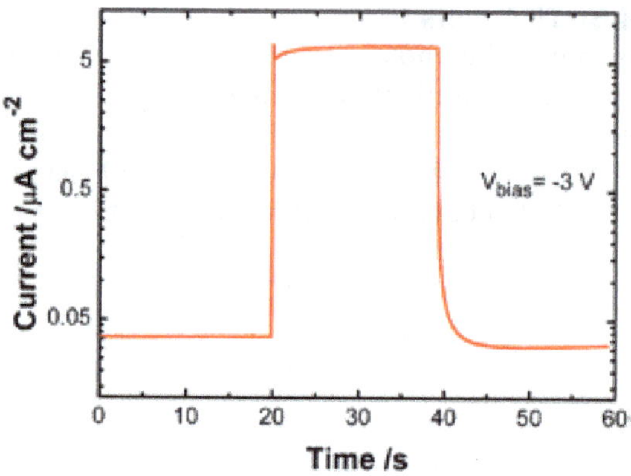

Fig. 6. On/off switching properties of the BDTDA photocell.

It is notable that the ITO/BDTDA/Al cells produce a photocurrent even at V_{bias}= 0 V, due to the potential difference of the electrodes, specifically ITO (4.8 eV) and Al (4.3 eV). This photovoltaic behaviour is consistent with the energy scheme in Fig. 3 taken by UPS/IPES measurements (Iwasaki et al., 2009). It is possible that the charge separation character in the photoexcited state, namely R+R-, contributes to this photovoltaic behaviour.

2.2.2 Anomalous transient photocurrent of BDTDA films

Figure 7(a) shows the photoresponses of an ITO/BDTDA/Al photocell with a bias voltage of 0 V. Upon illumination, a large anomalous transient photocurrent followed by a steady-state photocurrent was observed. Upon removal of illumination, a negative anomalous transient photocurrent was detected. Both the anomalous transient photocurrent and steady-state photocurrent increase with increases in light intensity. Figure 7(b) demonstrates the short circuit photoresponses under a reverse bias voltage of -2 V. Note that the anomalous transient photocurrent can be dramatically suppressed by applying a bias voltage. In particular, the negative current is nearly eliminated, while the steady-state current is increased. It is notable that anomalous transient photocurrent values under the zero bias can be comparable to those of the steady-state photocurrent under a bias voltage *V*. Positive anomalous transient photocurrent with weak excitation light intensity (≤ 0.57 μW/cm^2) decreases exponentially with time, and decay time of the positive anomalous transient photocurrent shows light-intensity dependence. As shown in Fig. 7(a), a stronger illumination causes faster decay. Meanwhile, for the light intensity of > 0.57 μW/cm^2, positive anomalous transient photocurrent cannot fit well with a single exponential simulation. This indicates that anomalous transient photocurrent is a superposed signal with different mechanisms.

Quantum efficiencies for steady-state photocurrent and anomalous transient photocurrent were calculated by neglecting reflection losses at the device surfaces. Figure 8(a) shows the intenal quantum efficiency (Pettersson et al., 2001) versus photon energy plots for the peak values of the positive (red curve) and negative anomalous transient photocurrent (blue curve) and for steady-state photocurrent under monochromatic illumination with weak intensity from a halogen lamp. Intenal quantum efficiency values for the positive and negative anomalous transient photocurrent show increases with an increase in photon energy, and their values are considerably higher than that of the steady-state photocurrent (black curve). It is notable that the transient intenal quantum efficiency for the positive anomalous transient photocurrent reaches an extremely high value of 65% at the photon energy of 2.8 eV, and its root mean square value is estimated to be ~30%; intenal quantum efficiency values of steady-state photocurrent are ~6%, corresponding to an external quantum efficiency of ~2%.

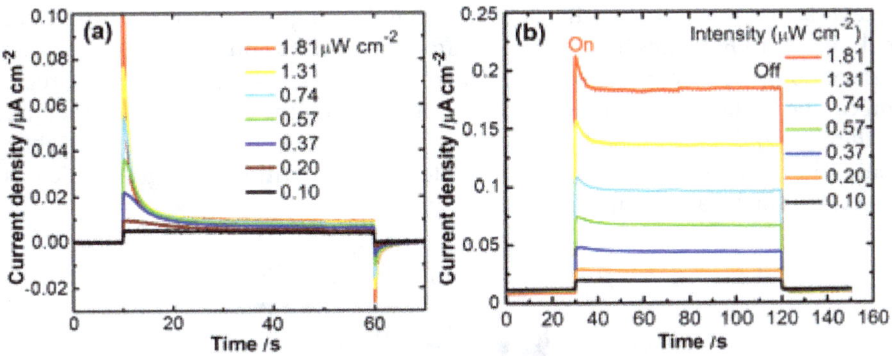

Fig. 7. Photoresponses of a BDTDA photocell with an illumination of 560 nm; (a) photoresponses under different light intensities with a zero bias voltage; (b) photoresponses under different light intensities with a bias voltage of -2 V.

To explore the recombination processes and mechanisms for anomalous transient photocurrent, we examined the light intensity dependence of the positive anomalous transient photocurrent and steady-state photocurrent. The results are shown in Fig. 8(b), where both axes are in a logarithmic scale. Both anomalous transient photocurrent and steady-state photocurrent obey a power law: $J \propto I^{\alpha}$, with α = 0.93 for the former or α = 0.28 for the latter. The former value suggests that monomolecular or geminate recombination (Binet et al., 1996) plays a role in the process. The latter α value suggests that the steady state suffers from higher order recombination processes, such as Auger (Wagner & Mandelis, 1996) and quadrimolecular recombinations (Marumoto et al., 2004). Considering that the α value is close to 0.25, quadrimolecular recombinations are more likely; adjacent photogenerated $R^{+}R^{-}$ pairs may interact with each other and recombine simultaneously.

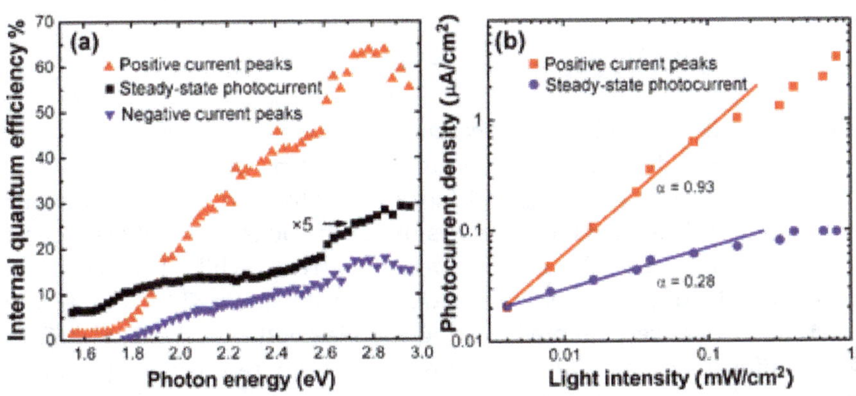

Fig. 8. (a) Intenal quantum efficiency values of the anomalous transient photocurrent and steady-state photocurrent for a BDTDA photocell; (b) light intensity dependence of the positive anomalous transient photocurrent (red points) and the steady-state photocurrent (blue points) induced by the green laser.

3. Mechanisms of anomalous photocurrent in BDTDA

Due to imbalance of carrier transports and the energy scheme of photocells, the junction at the Al/BDTDA interface plays the dominant role for the transient photoresponse (Hu et al., 2010a) if the BDTDA film is fully depleted. On the contrary, ITO/BDTDA with a larger barrier plays a main role if the film is not depleted. The junction acts as an active region (dark pink region in Fig. 9), which makes a different contribution to the anomalous transient photocurrent compared with the bulk region as blocking region (shallow pink region). The thick film can be treated as a double-layer system with widths of d_a and d_b (Hu et al., 2010b). Due to the large thickness and an imbalance of carrier transport, space charges are accumulated in the active layer. The built-in electric field will be changed, which may lead to the generation of polarisation current in the film.

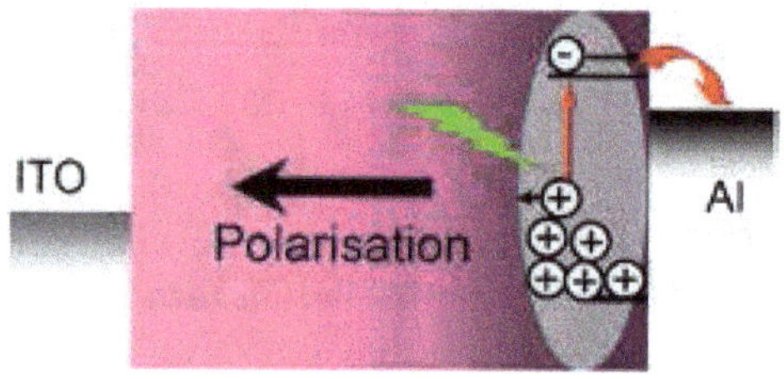

Fig. 9. A schematic view of BDTDA photocells.

3.1 Total current in a double-layer model

Theoretic analyses were performed to explore the mechanisms. To simplify the related theoretic analyses, electric fields in both regions are regarded as uniform and thus the BDTDA films can be separated into a double-layer film. Moreover, the thickness of both layers is assumed to be constant.

3.1.1 Theoretic analyses for a double-layer model

Based on the total current equation (Guru & Hiziroğlu, 2004), the current density j through the double layers is written as follows:

$$j=(\sigma_{b0}+\sigma_b)E_b(t)+\varepsilon_0\varepsilon_b\frac{dE_b(t)}{dt}=(\sigma_{a0}+\sigma_a)E_a(t)+\varepsilon_0\varepsilon_a\frac{dE_a(t)}{dt}, \tag{1}$$

where σ_{b0}, σ_b, ε_b, and $E_b(t)$ pertain to dark conductivity, photoconductivity, relative dielectric constant, and the time t dependence of the uniform electric field, respectively, in the bulk blocking region. Meanwhile, σ_{a0}, σ_a, ε_a, and $E_a(t)$ express the corresponding quantities in the active junction region. The constant ε_0 denotes the dielectric constant of free space. The first terms in both sides of Eq. (1) are conduction current, while the second terms are displacement current. All parameters for conductivity are assumed to be time independent and dark conductivities were ignored. Taking the bias voltage ($V = E_b d_b + E_a d_a$) and the boundary condition $\varepsilon_b E_b(0)=\varepsilon_a E_a(0)$ into account, we can resolve Eq. (1), and the time dependence of E_b, E_a and j can be written as:

$$E_b(t)=\frac{\sigma_a V}{d_b\sigma_a+d_a\sigma_b}+\left(\frac{\varepsilon_a}{d_b\varepsilon_a+d_a\varepsilon_b}-\frac{\sigma_a}{d_b\sigma_a+d_a\sigma_b}\right)Ve^{-t/\tau}, \tag{2a}$$

$$E_a(t)=\frac{\sigma_b V}{d_b\sigma_a+d_a\sigma_b}+\left(\frac{\varepsilon_b}{d_b\varepsilon_a+d_a\varepsilon_b}-\frac{\sigma_b}{d_b\sigma_a+d_a\sigma_b}\right)Ve^{-t/\tau}, \tag{2b}$$

$$j=\frac{\sigma_b\sigma_a}{d_b\sigma_a+d_a\sigma_b}V+\frac{(\varepsilon_a\sigma_b-\varepsilon_b\sigma_a)^2 d_b d_a V}{(d_b\varepsilon_a+d_a\varepsilon_b)^2(d_b\sigma_a+d_a\sigma_b)}e^{-t/\tau}, \tag{3}$$

where

$$\tau=\frac{\varepsilon_0(d_b\varepsilon_a+d_a\varepsilon_b)}{d_b\sigma_a+d_a\sigma_b}. \tag{4}$$

As shown in Eq. (4), the physics of decay time τ relates to the extraction speed for the free carriers by electrodes and dielectric property/polarisation in the films. Subsequently, we can estimate the total collected charges at time t in the active side electrode, which is given by the following:

$$Q(t)=\int_0^t Aj(t)dt-\int_0^t i(t)dt, \tag{5}$$

where $i(t)$ is the discharging current in the external circuit. We take into account that the voltage drop across load resistor R is equal to that across the photocell, particularly the following:

$$i(t)R=\frac{\int_0^t Aj(t)dt-\int_0^t i(t)dt}{C}, \tag{6}$$

where C is the capacitance of the photocell. Since the initial current $i(0)$ = 0, Eq. (6) can be resolved and the external discharging current $i(t)$ (i.e., anomalous transient photocurrent) is expressed as follows:

$$i(t)=\frac{S\xi_0}{(\tau-RC)}\left(e^{-\frac{t}{\tau}}-e^{-\frac{t}{RC}}\right)+S\varsigma(1-e^{-\frac{t}{RC}}), \tag{7}$$

where

$$\xi_0=\frac{\varepsilon_0(\varepsilon_a\sigma_b-\varepsilon_b\sigma_a)^2 d_b d_a V}{(d_b\varepsilon_a+d_a\varepsilon_b)(d_b\sigma_a+d_a\sigma_b)^2},\ \varsigma=\frac{\sigma_b\sigma_a}{d_b\sigma_a+d_a\sigma_b}V$$

3.1.2 Simplified analyses for BDTDA photocells

In general, photogenerated excitons in organic materials can be dissociated only at donor-acceptor interfaces, or by a strong local electric field (Nicholson & Castro, 2010). If the film thickness is considerably larger tthan exciton diffusion length and carrier drift length, the excitons and carriers far from the electrodes cannot contribute to the photocurrent. In other words, the photoconductivity σ_a, which is proportional to carrier mobility μ and density n in the junction (active region), is considerably larger than that in the bulk region, as well as the dark conductivity. Therefore, other conductivities (σ_{b0}, σ_b, and σ_{a0}) can be ignored. Eqs. (2) and (7) can therefore be expressed as follows:

$$E_b(t) = \frac{V}{d_b} + \left(\frac{\varepsilon_a}{d_b\varepsilon_a + d_a\varepsilon_b} - \frac{1}{d_b}\right)Ve^{-t/\tau}, \tag{8a}$$

$$E_a(t) = \frac{\varepsilon_b}{d_b\varepsilon_a + d_a\varepsilon_b}Ve^{-t/\tau}, \tag{8b}$$

$$i(t) = \frac{S\xi}{(\tau - RC)}\left(e^{-\frac{t}{\tau}} - e^{-\frac{t}{RC}}\right) \tag{9}$$

with

$$\xi = \frac{\varepsilon_0\varepsilon_b^2 d_a V}{(d_b\varepsilon_a + d_a\varepsilon_b)d_b} \tag{10}$$

and

$$\tau = \frac{\varepsilon_0(d_b\varepsilon_a + d_a\varepsilon_b)}{d_b\sigma_a}. \tag{11}$$

Obviously, photogenerated carriers in the junction region that can be collected by electrodes will be exhausted if photoconductivity of the blocking region is extremely small. The space charge will be accumulated in the film and thus the electric field can be changed, as shown in Eq. (9). This naturally leads to a polarisation current. Two mechanisms, including τ and RC time constant, are responsible for the decay of the anomalous transient photocurrent.
The derivative calculation was performed for Eq. (9), and a rise time τ_R can be obtained. In particular, after a time

$$t = \tau_R = \frac{RC\kappa}{RCI - \kappa}\ln\frac{RCI}{\kappa}, \tag{12}$$

with $\kappa = \tau I = \varepsilon_0(d_b\varepsilon_a + d_a\varepsilon_b)/d_b e a\mu$, the largest current density J_m can be achieved, which is expressed as follows:

$$J_m = \frac{\xi}{(\tau - RC)}\left(\frac{RC}{\tau}\right)^{-\frac{RC}{RC-\tau}}\left(1 - \frac{RC}{\tau}\right). \tag{13}$$

Since $\sigma_a = e\alpha I\mu$, where a is quantum efficiency and aI means carrier density with a light intensity of I, and e means the elementary charge, the decay time τ can be written as follows:

$$\tau = \frac{\varepsilon_0(d_b\varepsilon_a + d_a\varepsilon_b)}{ed_b\alpha I\mu}, \tag{14}$$

which suggests a relationship of $\tau \propto I^{-1}$. This relation fits the experimental data well. We consider the situation of a weak illumination, which will lead to a large τ. If $\tau >> RC$, the discharging current density in Eq. (9) will be

$$J \approx J_m e^{-\frac{t}{\tau}}, \tag{15}$$

with a maxima J(t) value J_m

$$J_m = \frac{d_a Ve\alpha I\mu}{\left(d_b\, \varepsilon_a/\varepsilon_b + d_a\right)^2} \propto I\,. \tag{16}$$

Equation (15) suggests that the anomalous transient photocurrent exhibits exponential decay under weak irradiation and/or with a very small RC time constant in the circuit, which fits well with the experimental behaviour in Fig. 7(a). It is notable that $J_m \propto \varepsilon_b^{\,2}$ in Eq. (16), indicating the effects from the dielectric constant of the bulk region. On the contrary, stronger illumination triggers a smaller τ, which is related to the dielectric constants of the materials and photoconductivity in the junction region. If $\tau << RC$, the time constant in the circuits will dominate the decay, and shows resistance dependence as well as an exponential relationship.

3.2 Discussions

Both Eqs. (7) and (9) indicate that the anomalous transient photocurrent is a superposed signal with two mechanisms, namely, electron extraction from the junction region, and discharging process in the external circuit with a time constant of RC. It is clear that the thickness of our BDTDA films (300 nm) is excessively large, exceeding the exciton diffusion length and carrier drift length. Upon illumination, photogenerated electrons near the cathode are extracted as conduction current, while electrons on the other side cannot move across the thick film to compensate. This induces the transient conduction current.

The capacitance and dielectric constant in the equations involve polarization mainly in the bulk region triggered by photogenerated space charges in the films. The dielectric property of BDTDA strongly influences the anomalous transient photocurrent. Based on theoretic analyses, it is natural that the anomalous behaviour is universal for the thin films with large polarity, poor mobility and relatively large thickness. Though the carriers in organic materials cannot withstand a long trip due to various means of dissipation, including traps and recombination, displacement or a polarisation current can generate a large anomalous transient photocurrent without experiencing a long trip. Fast generation of this photocurrent is possible because the photoinduced polarisation current allows localised charges to oscillate around their equilibrium states. This is promising for high-speed organic photodetectors.

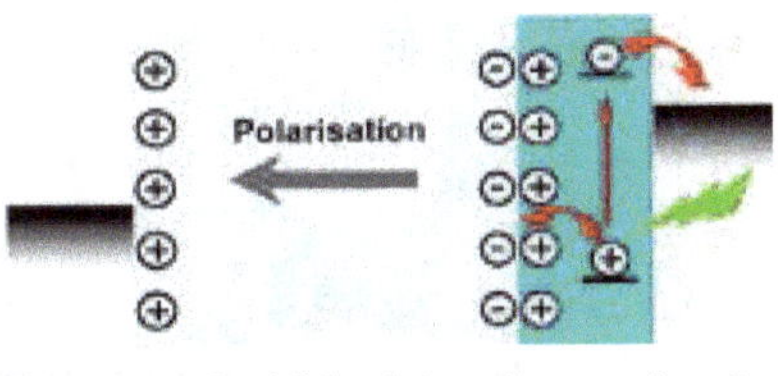

Fig. 10. A schematic display of an anode/ blocking layer / active layer/ cathode photosensor.

4. Metal/insulator/semiconductor/metal type photocells

Based on the double-layer model, we developed a device to confirm the theoretic analyses in Section 3. A transparent thick organic insulator layer as a blocking layer was adopted to substitute the bulk region in BDTDA photocells, and an organic semiconductor thin layer as an active layer was chosen to substitute the junction region. Figure 10 demonstrates the photocell with a structure of metal/organic insulator/organic semiconductor/metal, which may be utilised for light detection as well. The thickness of the semiconductor layer is targeted around 20 nm, which is equivalent to the carrier drift length. The organic double layers between the metals induce an imbalance of carrier transports; in particular, only one type of carrier can be collected by the electrodes. These will facilitate accumulation of the other type of carriers as space charges at the interface of the blocking layer and active layer. In this structure, the dielectric property of the insulator layer will strongly influence the signals.

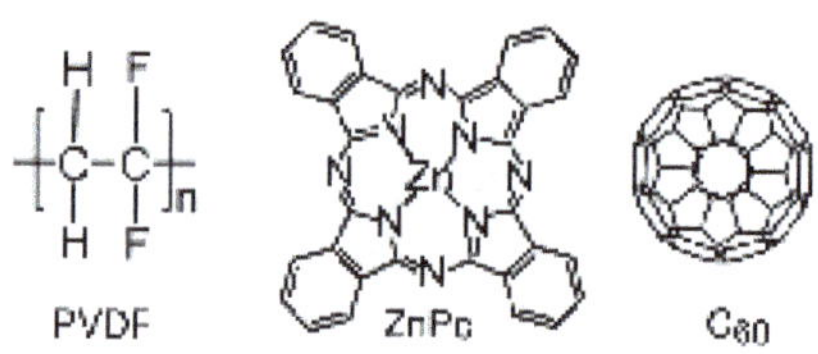

Fig. 11. Chemical structures of PVDF and ZnPc:C_{60} donor-acceptor systems.

4.1 Photoresponses of ITO/PVDF/ZnPc:C_{60}/Al

To check the photoresponse of this kind of photocell, an equivalent metal/blocking layer/semiconductor layer/metal photocell was fabricated with ITO and Al electrodes. A well-known transparent polymer, polyvinylidene fluoride (PVDF, 8 wt% in dimethyl formamide), was adopted for the blocking layer and spin-coated onto a hot ITO glass slide (100 °C). Thickness was estimated to be ~1 μm by cross-sectional SEM images. At the top of the blocking layer, a 30-nm active layer with a high charge-separation efficiency was prepared with zinc phthalocyanine (ZnPc) and fullerene (C_{60}) (molar ratio: 1:1, see Fig. 11 for their molecular structures) by co-deposition. Subsequently, the Al cathode was thermally evaporated onto the blend film. Photocurrent measurements were conducted under an illumination from a green laser (532 nm) controlled by a multifunction synthesiser. Photoresponses across a load resistor of 10^5 Ω were recorded on an oscilloscope.

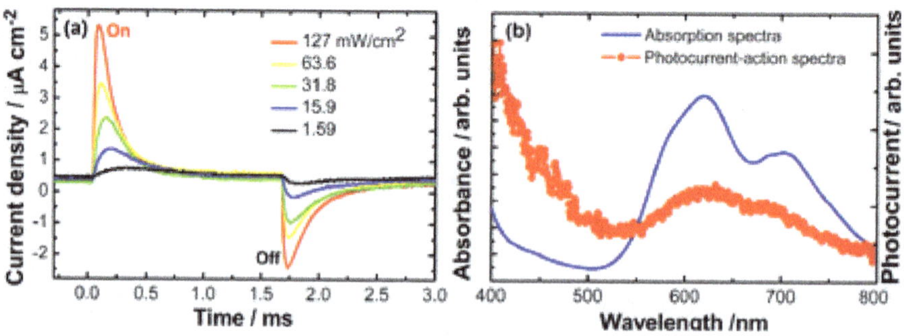

Fig. 12. (a) Photoresponses of an ITO/PVDF/ZnPc:C_{60}/Al photocell under an illumination of different intensities; (b) a comparison between the absorption spectra of the blend films (blue curve) and photocurrent-action spectra of the photocell (red curve).

4.1.1 Photoresponses

Figure 12 shows the photoresponses with various light intensities. Upon laser illumination, a large anomalous transient photocurrent similar to that in the BDTDA photocells is observed, and a negative anomalous transient photocurrent appears just after the illumination. Both the positive and negative anomalous transient photocurrent increase with increases in light intensity, and a faster decay can be obtained under a stronger illumination, which fits the expectation of Eq. (12). Absorption spectra of the blend films and photocurrent-action spectra (Fig. 12(b)) were collected for comparison. The peaks in these spectra are in agreement, indicating that the active layer does play a primary role in the production of this anomalous transient photocurrent. It is notable that no signals were obtained in the ITO/PVDF/Al structure, suggesting that only the active layer was the sensitive component. In addition, the relationship between anomalous transient photocurrent and weak light intensity was observed to exhibit linearity.

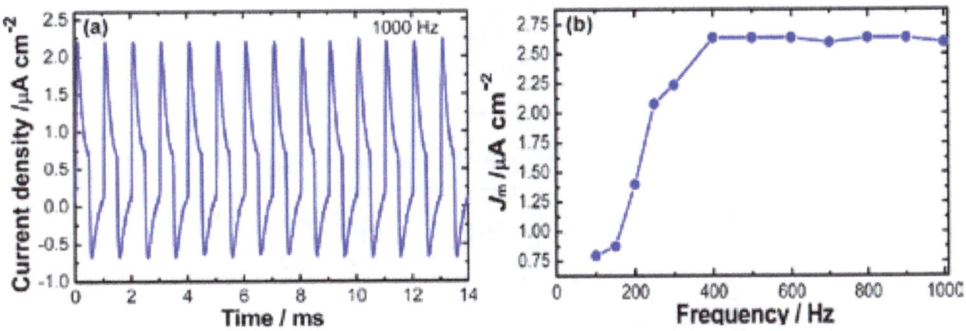

Fig. 13. (a) Photoresponses of an ITO/PVDF (1 μm)/ZnPc:C_{60} (30 nm)/Al photocell with a light modulation of 1 kHz (31.8 mW/cm^2). (b) Frequency dependence of the photoresponses.

We examined the reproducibility of the anomalous transient photocurrent as well. Continuous current oscillation induced by frequency modulation is stably observed without degeneration (Fig. 13(a)). Evidently, the effective current will be increased as modulation frequency increases, as more current peaks can be generated in a fixed time period. It is notable that the values of the anomalous transient photocurrent peaks increase with increases in modulation frequency, and saturation is subsequently achieved after a certain modulation frequency, as shown in Fig. 13(b).

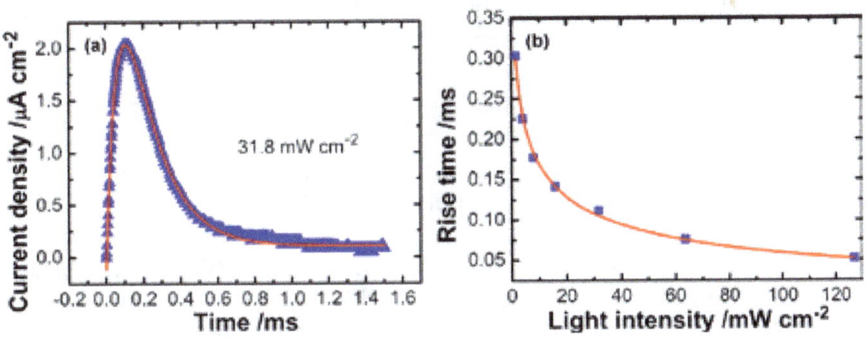

Fig. 14. (a) Simulations for the positive anomalous transient photocurrent based on (a) Eq. (7) and (b) Eq. (12) at 100 Hz.

4.1.2 Theoretic analyses for the transient photocurrent

We performed theoretic simulations for the anomalous transient photocurrent from the metal/blocking layer/semiconductor layer/metal photocells based on Eqs. (7) and (12), as shown in Fig. 14. The blue triangles in Fig. 14(a) show the time dependence of the current density of positive anomalous transient photocurrent obtained under an illumination of 31.8 mW/cm^2. The solid red curve in this figure shows the theoretical simulations from Eq. (7). The *RC* time constant was extracted from the simulation to be 6.8×10^{-5} s, which is considerably close to the experimental value (4.2×10^{-5} s, experimentally determined for the present circuit by an LCR meter at 100 Hz). τ was estimated to be $\sim1.6\times10^{-4}$ s, during which 1-(1/e) of the photogenerated carriers that can be extracted will be collected by electrodes. The blue squares in Fig. 14(b) depict the dependence of the rise time τ_R on light intensity. This behaviour is reproduced by Eq. (12) (solid curve) as well. The *RC* time constant and τ are estimated to be 8.3×10^{-5} s and 1.4×10^{-4} s under an illumination of 31.8 mW/cm^2, respectively. Both simulated values from Eqs. (7) and (12) are in approximate agreement with each other, suggesting that the established double-layer model is reasonable for the explanation of anomalous transient photocurrent.

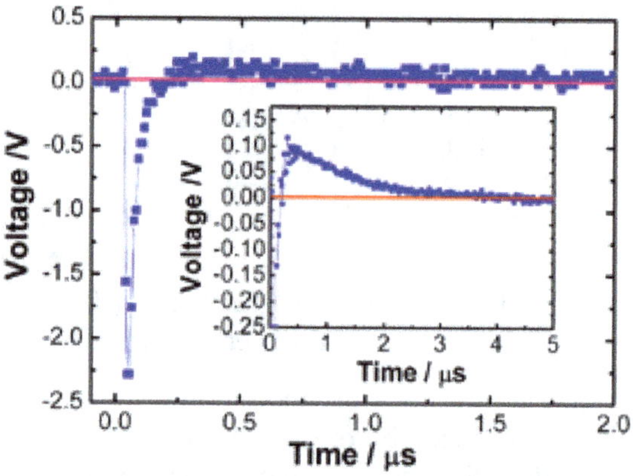

Fig. 15. Impulse response of the ITO/PVDF/ZnPc:C_{60}/Al photocell under a zero bias voltage; the inset is a magnified version of the recovery process.

4.1.3 Impulse response

To evaluate the lifetime of anomalous transient photocurrent, an impulse response was examined with a nanosecond laser beam (600 nm) from an optical parametric oscillator pumped by a Nd:YAG laser (10 Hz; pulse width: ~6 ns; power: ~1.08 μJ/pulse). A digital oscilloscope and a dc 300-MHz amplifier were used to collect voltage response with an input resistance of 50 Ω. A photocell with a structure of ITO/polystyrene (1 μm)/ZnPc:C_{60} (20 nm)/Al was prepared for comparison with the ITO/PVDF (1 μm)/ZnPc:C_{60} (20 nm)/Al photocells. The fabrication method for the polystyrene blocking layer was the same as that for PVDF.

Figure 15 shows the impulse response of the photocell with a PVDF blocking layer, which consists of rise, decay, and recovery processes. This behaviour is similar to that of the pyroelectric detectors with slower rise, decay, and recovery times (Odon, 2005; Polla et al., 1991), though their mechanisms are quite different. The *RC* constant in this circuit was estimated to be ~5 ns. Rise and decay time of the PVDF photocell can be observed as ~15 and 100 ns, respectively. Both the rise and decay times show an *RC* constant dependence; they increase along with increases in the *RC* constant (not shown). However, the recovery time exhibits a long time scale of ~2.5 μs (see inset of Fig. 15) and is independent of the *RC* constant.

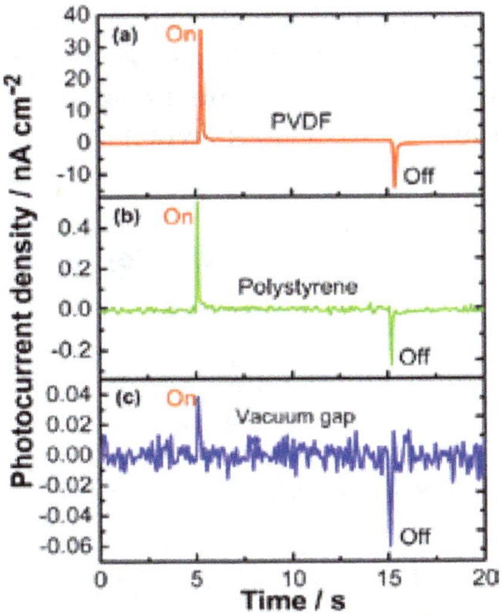

Fig. 16. Dielectric constant dependence of the anomalous transient photocurrent under an illumination (532 nm) of 160 mW/cm^2; (a), (b) and (c) show the short-circuit anomalous transient photocurrent in the metal/blocking layer/semiconductor layer/metal photosensor with PVDF, polystyrene, and vacuum gap as blocking layers, respectively.

Faster response can be achieved by decreasing the dielectric constant ε of blocking layer. For example, substitution of the PVDF layer (ε ≈ 7-13) (Kerbow & Sperati, 1999) by polystyrene (ε ≈ 2.6) (Cullen & Yu, 1971) brings about a considerably faster rise (~5 ns) and decay (~8 ns) times at 0 V, but recover time remains to be ~1 μs. Slow recovery time could be ascribed to an energy barrier between the donor-acceptor and/or semiconductor-metal interfaces. Considering that the polarisation current is proportional to the variation rate of E_b triggered by the photogenerated space charges, a faster generation of space charges by a sharper light pulse can bring about a larger anomalous transient photocurrent, even when only a small number of space charges are generated. Therefore, device speed is mainly determined by the rise and decay time, even though the system does not completely recover.

4.1.4 Dielectric influences

We examined the relation between the dielectric constant ε_b of the blocking layer and the quantum efficiency of anomalous transient photocurrent. Photocells with three different blocking layers (1 μm), namely, with vacuum gap (ε = 1), polystyrene, and PVDF were prepared. Thickness of all the active layers is approximately 20 nm. Figure 16 demonstrates the short-circuit photoresponses of the three photocells against a strong illumination (160 mW/cm^2). The values of the anomalous transient photocurrent dramatically increase with ε_b as predicted in Eq. (16). As such, we can control the transient conversion efficiency by changing the ε value of blocking layer. It is notable that the positive anomalous transient photocurrent of the PVDF photocell is ~8×10^2 times larger than that of the vacuum-gap photocell, though a rough estimation based on Eq. (16) suggests a difference of two orders of magnitude. The internal quantum efficiency of anomalous transient photocurrent in this PVDF cell under a weak illumination (0.2 μW/cm^2; 560 nm) from a halogen lamp is calculated to be approximately 34% (root mean square, rms). The photoresponsivity at 560 nm (0.2 μW/cm^2) reaches 10 mA/W (rms) even without applying a bias voltage, which is comparable to those of conventional organic photodetectors operated by a bias voltage (Iwasaki et al., 2009; Narayan & Singh, 1999; O'Brien et al., 2006). It is believed that more charges stored in the PVDF photocell with a larger ε were released upon illumination, when compared with the polystyrene and vaccum gap photocells.

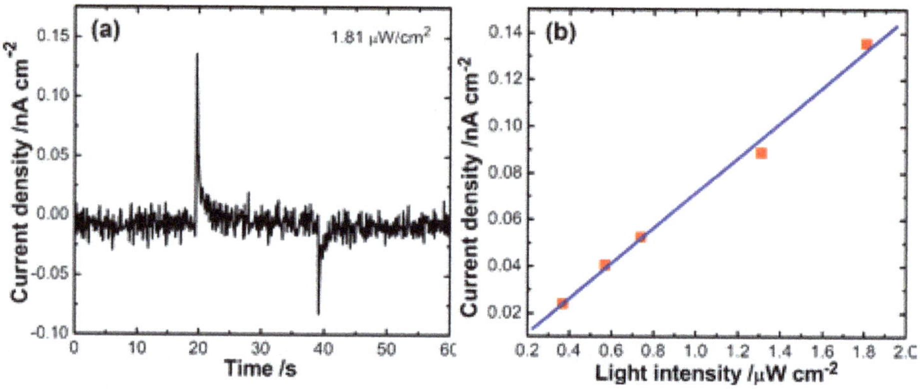

Fig. 17. (a) Photoresponse of ITO/ ZnPc/polystyrene (1 μm)/Al photocells under a zero bias voltage; (b) light intensity dependence of peak anomalous transient photocurrent value.

4.2 Photoconductivity dependence

Based on Eq. (14), decay time τ is inversely proportional to photoconductivity $\sigma_a = en\mu$, where $n = aI$ is the photogenerated free carrier density. Therefore, larger carrier mobility and density will induce faster decay and a larger signal if the RC time constant in a circuit is very small. To check this relationship, ZnPc (30 nm) as the active layer was utilised in the metal/blocking layer/semiconductor layer/metal structure by thermal evaporation with a speed of 1 Å/s. Polystyrene layer (1 μm) by spin coating was adopted as the blocking layer. Two types of photocells with different structures for this material were produced, namely, ITO/polystyrene/ZnPc/Al and ITO/ZnPc/polystyrene/Al. As expected, the former does not exhibit signals since ZnPc is an excellent donor material. The latter shows a signal (see Fig. 17) and only holes are collected by the ITO electrode, which can be judged by the current direction. However, compared to those from the blend film (or bulk-heterojunction) devices, the signal from ZnPc photocells is considerably weaker due to a lower charge separation efficiency, which leads to a smaller carrier density. We likewise examined the light intensity dependence of anomalous transient photocurrent. As predicted in Section 3, intensity dependence of anomalous transient photocurrent does exhibit linearity (Fig. 17(b)) under weak illumination with a monomolecular or geminate recombination.

4.3 Discussions

Photoresponses from the metal/blocking layer/semiconductor layer/metal structure even with a vacuum gap is promising, indicating potential for pulse light detection. As we know, metal/semiconductor/metal type organic thin film device usually exhibits a large dark current due to pin holes, which leads to a small photocurrent. The employment of a blocking layer hampers the formation of pin holes and results in an extremely small dark current. It is possible now to utilise ultrathin films only with the highest internal quantum efficiency for light detection. Compared with ideal metal/semiconductor/metal photocells with the same thickness in which conduction photocurrent J_{ph} can be written as follows:

$$J_{ph} = \frac{Ve\alpha I\mu}{d_b + d_a}, \tag{17}$$

the metal/blocking layer/semiconductor layer/metal structure may exhibit a larger anomalous transient photocurrent current if we select an appropriate blocking layer and active layer with a small value of d_b/d_a, and small dielectric constant ratio ($\varepsilon_a/\varepsilon_b$). These can be judged from Eq. (16) [1]. Therefore, it is possible to release more stored charges in the capacitor type photocells upon illumination in addition to the photogenerated free carriers. Equation (14) likewise indicates a promising method for characterisation of carrier transport. As we can see from this equation, the anomalous transient photocurrent signal is related to carrier mobility and density. Therefore, mobility can be estimated if we extract the decay time τ from theoretic simulations, and obtain the photogenerated carrier density by other methods, such as light-induced electron paramagnetic resonance technique. In addition, we can utilise the metal/blocking layer/semiconductor layer/metal structure to determine the carrier type in an organic semiconductor, as described in Section 4.2.

5. Conclusion

In summary, we analysed the anomalous transient photocurrent in the BDTDA photocells based on a double-layer model. Resuts indicate that the dielectric property of organic materials will strongly influence the anomalous behaviour. For instance, a large dielectric constant will induce a larger anomalous photocurrent. This was confirmed in equivalent devices, such as ITO/PVDF or polystyrene/ZnPc:C_{60}/Al, in which the PVDF and polystyrene layer act as the bulk region, and the blend film acts as the junction region in the BDTDA photocells. Both theoretic and experimental results fit well with each other, suggesting that polarisation and fast extraction of photogenerated carriers in the blend film play significant roles in this behaviour. The theoretic analyses likewise indicate that the anomalous transient photocurrent may achieve a larger value if proper conditions are satisfied, compared to the conventional metal/semiconductor/metal photocells with the same total thickness. It is notable that the metal/blocking layer/semiconductor layer/metal structure is immune from pin-hole effects which usually exist in ultrathin conventional devices. Stored charges in the metal/blocking layer/semiconductor layer/metal capacitor photocells can be released upon illumination, which is quite different from the conventional principles for light detection or harvesting. These indicate potential optoelectronic conversion for pulse light detection in various fields, including communications, remote control, and image sensors. The obtained theory may also play a role in the characterisation of carrier transport along the directions perpendicular to the electrodes in the metal/blocking layer/semiconductor layer/metal photocells.

---

[1]For comparison, Eq. (16) can be changed to the following:

$$J_m = Ve\alpha I\mu \,/\left(d_a + 2d_b\,\varepsilon_a/\varepsilon_b + d_b^{\,2}\,\varepsilon_a^{\,2}/d_a^{\,2}\varepsilon_b^{\,2}\right).$$

6. Acknowledgments

The authors are indebted to Prof. Hiroshi Ito for his technical supports and constructive suggestions. Also acknowledged are the research group members who contributed to this work through useful discussions and provision of materials. This research was supported by a Grant-in-Aid for Scientific Research from the Ministry of Education, Culture, Sports, Science, and Technology (MEXT) and by CREST, JST. Dr. Hu also thanks the National Natural Science Foundation of China (No. 11004172 and 10804098) and the Zhejiang Provincial Natural Science Foundation of China (No. Y607472).

7. References

Andriesh, A. M. *et al.* (1983), Anomalous Transient Photocurrent in Disordered Semiconductors - Theory and Experiment. *Solid State Communications*, Vol. 48, No. 12, pp. 1041-1043, ISSN 0038-1098.

Binet, F. *et al.* (1996), Mechanisms of recombination in GaN photodetectors. *Applied Physics Letters*, Vol. 69, No. 9, pp. 1202-1204, ISSN 0003-6951.

Blanchet, G. B. *et al.* (2003), Large area, high resolution, dry printing of conducting polymers for organic electronics. *Applied Physics Letters*, Vol. 82, No. 3, pp. 463-465, ISSN 0003-6951.

Blom, P. W. M. *et al.* (1997), Electric-field and temperature dependence of the hole mobility in poly(p-phenylene vinylene). *Physical Review B*, Vol. 55, No. 2, pp. R656-R659, ISSN 0163-1829.

Bryan, C. D. *et al.* (1996), Preparation and characterization of the disjoint diradical 4,4'-bis(1,2,3,5-dithiadiazolyl) [S_2N_2C-CN_2S_2] and its iodine charge transfer salt [S_2N_2C-CN_2S_2]. *Journal of the American Chemical Society*, Vol. 118, No. 2, pp. 330-338, ISSN 0002-7863.

Chakraborty, A. K. & Mallik, B. (2009), Photoinduced anomalous current changes in some organometallic materials. *Current Applied Physics*, Vol. 9, No. 5, pp. 1079-1087, ISSN 1567-1739.

Coropceanu, V. *et al.* (2007), Charge transport in organic semiconductors. *Chemical Reviews*, Vol. 107, No. 4, pp. 926-952, ISSN 0009-2665.

Cullen, A. L. & Yu, P. K. (1971), Accurate Measurement of Permittivity by Means of an Open Resonator. *Proceedings of the Royal Society of London Series a-Mathematical and Physical Sciences*, Vol. 325, No. 1563, pp. 493-509, ISSN 0950–1207

Forrest, S. R. & Thompson, M. E. (2007), Introduction: Organic electronics and optoelectronics. *Chemical Reviews*, Vol. 107, No. 4, pp. 923-925, ISSN 0009-2665.

Gunes, S. *et al.* (2007), Conjugated polymer-based organic solar cells. *Chemical Reviews*, Vol. 107, No. 4, pp. 1324-1338, ISSN 0009-2665.

Guru, B. S. & Hiziroğlu, H. R. (2004), *Electromagnetic Field Theory Fundamentals*, Cambridge Univ. Press, Cambridge. ISBN 0521830168.

Hamilton, M. C. & Kanicki, J. (2004), Organic polymer thin-film transistor photosensors. *Ieee Journal of Selected Topics in Quantum Electronics*, Vol. 10, No. 4, pp. 840-848, ISSN 1077-260X.

Hu, L. G. *et al.* (2010), Highly efficient alternating photocurrent from interactive organic-radical dimers: A novel light-harvesting mechanism for optoelectronic conversion. *Chemical Physics Letters*, Vol. 484, No. 4-6, pp. 177-180, ISSN 0009-2614.

Hu, L. G. *et al.* (2010), Optoelectronic conversion by polarization current, triggered by space charges at organic-based interfaces. *Applied Physics Letters*, Vol. 96, No. 24, pp. 243303, ISSN 0003-6951.

Ito, H. *et al.* (2008), Photocurrent of regioregular poly(3-alkylthiophene)/fullerene composites in surface-type photocells. *Thin Solid Films*, Vol. 516, No. 9, pp. 2743-2746, ISSN 0040-6090.

Iwamoto, M. (1996), Transient current across insulating films with long-range movements of charge carriers. *Journal of Applied Physics*, Vol. 79, No. 10, pp. 7936-7943, ISSN 0021-8979.

Iwasaki, A. *et al.* (2009), Interactive Radical Dimers in Photoconductive Organic Thin Films. *Angewandte Chemie-International Edition*, Vol. 48, No. 22, pp. 4022-4024, ISSN 1433-7851.

Kao, K.-C. (2004), *Dielectric Phenomena in Solids* Elsevier Academic Press, Amsterdam. ISBN 978-0-12-396561-5.

Karl, N. (2003), Charge carrier transport in organic semiconductors. *Synthetic Metals*, Vol. 133, No. pp. 649-657, ISSN 0379-6779.

Kumar, A. *et al.* (1987), Anomalous Decay of Photocurrent in Amorphous Thin-Films of $Ge_{22}Se_{78}$. *Physical Review B*, Vol. 35, No. 11, pp. 5635-5638, ISSN 0163-1829.

Marumoto, K. *et al.* (2004), Quadrimolecular recombination kinetics of photogenerated charge carriers in regioregular poly (3-alkylthiophene) / fullerene composites. *Applied Physics Letters*, Vol. 84, No. 8, pp. 1317-1319, ISSN 0003-6951.

Mihailetchi, V. D. *et al.* (2005), Space-charge limited photocurrent. *Physical Review Letters*, Vol. 94, No. 12, pp. 126602, ISSN 0031-9007.

Narayan, K. S. & Singh, T. B. (1999), Nanocrystalline titanium dioxide-dispersed semiconducting polymer photodetectors. *Applied Physics Letters*, Vol. 74, No. 23, pp. 3456-3458, ISSN 0003-6951.

Nicholson, P. G. & Castro, F. A. (2010), Organic photovoltaics: principles and techniques for nanometre scale characterization. *Nanotechnology*, Vol. 21, No. 49, pp. 492001, ISSN 0957-4484.

O'Brien, G. A. *et al.* (2006), A single polymer nanowire photodetector. *Advanced Materials*, Vol. 18, No. 18, pp. 2379-2383, ISSN 0935-9648.

Odon, A. (2005), Voltage Response of Pyroelectric PVDF Detector to Pulse Source of Optical Radiation. *Measurement Science Review*, Vol. 5, No. pp. 55-58, ISSN 1335-8871

Pandey, A. K. *et al.* (2008), Size effect on organic optoelectronics devices: Example of photovoltaic cell efficiency. *Physics Letters A*, Vol. 372, No. 8, pp. 1333-1336, ISSN 0375-9601.

Pettersson, L. A. A. *et al.* (2001), Quantum efficiency of exciton-to-charge generation in organic photovoltaic devices. *Journal of Applied Physics*, Vol. 89, No. 10, pp. 5564-5569, ISSN 0021-8979.

Polla, D. L. *et al.* (1991), Surface-Micromachined Pbtio3 Pyroelectric Detectors. *Applied Physics Letters*, Vol. 59, No. 27, pp. 3539-3541, ISSN 0003-6951.

Saragi, T. P. I. *et al.* (2007), Spiro compounds for organic optoelectronics. *Chemical Reviews*, Vol. 107, No. 4, pp. 1011-1065, ISSN 0009-2665.

Sharma, G. D. (1995), Electrical and photoelectrical properties of Schottky barrier devices using the chloro aluminium phthalocyanines. *Synthetic Metals*, Vol. 74, No. 3, pp. 227-234, ISSN 0379-6779.

Slooff, L. H. *et al.* (2007), Determining the internal quantum efficiency of highly efficient polymer solar cells through optical modeling. *Applied Physics Letters*, Vol. 90, No. 14, pp. 143506, ISSN 0003-6951.

Spanggaard, H. & Krebs, F. C. (2004), A brief history of the development of organic and polymeric photovoltaics. *Solar Energy Materials and Solar Cells*, Vol. 83, No. 2-3, pp. 125-146, ISSN 0927-0248.

Sugimura, A. *et al.* (1989), Anomalous Photoinduced Current Transients in Nematic Liquid-Crystals. *Physical Review Letters*, Vol. 63, No. 5, pp. 555-557, ISSN 0031-9007.

Tahira, K.&Kao, K. C. (1985), Anomalous Photocurrent Transient in Polyethylene. *Journal of Physics D-Applied Physics*, Vol. 18, No. 11, pp. 2247-2259, ISSN 0022-3727.

van Duren, J. K. J. *et al.* (2003), Injection-limited electron current in a methano-fullerene. *Journal of Applied Physics*, Vol. 94, No. 7, pp. 4477-4479, ISSN 0021-8979.

Wagner, R. E. & Mandelis, A. (1996), Nonlinear photothermal modulated optical reflectance and photocurrent phenomena in crystalline semiconductors .1. Theoretical. *Semiconductor Science and Technology*, Vol. 11, No. 3, pp. 289-299, ISSN 0268-1242.

Xu, Y. F. *et al.* (2004), Photoresponsivity of polymer thin-film transistors based on polyphenyleneethynylene derivative with improved hole injection. *Applied Physics Letters*, Vol. 85, No. 18, pp. 4219-4221, ISSN 0003-6951.

Xue, J. G. (2010), Perspectives on Organic Photovoltaics. *Polymer Reviews*, Vol. 50, No. 4, pp. 411-419, ISSN 1558-3724.

8

Optical Transmission Systems using Polymeric Fibers

U. H. P. Fischer, M. Haupt and M. Joncic
Harz University of Applied Sciences
Germany

1. Introduction

Polymer Optical Fibers (POFs) offer many advantages compared to alternate data communication solutions such as glass fibers, copper cables and wireless communication systems. In comparison with glass fibers, POFs offer easy and cost-efficient processing and are more flexible for plug interconnections. POFs can be passed with smaller radius of curvature and without any mechanical disruption because of the larger diameter in comparison with glass fibers.

The clear advantage of using glass fibers is their low attenuation, which is below 0.5 dB/km in the infrared range (Fischer, 2002; Keiser, 2000). In comparison, POF can only provide acceptable attenuation in the visible spectrum from 450 nm up to 750 nm (Fig. 1). The attenuation has its minimum with about 85 dB/km at approximately 570 nm, which is due to absorption bands of the used Polymethylmetacrylat (PMMA) material (Daum, 2002). For this reason, POF can only be efficiently used for short distance communication up to 100 m. The large core diameter combined with higher Numerical Aperture (NA) results in strong optical mode dispersion, see Fig. 2.

Sources both LEDs and laser diodes in the 650 nm window have been available for some time. It is only recently that LED and Resonant Cavity LEDs (RC-LEDs) sources have become available in the 520 nm and 580 nm windows.

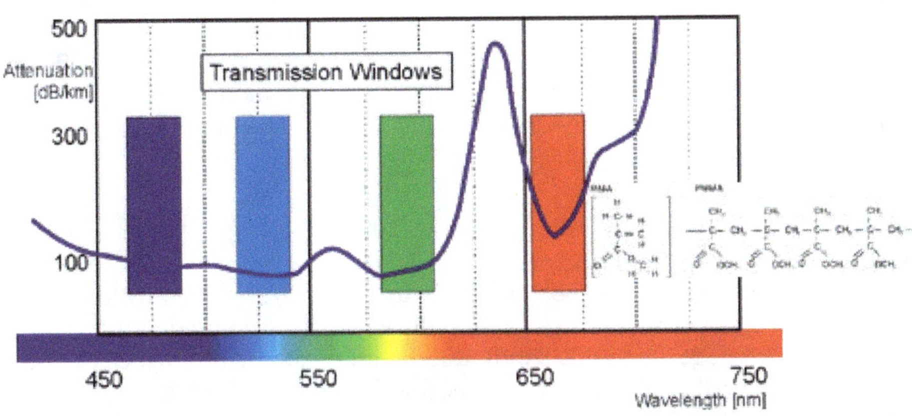

Fig. 1. Attenuation of POF in the visible range, insert: structure of PMMA.

The Numerical Apertur is directly given by the difference of the refractive indices of core and cladding material of the waveguide.

$$NA = (n_1^2 - n_2^2)^{1/2} \quad (1)$$

$$\Theta = \arcsin(NA) \quad (2)$$

The aperture angle of the waveguide is defined by the arcsin of the NA, which is the amount of input light that can be transferred by the waveguide by total reflection (Senior, 1992). For polymeric fiber systems, the NA calculates to 0.5, which results in the aperture angle of 30°. The difference of the core and cladding refractive indices is in comparison to glass fibers very high : 5%. The numerical aperture NA is correlated to the so-called V-parameter, which gives a correlation to the number of optical modes in the fiber waveguide. The number of the modes allowed in a given fiber is determined by a relationship between the wavelength of the light passing through the fiber, the core diameter of the fiber, and the material of the fiber. This relationship is known as the Normalized Frequency Parameter, or V number. The mathematical description is:

$$V = 2\pi \, NA \, a / \lambda \quad (3)$$

where NA is the Numerical Aperture, a is the fiber radius , and λ is wavelength.

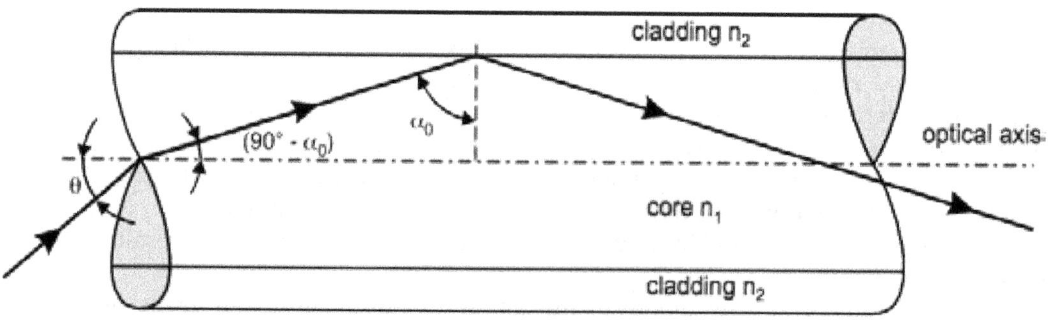

Fig. 2. Optical fiber waveguide.

A single-mode fiber has a V number that is less than 2.405, for most optical wavelengths. It will propagate light in a single guided mode. The multi-mode step index POF has a V number of 2,799, by a given optical wavelength of 550 nm, core radius of 490 μm, and NA of 0.5. This is more than 1000 times larger than for single-mode fiber Therefore the light will propagate in many paths or modes through the fiber. The number of optical modes can be calculated by:

$$N = 0.5 \, V^2 \, g/(g+2) \quad (4)$$

where g is the index profile exponent, which is infinity for step index fibers. For step index POF the mode number can be calculated to $N \cong V^2/2$ = 3.917 Mio modes. For longer wavelengths the number of modes will reduce to 2.804 Mio modes at 650 nm. The number of modes will reduce the usable bandwidth by mode dispersion, which can be calculated by the difference of the optical path of the mode which is lead through the fiber without reflection t_1 at the core/cladding interface and the path of the mode t_2 which is most reflected due to a high aperture angle of 30°.

$$\Delta t_{mod} = t_1 - t_2 = L_1 \, NA^2/(2 \, c \, n_2) \tag{5}$$

The skew between the two modes in a POF step index fiber can be calculated to $\Delta t_{mod} \cong 25$ ns for $L_1 = 100$ m and c = velocity of light in vacuum. The bandwidth length product for uniform Gaussian pulses (Ziemann, 2008b)

$$B\,L \cong (0.44/\Delta t_{mod})\, L_1 \tag{6}$$

will result in a theoretical bandwidth of 14 MHz for 100 m fiber length. A reduced NA will magnify the bandwidth length product BL up to 100 MHz for a step index POF with a NA of 0.19. To increase the BL product, other types of POF, which are described in detail in chapter 3., are introduced

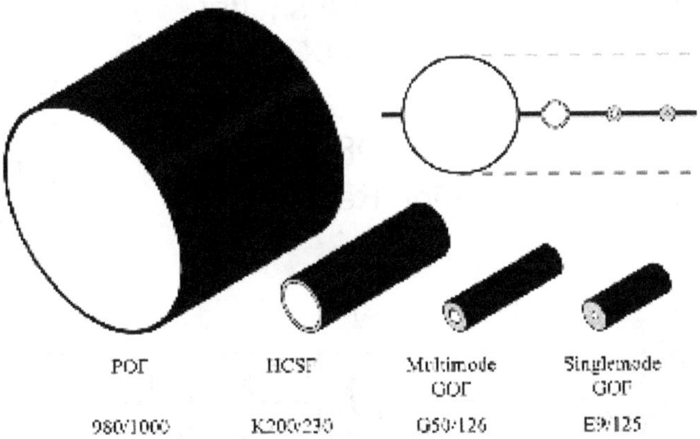

Fig. 3a. Polymeric step index fiber, b. Comparison of the dimension of different optical fiber types.

Like all optical transmission systems, at the beginning of the transmission an electro-optical conversion in a transmitter turns the electrical modulated signals into optical signals (see Fig. 4). This is typically performed by the use of a LED for data speeds up to 150 Mbit/s. For higher data speeds the use of a Laser diode like a VCSEL or edge emitter is necessary.
Modulation format in the existing Fast Ethernet systems is direct modulation by ASK: Non-Return-to-Zero (NRZ). NRZ means that the transmitter switches from maximum level to zero switching with the bit pattern. The advantage is the very easy system set-up. The disadvantage is the large required bandwidth. Usually a minimum bandwidth corresponding to the half of the transmitted bit rate is needed (e.g. 50 MHz for a bit rate of 100 Mbit/s).
For 1 Gbit/s Ethernet direct modulation techniques are not possible for use in POF systems, because of the high mode dispersion of the SI POF. Here, different higher modulation techniques must be implemented:

2. Pulse Amplitude Modulation (PAM)

In pulse-amplitude modulation there are more than two levels possible. Usually 2^n levels are used, with $4 < n < 12$. Due to every symbol transmitting n bits, the required bandwidth and the noise is reduced by $1/n$. A great advantage of PAM is its flexibility and adaptability to the actual signal to noise ratio (Gaudino et al., 2007a, 2007b; Loquai et al., 2010).

2.1 Discrete Multi Tone (DMT)

At DMT the used spectrum is cut into many sub-carriers. Each sub-carrier can now be modulated discrete by quadrature amplitude modulation QAM. Strong signal processing must be implemented with a fast analog-to- digital converter and forward error correction, which makes the overall system expensive. Nowadays, many communication systems like DSL, LTE or WLAN use this method (Ziemann, 2010).

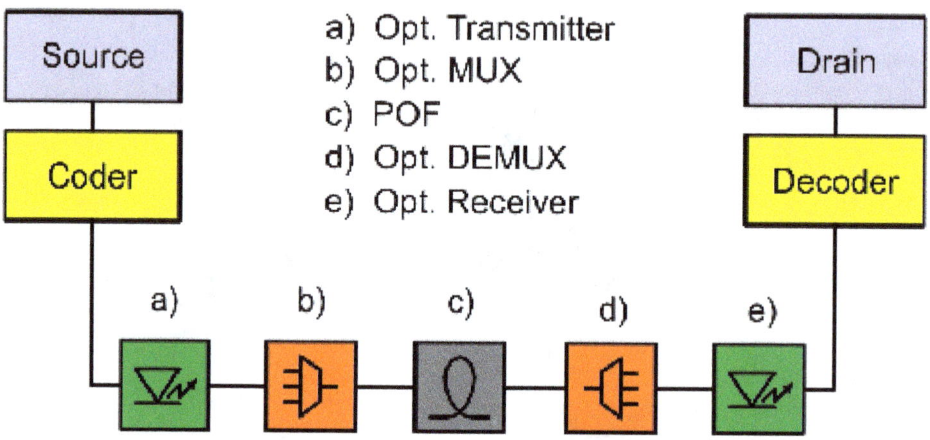

Fig. 4. Basic key elements of an optical transmission line.

At the end of the optical transmission path, an optical/electrical converter must be used. Typically, pin-photo diodes with large active areas are used. In between, the POF medium is situated using multiplexers (MUX) and demultiplexers (DEMUX) for higher effective data rates in the optical pathway. In this paper special optical DEMUX und MUX for wavelength multiplexing are described to extend the data rate of the whole systems for a factor of 4 - 10 in comparison to todays one channel transmission.

The use of copper as communication medium is technically out-dated, but still the standard for short distance communication. In comparison, POF offers lower weight, 1/10 of the volume of CAT cables and very low bending losses down to 20 mm radius. Another reason is the non-existent susceptibility to any kind of electromagnetic interference.

Wireless communication is afflicted with two main disadvantages:

- electromagnetic fields can disturb each other and probably other electronic device,
- wireless communication technologies provide almost no safeguards against unwarranted eavesdropping by third parties, which makes this technology unsuitable for the secure transmission of volatile and sensitive business information.

For these reasons, POF is already applied in various applications sectors. Two of these fields should be described in more detail in the next sections: the automotive sector and the in house communication sector.

2.2 Application areas of POF

2.2.1 Automotive

Since 2000 POF displaces copper in the passenger compartment for multimedia applications, see Fig. 5. The benefits for the automobile manufacturers are clear: POF offers a high operating bandwidth, increased transmission security, low weight, immunity to electromagnetic interference, and ease of handing and installation (Daishing POF Co., Ltd, n.d.). This vehicle bus standard is called Media Oriented Systems Transport (MOST). It is based on synchronous data communication and is used for transmission of multimedia signals over polymer optical fiber (MOST25, MOST50, MOST150) or via electrical conductors (MOST50). The technology was developed, standardized and up to date regularly refined by the MOST Cooperation founded in 1998. MOST was first introduced by BMW in the 7er series in 2001. Since then, MOST technology is used in almost all major car manufacturers in the world, such as VAG Group, Toyota, BMW, Mercedes-Benz, Ford, Hyundai, Jaguar and Land Rover (Wikipedia, 2011). In 2011 there are more than 50 different car types on the market which use the POF in the passenger cabin network structure for multi media data services.

The MOST specification covers all seven layers of the ISO/OSI Reference Model for data communication. On a physical layer polymer optical fiber is used as a media. A light emitting diode (LED) is used for transmission in red wavelength area at 650 nm. PIN photo diode is used as receiver (Grzemba, 2008).

The basic architecture of a MOST network is a logical ring, which consists of up to 64 devices (nods). The logical ring structure is usually implemented on a physical ring, which is however not mandatory. Combined ring, star network or double ring (for critical applications) can also be realised. Plug and play functionality enables easy adding or removing of devices.

In a MOST network one MOST device handles the role of the Timing Master which feeds MOST frames into the ring at a sampling rate of 44.1 kHz (frame is transmitted 44,100 times a second) or 48 kHz. The latest MOST specification recommends sampling rate of 48 kHz. The exact data rate depends on the sampling rate of the system. One after another Timing Slaves on the logical ring receive the signal, synchronize themselves with the preamble, parse the frame, process the desired information, add information to the free slots in the frame and transmits the frame to their successor. Since the MOST system is fully synchronous, with all devices connected to the bus being synchronized, no memory buffering is needed. Each Time Slave contain a fiber optic transceiver - received light signals are converted into electrical domain, processed, converted back into the optical domain and forwarded further.

A MOST frame includes one area for the synchronous transmission of streaming data (audio and video data), one area for the asynchronous transmission of packet data (TCP/IP packets or configuration data for a navigation system), and one area for the transmission of control data. MOST25 frame consists of 512 bits (64 bytes). 60 bytes are used for transmission of data. 6 – 15 quadlets (qualet consists of 4 bytes) of the data can be synchronous data, while the rest of the 60 bytes (0 – 9 quadlets) hold asynchronous data. Two bytes transport the part of the control message which spreads over 16 frames (one block). The first and the last byte of the frame contain the control information for the frame. MOST25 provides a data rate of 22.58 Mbit/s at a sampling rate of 44.1 kHz. This allows up to 15 uncompressed stereo audio channels in CD quality (2x16 bits per channel) / 15 MPEG1 channels for audio-video transmission or up to 60 1-byte connections to be established simultaneously. Maximal data rate is 24.58 Mbit/s at a sampling frequency of 48 kHz.

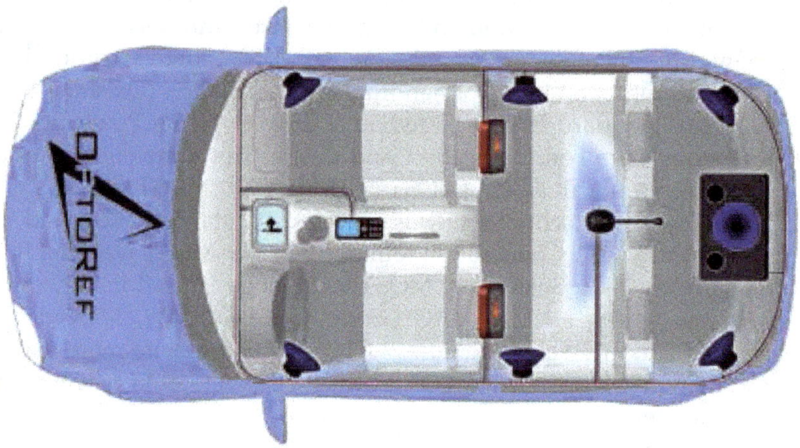

Fig. 5. Multimedia Bus System (MOST-Bus) with POF.

Next MOST generation uses a bit rate of just under 50 Mbit/s for doubling the bandwidth. The name MOST50 derives from this fact. Each frame consists of 1024 bits (128 bytes): 11 bytes for header, which also includes the control channel, and 117 bytes for the payload. The border between synchronous and asynchronous data can be adapted dynamically to the current requirements. The synchronous area can have a width of 0 to 29 quadlets plus one byte (0 to 117 bytes) and the asynchronous area can have a width of 0 to 29 quadlets (116 bytes). Control message consists of 64 bytes.

The latest MOST version (MOST150) was presented in October 2007. MOST150 is designed for high data rate of just under 150 Mbit/s and has a frame of 3027 bits (384 bytes): 12 bytes for header, which also includes the control channel, and 372 bytes for streaming and packet data transfer. It also has access to the dynamic boundary. Both, synchronous and asynchronous areas can have a width in between of 0 and 372 bytes. Besides the three known channels, an Ethernet channel with adjustable bandwidth and isochronous transfer on the synchronous channel for HDTV were introduced. This enables the transmission of synchronous data that require a different frequency than that given by the frame rate of the MOST. MOST150 thus a physical layer for Ethernet in the vehicle (MOST Cooperation, 2010).

Not just multimedia functions can exploit POF. For example, BMW has developed a 10 Mbit/s protocol called ByteFlight, which it uses to support the rapidly growing number of sensors, actuators and electronic control units within cars. Unlike MOST, which employs real-time data transfer, ByteFlight is a deterministic system in which the focus is on making sure that no data is lost (BMW, n.d.). The glass temperature of POF (below 85°C) makes using the fiber in the engine compartment impossible, although this problem might be solved in the foreseeable future. Up to date, a number of different in-car networks for multimedia and security applications has been developed, see Fig. 7.

Fig. 6. MOST applications in the multimedia bus.

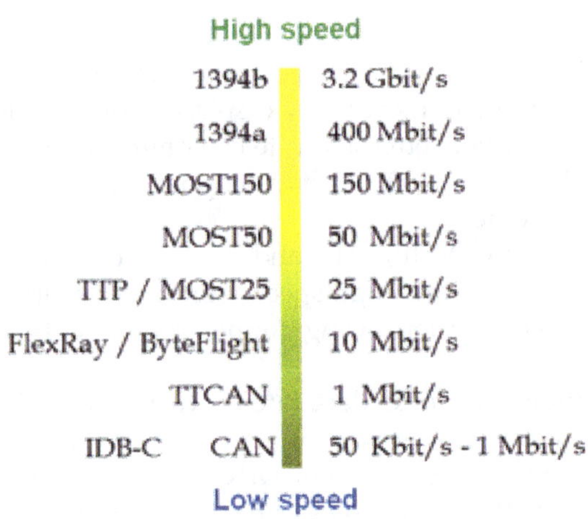

Fig. 7. In-Car network data rates.

2.3 Use of POF in aircraft

To use POF as the transmission media for aircrafts is under the research of different R&D groups due to its specific advantages. The DLR (German Aerospace Center) researches this kind of fiber under the conditions in civil aircrafts. They concluded that "the use of POF multimedia fibers appears to be possible for future aircraft applications" (Cherian et al., 2010). The Boeing Company develops special measurement setups to investigate and analyze POFs for the application under the conditions of daily use in aircrafts. Especially the low weight and the easy and economic handling make this kind of fiber the first choice. But for now the data rates and the temperature range are too low to replace copper for multimedia purposes.

To build aircraft with less weight, all big aircraft manufacturers will use carbon fibers for the aircraft body in all the new aircraft models. Because of its better weight performance, the aviation will loose a lot of its resistance against EMV and outer space radiation. To use optical cables like glass fibers or polymeric fibers is a good approach to bypass the problems of EMV in signal transmission. One coming solution will be the replacement of the electrical copper cables by POF and the application of the bus protocols FlexRay or MOST, which is widely used in the automotive industry (Lubkol, 2008; Strobel, 2010).

In aviation, strong test procedures are introduced for high reliable operation of all system components. High and low temperature operation starting from -60°C up to +130°C must be considered. Also high vibration stability in case of using optical connectors is required. For system relevant usage in the airplane, it is necessary to design the cable in the aircraft for POF use fire- and heat resistant and also waterproof, respectively. Additionally, high temperature POF must be implemented to force stable operation at temperatures in the aircraft up to +130°C, which can occur in the cockpit system unit.

To implement MOST technology in the airplane in the cabin for multimedia usage, the normal standard fiber can be used, because of the not relevant system impact of multimedia provision of the passengers. Up to now, the usage of POF in the airplane is focused in the research area and it will take years to test the reliability for everyday use in the airplane industry.

2.4 In-house

Another sector where POF displaces the traditional communication medium is in-house communication, although the possibilities of application are not confined to the inside of the house itself. In the future, POF will most likely displace copper cables for the so-called last mile between the last distribution box of the telecommunication company and the end-consumer (Koonen et al., 2005, 2009). Today, copper cables are the most significant bottleneck for high-speed Internet.

"Triple Play", the combination of VoIP, IPTV and the classical Internet, is being introduced to the market with force, therefore high-speed connections are essential. It is highly expensive to realize any VDSL system using copper components, thus the future will be FTTH (Fischer, 2007a).

For in-house communications networks data rates between 10 Mbit/s and 100 Mbit/s are typically in use. Copper-cables (Category 5/6) are most widely used in office networks in combination with structured wiring system of DIN EN 50173-1 and DIN EN 50173-2. The 8-core wire in combination with the RJ45 plug can transmit 100/1000 Mbit/s over distances up to 100 meters using Ethernet protocol. Due to the mass-market application of Ethernet (IEEE 802.3), this technique has become very cheap. Most broadband home networks today focus on the combination of Ethernet and RJ45 data cable interface. The disadvantage of this technique depends on the lack of structured cabling in most apartments. The possibilities for re-installation of the thick and inflexible CAT cables are very limited, while most of the wiring has no professional electrical grounding.

In the following the available in-house network technologies are depicted and compared in detail with their specific advantages and disadvantages with POF applications in Table 1:

- Twisted-pair cables belong to the Ethernet standard CAT 5/6 with a star network topology and data rates up to 1 Gbit/s up to 100 m, but due to very thick cables (Ø 7 mm) wide cable channels and complex plug required. They have no electrical isolation, which also leads to a high EMC sensitivity. This disturbing especially in the industrial and automotive environment the transmission.

- Coaxial cables, as they are known from the TV connection, have a diameter of 5 mm and a much higher bandwidth up to 1 GHz for 30 m with large bend radii. However, the electrical isolation from the 230 V power is problematic, which can lead to problems. The EMC problem is related critical as the twisted-pair cable.
- Glass fibers are the media with the highest range and data rate, but expensive compared to alternative techniques, also because of expensive connector assembly and low possible bending radii. Additionally, the small core diameter of 9 microns for single mode fiber is highly vulnerable to pollution. This leads to significant problems in the industrial environment, but without EMC problems.
- Polymer fibers can be easily laid with small bend radii, are very tolerant in terms of buckling and pollution (large core cross-section), without the need of using connectors. It can be shown that POF have a high future potential for increased data rate without having to install additional fibers. Like the glass fiber, POF has a fiber optic to electrical isolation and has a very low EMC sensitivity.
- WLAN is a pure wireless technology with a possible range up to 20 m. Due to absorption by walls, and ceilings the effective range is poor. Furthermore due to interference by third parties, the transmission is not secure. In addition, neighbouring networks will reduce the data rate significantly. This leads especially in the industrial environment to a very large problem, if there are installed WLAN nodes in a very large number. Data rates from 2 up to 100 Mbit/s data rate are possible under optimal conditions, most of the achievable data rates remains well below it.
- Powerline uses the 230 V-house power grid. The range is very limited and depends on the power grid. However, there are only low installation costs, but the high electromagnetic radiation and the uncontrolled distribution over the network are major disadvantages, which makes this network technology for in-house use unattractive.

Technique	Data rate	Range	Security	Costs	Handling	Deployability	Total
Twisted-Pair cable	+	0	0	++	-	0	2+
Coax cable	0	0	0	+	0	0	1+
Glass fiber	++	++	++	--	--	-	1+
POF	0	-	++	+	+	+	4+
WLAN	--	-	--	++	++	++	1+
Powerline	-	-	--	+	+	++	0

Table 1. In-house networks in comparison, division between particularly poor -- and particularly well: ++

In Table 1 an overview is summarized to assess the respective qualities of the alternative networks in view of the most important criteria. It turns out that the most widely used networking technologies such as wireless or twisted pair are leader in the field in terms of costs, but in total the polymer fiber technology shows superior overall properties and combines many advantages of the other transmission media, without their main drawbacks. Keeping these reasons in mind, the further potential of POF seems to be very high.

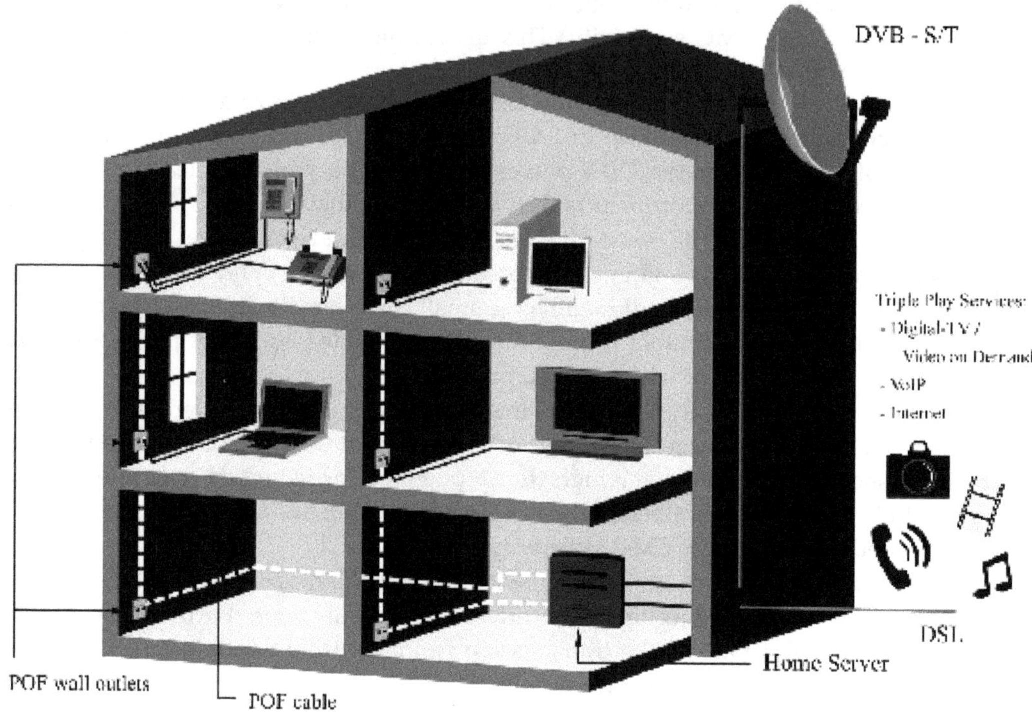

Fig. 8. In-house communication scenario with POF -Ethernet system.

3. State of the art in POF transmission systems

3.1 POF fiber types

Plastic optical fibers for data transmission until recently were limited to step index PMMA fibers that had bandwidths of 38 MHz-100 m (Mitsubishi Eska). More recent results by Mitsubishi with an Eska-Mega fiber shows a three fold increase in bandwidth to 105 MHz-100 m. Increases in bandwidth are also possible with the use of dual step index (DSI), multi-step index (MSI) profiles, multi-core (MC), or combinations of these (Poisel et al., 2003). Here, only the most relevant used POF types are described.
For POF there are in general three fiber types existing, which are on the market available (Table 2).

3.1.1 Step index fibers SI POF

The SI POF (Fig. 9) is already standardized in a IEC 60793-2-40: and IEC 60794-2-40: specification for A4 fiber cables and also in an ETSI recommendation TS 105 175-1. The SI POF fiber is called in the specification as A4.a1 and A4.a2, respectively. Optical properties and also mechanical ones are strictly defined to guaranty a international reliable high level of fabricated fibers. Optical specs are 980/1000 μm diameters, temperature range of -40 °C - +85 °C, and A4a1 is a fiber with attenuation 180 dB/km at 650 nm and bandwidth 4 MHz-km, which is of course the performance of the standard 1 mm, high NA, SI POF. A4a2 refers to the nominally 0.5 mm PF-GI-POF which has an attenuation of less than

100 dB/km in the red, but more significantly, transmits at 850 nm with an attenuation of 40 dB/km. The bandwidth of this fiber is very high, typically of that of GI POF fibers, ~200 MHz-km, but is not in use in cars or in-house networks.

3.1.2 Gradient index fibers GI POF

Here the core-refracting index is distributed in a quadratic behaviour (Fig. 10). This reduces the mode dispersion significantly and relates to a better BL than in step index fibers. BL products of more than 2 GHz-100 m are realized with OM Giga from Optimedia Inc. in Korea. Specifications of different POF types are shown in Table 2:

Specs/Fiber type		SI POF (IEC 60793-2-40)	GI POF (Park 2006)	MC POF (Asahi Chem.)
Diameter (core)	mm	1.0 (0.9)	1.0 (0.9)	1.0 (0.9)
Variation of Diameter	%	± 5	± 5	± 5
Tensile at break	N	> 65	> 65	> 65
Bending Radius	mm	> 20	> 35	> 15
Operating Temperature	°C	-40 ~ 85	-30 ~ 70	-40 ~ 85
Attenuation	dB/km	< 180 at 650 nm	< 200 at 650 nm	< 180 at 650 nm
Bandwidth	Mbps	~ 40 @ 100 m	~ 2000 @ 100 m	> 500 @ 100 m

Table 2. Specifications of different POF types.

3.1.3 Multicore fiber MC POF

In this approach, the fiber can be made of many tiny cores, a multi-core POF (Fig. 11). The partition of the core into many individual light guiding areas allows for very small bending radii, helping to ease the installation of the fiber. The numeric aperture and the bandwidth is nearly the same as of the SI-POF. Up to now, there is no international standard available for this new type of POF. In the market available is the fiber from Asahi Chemical, shown in Fig. 11. 19 multi cores of PMMA are introduced in one complete fiber. The NA of this fiber is 0.27 with 1 mm outer diameter and bandwidth length product of 500 Mbit/s-100 m. Operating temperature ranges from -40 °C up to +85 °C. This fiber is an excellent candidate to replace in the next future the SI POF in the mass market, because of its good bandwidth performance and comparable price. On the other hand, standardizing procedures must be realized to make this MC POF acceptable for international network markets.

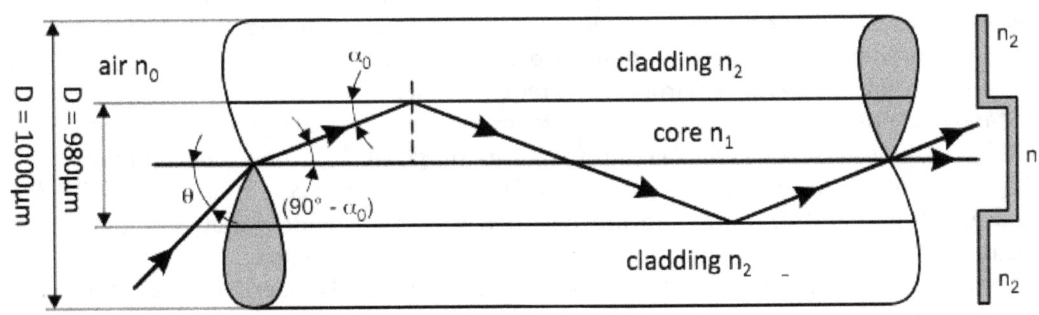

Fig. 9. Step index POF.

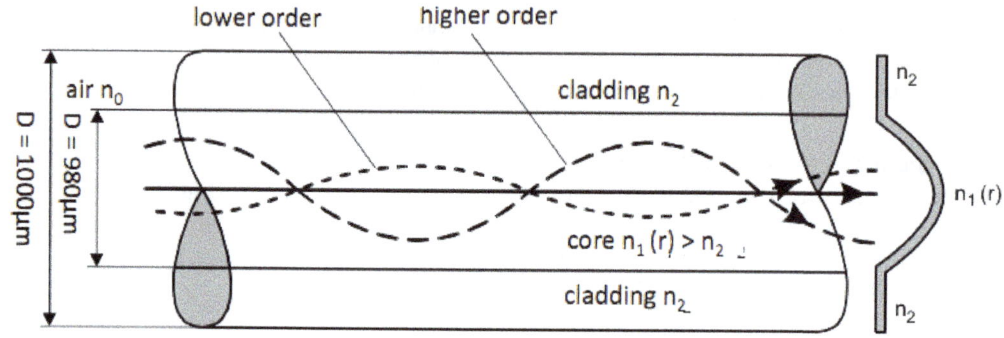

Fig. 10. Gradient index POF.

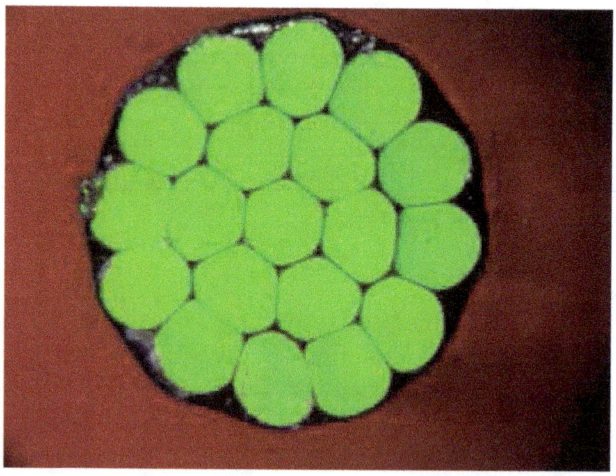

Fig. 11. Multi core POF (Asahi Chemical) 1 mm diameter, 500Mbit/s-100 m.

3.2 Standardization of POF

As shown in this chapter, SI POF is standardized by the International Electrotechnical Commission (IEC) as the A4 category of fibers. In completion, this category contains four types (families A4a-A4d) of SI POF having core diameters ranging from 490 microns to 980 microns for different applications in networks, multimedia sources and sensor systems. This standard also defines other dimensional requirements for these fibers, as well as minimum mechanical and transmission properties (Ziemann 2008). The existing IEC POF standards do not specify any environmental requirements, however.

OFS and Nexans have recently proposed to modify the A4 category fiber standards to include perfluorinated GI POF. According to this proposal, four new fiber families (A4e-A4h) are be added to the A4 category. These families will have core diameters of 500 µm, 200 µm, 120 µm, and 62.5 µm, and are intended to serve a wide variety of applications ranging from consumer electronics to multi-Gb/s data communication.

In Germany, the DKE as the Standardization Division of the VDE Germany has established a POF working group DKE 412.7.1, which is responsible for the international standardisation of Gbit/s POF transmission systems with active and passive elements.

4. WDM over POF

Several sectors will be introduced, where POF offers advantages when compared to the established technologies. Other possible industrial sectors include the aviation or the medical sector. All these applications have one thing in common - they all need high-speed communications systems. The standard communication over POF uses only one single channel. To increase bandwidth for this technology the only possibility is to increase the data rate, which lowers the signal-to-noise ratio and therefore can only be improved in small limitations.

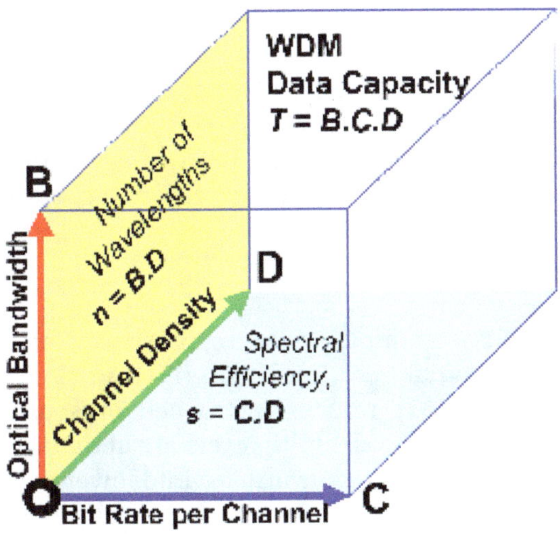

Fig. 12. Advantages of extending the overall transmission capacity using WDM.

Wavelength Division Multiplexing (WDM) is a technique that combines multiple, unique optical signals at different wavelengths (colours) onto a single strand of fiber. At the receiving location, these optical signals are split back out, or demultiplexed, into separate fibers. Essentially, the bandwidth capacity of the fiber is multiplied by the number of wavelengths multiplexed onto the fiber. Fig. 12 conceptually shows the principal benefits of the WDM technique. The WDM system capacity (T) depends on three basic parameters, which are open to modify:

Bit Rate per Channel (C)
Optical Bandwidth (B)
Channel Density (D)

$$T = B \cdot C \cdot D \tag{7}$$

In comparison to single channel transmission systems that only extend the capacity by the help of higher bitrates per channel, WDM will allow to boost up the overall transmission capacity by two additional factors:

- the channel density and
- the optical bandwidth of the system.

Both factors in combination will lead to the total number of wavelength channels which are possible to implement in the whole system. For glass fiber systems the optical bandwidth is characterised by the fibers attenuation curve between 1300 nm and 1650 nm. Here using

POF the bandwidth ist allocated beween 400 nm and 800 nm. Assuming a bandwidth of B = 380 nm and a channel density of D = 1/40 nm, a bit rate per channel of 1 Gbit/s the total capacity will be T = 1 Gbit/s x (380 nm/40 nm) = 9,5 Gbit/s. The application of fixed reference channels for POF systems will be described in chapter 4.6.
For the use of different channel densities, a international system was established for glass fiber systems which defines fixed channel spacing for long distance, metro and short haul networks: this variations of WDM that are commonly used for glass fiber systems: Broad WDM, Coarse WDM, and Dense WDM. Each variation has different capabilities, costs, and operational friendliness.

4.1 Broad WDM

Broad WDM (often just called WDM) utilizes two wavelengths with are parted by more than 200 nm. Broad WDM is very simple to implement. Off-the-shelf optical transmitters without tight control of wavelengths can be used. These applications also utilize low-cost optical multiplexers and demultiplexers with low insertion loss, but are not useful for higher speed systems.

4.2 Coarse WDM

Coarse WDM (CWDM) utilizes multiple wavelengths spaced at 20 nm in the infra red region. The International Telecommunication Union (ITU) in G 694-2 specifies 18 CWDM wavelengths from 1271 nm to 1611 nm for metro networks using optical glass fibers. Transmitters, optical multiplexers, and demultiplexers are at defined wavelengths, but they do not need to be tightly controlled, which translates into lower equipment costs compared to Dense WDM.

4.3 Dense WDM

Dense WDM (DWDM) utilizes many wavelengths spaced narrowly, and they are most commonly located in the C-band, the wavelength range from 1530 nm to 1565 nm. ITU G 694-1 in specifies the center of the DWDM wavelengths. Practical deployments of DWDM today are spaced at 100 GHz frequencies (or approximately 0.8 nm spacing), which allow about 40 wavelengths in the C-band. DWDM requires that the optical transmitters, multiplexers and demultiplexers have very tight control over the wavelength under all operating temperature conditions.
This contribution presents a possibility to open the WDM technique to the POF world. This basic concept can - in theory - also be assigned to POF. However POF shows different attenuation behavior, see Fig. 1. For this reason, only the visible spectrum between 400 nm - 780 nm can be applied when using POF for communication.
For WDM, two key-elements are indispensable: a multiplexer and a demultiplexer. The multiplexer is placed before the single fiber to integrate every wavelength to a single waveguide. The second element, the demultiplexer, is placed behind the fiber to regain every discrete wavelength. Therefore, the polychromatic light must be split in its monochromatic parts to regain the information. These two components are well known for infrared telecom systems, but must be re-developed for POF, because of the different transmission windows.
One technical solution for this problem is available, but it cannot be efficiently utilized in the POF application scenario described here, mostly because this solution is afflicted with high costs and therefore not applicable for any mass production.

4.4 Basic concept of the demultiplexer

As mentioned before, a demultiplexer is essential for WDM (Daum et al., 2008; Chen & Lipscomb, 2000). Several preconditions must be fulfilled to create a functional demultiplexer for POF. First of all, the divergent light beam, which escapes the POF, must be focused. This is done by an on-axis mirror. In the first attempt, a spherical mirror is used. To get perfect results without any spherical aberrations, an ellipsoid mirror should be used.

The second function is the separation of the different transmitted wavelengths (Fischer et al., 2007b). In Fig. 13, this principle is illustrated for three wavelengths (red, green, blue). This is not a limitation for possible future developments, but rather an experimental basis from where to run the various simulations described below. The diffraction is done by a diffraction grating. The diffraction is split into different orders of diffraction. The first order is the important one to regain all information. There a detector line can be installed to detect the signals.

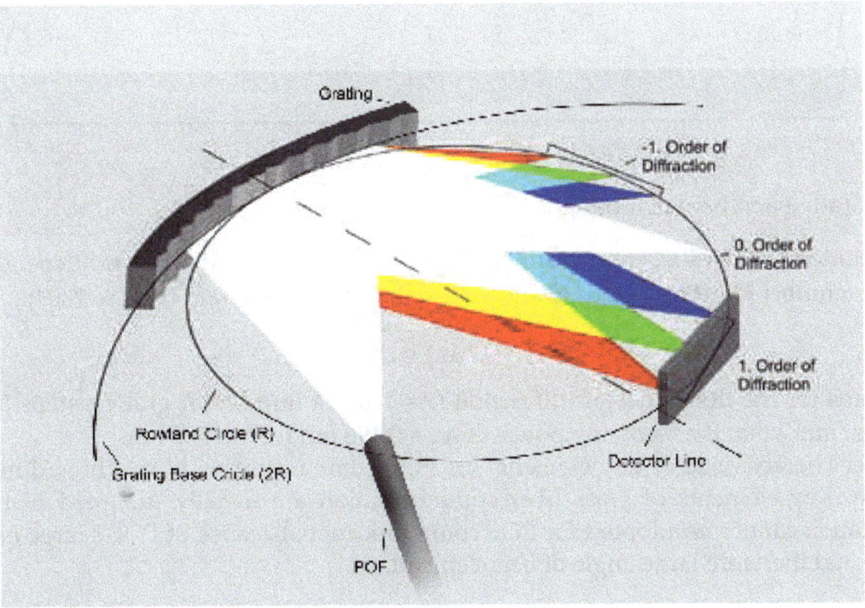

Fig. 13. Principle sketch of a Rowland Spectrometer.

Because the grating is attached to a bended basement only one element can cover both functions, the focusing and the diffracting. Hence the light is not afflicted with any aberrations or attenuations of a focusing lens or other refractive elements, which are necessary for any other setup. The diffraction is split into different orders of diffraction. The first order (z = 1) is the important one for data transmission. The higher the grating constant g of the diffraction grating, the more accurate the shape of the maxima of the different diffraction orders. The diffracted light interferes positively on the detection layer for (Demtröder, 2008):

$$z\lambda = g(\sin\alpha + \sin\beta) \quad (8)$$

with α angle of incidence, β emergent angle and g the grating constant. The following figure illustrates this formula (Fig 14).

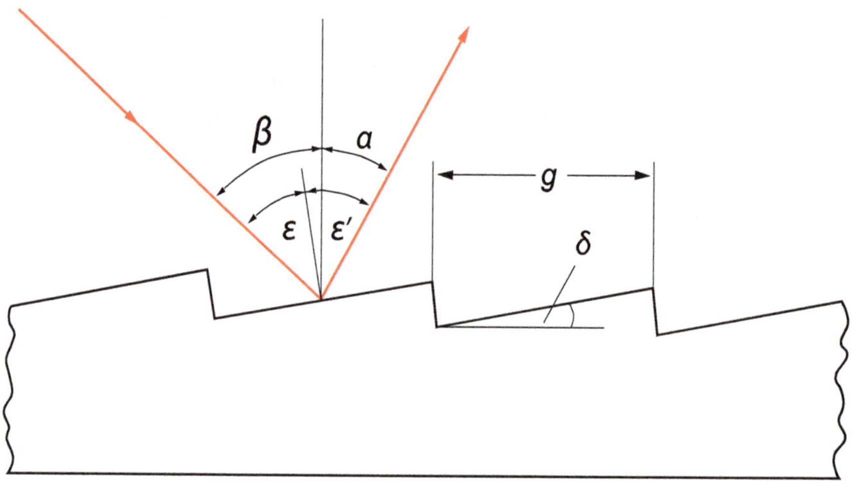

Fig. 14. Grating and behavior of light.

The resolution of the diffraction grating follows the Rayleigh Criterion and depends on the complete number of grating steps N and not on the grating constant (Hecht, 2009):

$$(\lambda/\Delta\lambda) \leq zN \tag{9}$$

This means for the first order of diffraction (z=1) and a number of grating steps N = 3000 (300 lines/mm) that the resolving power is $\Delta\lambda = 0.196$ nm for $\lambda = 589$ nm.
One other characteristic of key elements for POF communication is the three dimensional approach. Key elements of glass fiber communication are usually designed planar. This simplification cannot be adopted for POF communication, because of POF's large numerical aperture and therefore large angle of beam spread.

4.4.1 Results of the simulation

In the following steps, a software program is used to design a demultiplexer based on the general concept outlined above. For the current task, the software OpTaLiX, which is based on the raytracing method, provides all needed functionalities (Blechinger, 2008; Hecht, 2009; Demtröder, 2008). This approach offers different advantages, it is easy to design, analyze and evaluate the simulated results. Also, effective improvements of the configuration can be simulated fast.

4.4.2 Results of the simulation for different line densities

In figure 15, the 2D plot for the reference wavelength (520 nm) of the demultiplexer with an ellipsoid mirror and grating is shown. The multicolored light is emitted by a polymeric fiber. It hits the mirror, where it is focused and diffracted in its monochromatic parts. The light is focused onto a POF- or detector-array.

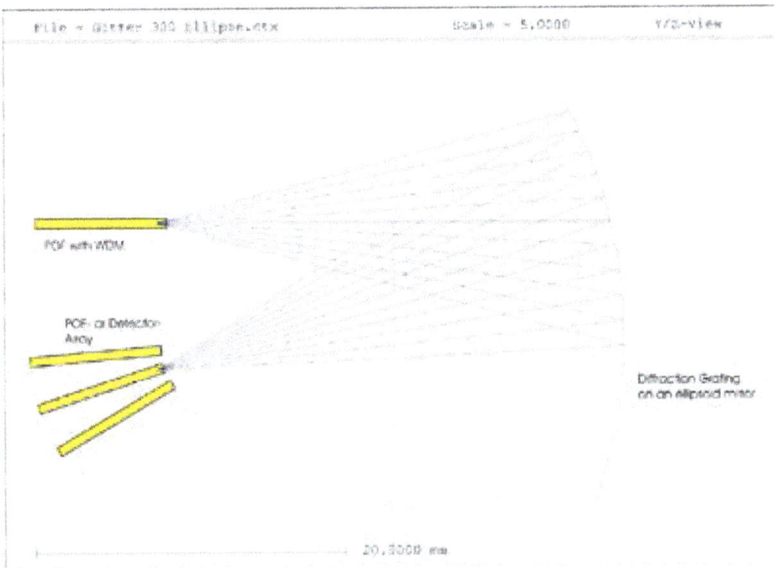

Fig. 15. 2D plot of the demultiplexer.

Without a grating, a perfect point to point mapping (without any aberrations) is possible with an ellipsoid mirror because of the two foci, but there is no separation of the different channels. With a grating stamped on the mirror, the separation of the multicolored light in its monochromatic parts is possible. But this grating distorts the optical path of light dramatically. The first change is the gap of the different colors in the image layer (here the POF- or Detector Array) increases with the line density of the grating. This can be noticed for an ellipsoid mirror (Fig. 16) and for a spherical mirror (Fig. 17) as well. The spherical mirror has the advantage, that the shape can be produced for injection molding easier.

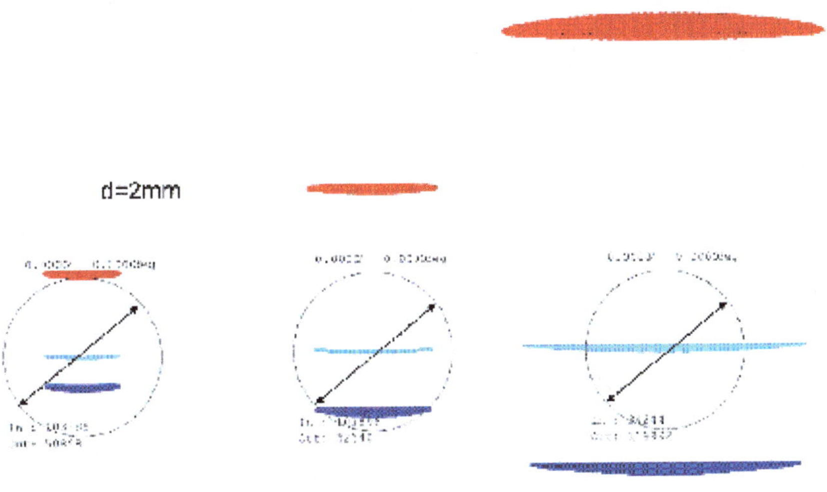

300 lines/mm 600 lines/mm 1200 lines/mm

Fig. 16. 2D Plot of the demultiplexer with an ellipsoid mirror and different line densities.

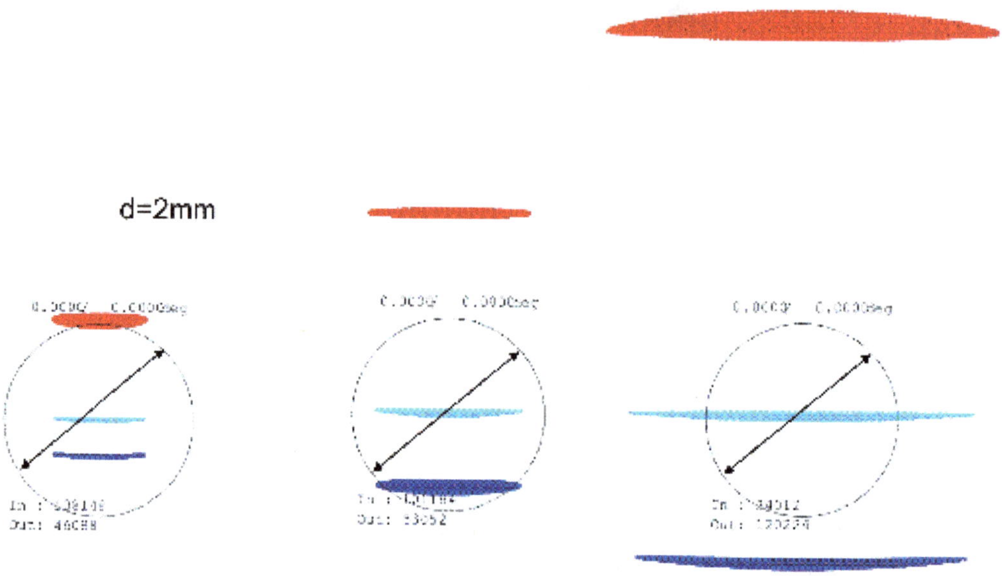

300 lines/mm 600 lines/mm 1200 lines/mm

Fig. 17. 2D Plot of the demultiplexer with a spherical mirror and different line densities.

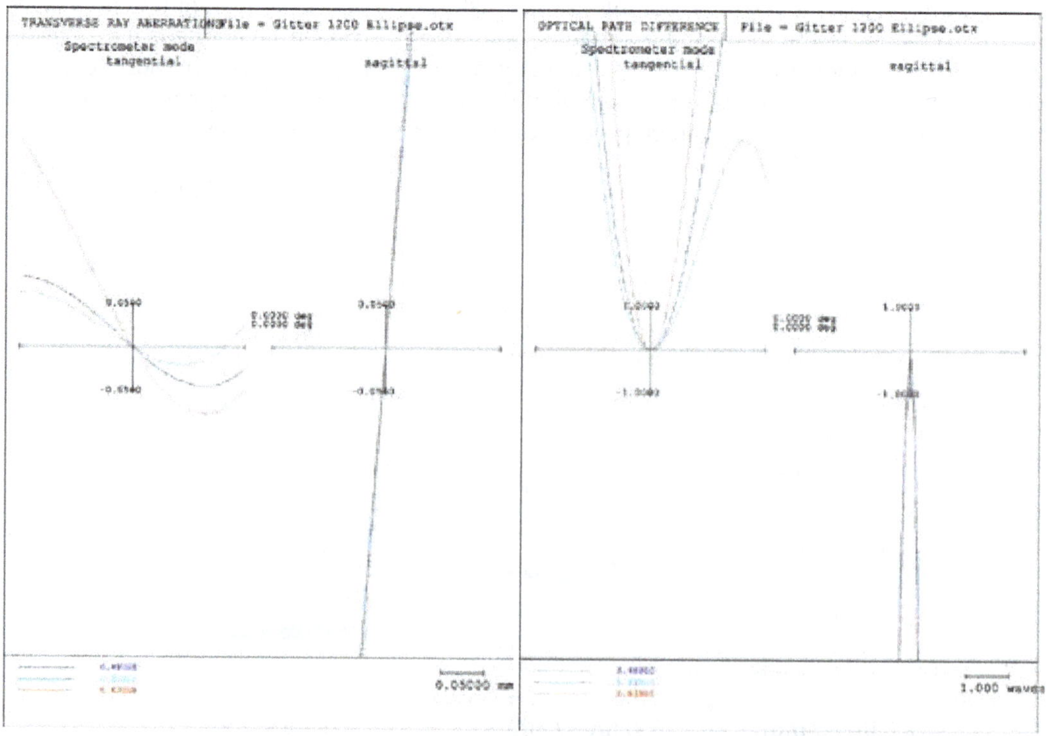

Fig. 18. TRA and OPD for the ellipsoid demultiplexer with 1200 lines/mm.

The second changes are the great aberrations especially for the demultiplexer high line density. To underline this result and to analyze the aberrations in detail, the transverse ray aberration (TRA) and the optical path difference (OPD) in spectrometer mode are shown in figure 18 for the demultiplexer with an ellipsoid mirror and 1200 lines/mm. The chief ray coordinates are irrespective for the TRA and OPD to overlap the different colors.
The TRA shows a slight defocusing for the meridional section, but a very strong defocusing for the sagittal section. The graph of the function in the meridional section exhibits a predominant third order Seidel coefficient. Therefore the slight defocusing in the meridional section compensates the astigmatism. The OPD shows as expected strong deviation from the ideal waveform especially in the sagittal section. This defocusing leads to high losses for the coupling efficiency for the POF- or detector- array in the image layer.
It is obvious that the grating changes the focal length especially of the sagittal section; therefore the shape of the mirror must be improved. It is necessary to change the radius of curvature notable in the sagittal section. Hence the basic shape of the mirror is not longer a sphere or ellipsoid. To meet the demands a higher order shape, which is nearly cylindrical, is used.

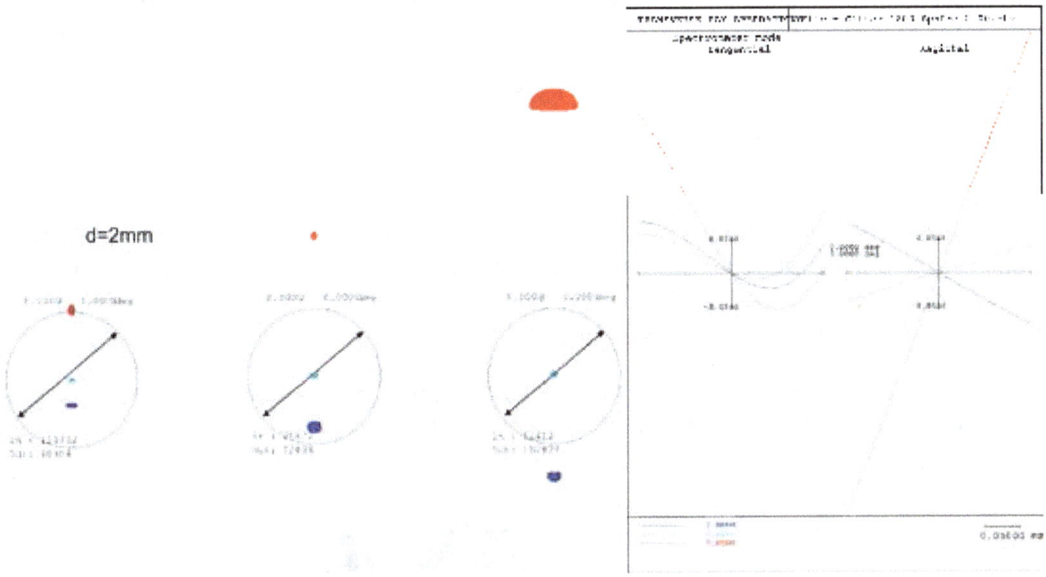

300 lines/mm 600 lines/mm 1200 lines/mm

Fig. 19. Spot Diagram and TRA for the improved DEMUX.

The change of the mirror shape improves the imaging quality substantial. The spot diagram and the TRA for the improved demultiplexer are shown in Fig. 19.
The spot diagram shows three dividable colors. The gap between every color is larger than 2 mm. The TRA shows a marginal shift of the focus of all wavelengths to offset the astigmatism in the meridional section. Because of the spectrometric function of the demultiplexer it is not possible to focus all three colors simultaneously. There is always a combination of over and under correction for the different colors. Hence the radius of the mirror in the sagittal section is optimized to focus the colors completely as much as possible.

This improved demultiplexer can separate three colors with enough space between them to regain the information with a POF- or detector-array. The shapes of the foci feature low coupling losses and the shape of the mirror is easy to produce in injection molding.

4.5 WDM reference comb

Different analyses will be shown in the full chapter including TRA and OPD. The way to the optimized setup will described in detail. Further on a first attempt to standardize the different wavelengths in visible spectrum will be discussed. As described in the previous chapter, WDM has a great chance to expand the overall bandwidth of POF transmission systems. Therefore it is necessary to standardize the WDM channels in frequency or wavelength, based on proven the glass fiber system channel allocation map of ITU recommendation G.694.2-1/2.

4.6 Optical channel allocation map proposal for POF

The usable transmission window in the visible spectrum of POF is located between 400 - 700 nm, which leads to the possible optical bandwidth of 300 nm for POF. Now the ITU proposes a frequency allocation map for WDM in its recommendation G 694.2. Assuming the correlation between wavelength and frequency of electro magnetic waves

$$\lambda f = c_{vac} \tag{10}$$

where f = frequency λ = wavelength and c_{vac} = speed of light in vacuum. Calculating the equivalent frequencies to 400 nm and 700 nm will result in 750 THz and 461.5 THz combined with a bandwidth of 320 THz for true WDM transmission. Additionally, a so-called anchor frequency (for glass fiber systems 193.1 THz) will be proposed as 750 THz. The possible transmission windows for WDM channels are dependent on the attenuation of the PMMA based standard SI POF. A possible transmission channel at 490 THz (610 nm) must be omitted because of the attenuation of the OH-Peak at 610 nm at that frequency (see Fig. 20).

The region of low attenuation of less than 90 dB/km is apportioned between 510 THz and 750 THz. In total 9 WDM channels, which are listed in Tab. 3, can be fixed with channel spacing of 40 THz.

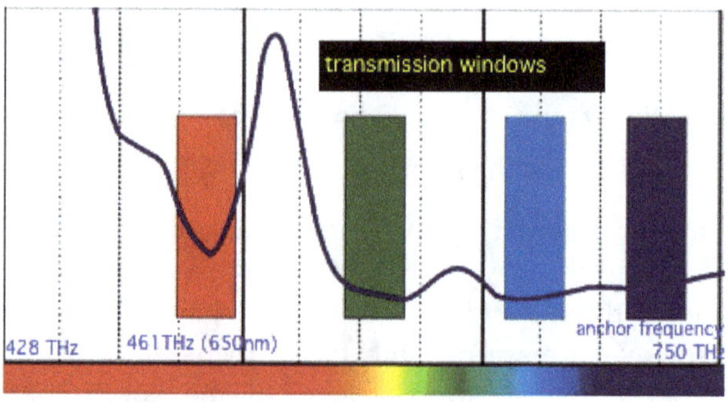

Fig. 20. Proposed optical frequency comb window.

The channel frequencies are calculated by:

$$f_{ch} = 750\text{THz} - n\,(40\text{THz}) \tag{11}$$

where n is the number of the channel.
In total a WDM system data rate of 9 x 2.5 Gbit/s = 22.5 Gbit/s seems to be possible assuming the todays data rates of POF systems using GI POF for transmission medium for a transmission length of 100 m. In this proposal for a new international POF WDM grid, most of the channels are located in the short wavelength region where the attenuation of POF is lower than in the long wavelength region. On the other hand, the "old red" window is already included at channel no 7: $f_7 = 470$ THz (638 nm).
In Fig. 21 a schematic view of the optical band pass behavior of the WDM filters of the DEMUX/MUX devices are depicted. Supposing a typical X-talk suppression of 30 dB for optical channel separation, a 3 dB filter width of 20 THz for each filter is needed.

Channel	Frequency / THz	Wavelength / nm
0	750	400
1	710	423
2	670	448
3	630	476
4	590	508
5	550	545
6	510	588
7	470	638
8	430	698

Table 3. Proposed optical frequency channels.

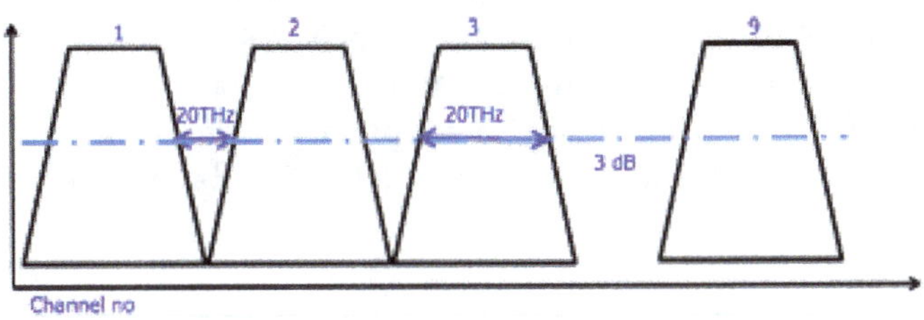

Fig. 21. Proposed optical frequency comb window.

5. Outlook

The simulation results show, that it is possible to build up a mass production convenient demultiplexer for polymeric fiber systems by means of a diffraction grating. A special shape of the mirror is needed to suppress most of the aberrations which results of the grating. The improved demultiplexer can separate all three colors with a gap of 2 mm and crosstalk lower than 30 dB. This demultiplexer has the chance to break through the limitation of standard POF communication also with broad range of usability in optical spectroscopy for sensor systems in automotive and medical applications due to its low cost realization. It can be implemented in combination with all in the market existing POF types like SI POF, MC POF or GI POF with 1 mm outer diameter and a Numerical Aperture of 0.3 - 0.5. The high number of modes in the fibers gives no restriction to optimal function the developed multiplexer for WDM transmission of minimum three different wavelength channels. In the future the device will be extended to multiplex 8 channels.

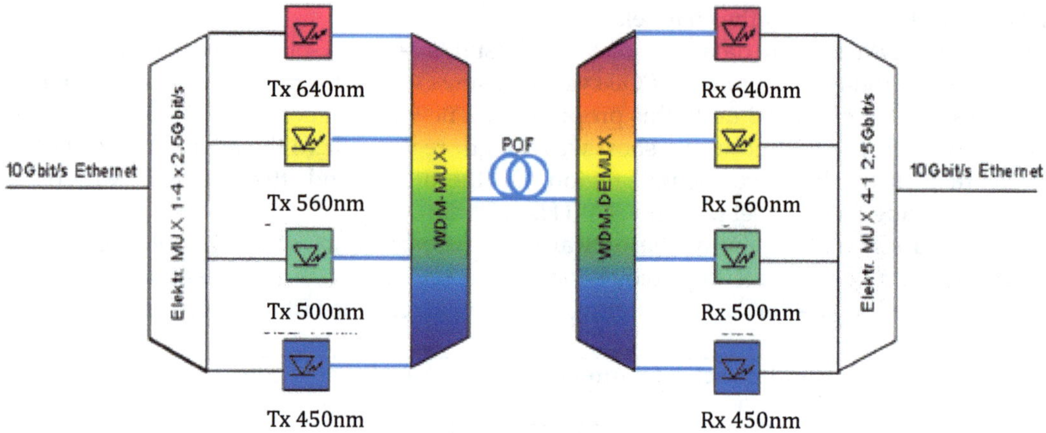

Fig. 22. Possible 10 Gbit/s via POF using WDM technique.

In the next years WDM over POF will expand the total bandwidth of POF transmission systems up to more than 20 Gbit/s. A channel allocation map for 9 WDM channels in the visible range is proposed as an input for the international standardisation organisations IEC and ITU to define a new optical reference standard for POF WDM systems like the ITU recommendation for glass fiber systems G 594.

Soon it would be possible to transmit 10 GbEthernet data via SI POF with the help of the here described WDM over POF technology, shown in Fig. 22. The electrical data stream of 10 GbEthernet will be electrically demultiplexed to four sub data streams of 2.5 Gbit/s. Each of this sub streams will modulate an optical laser diode source with different WDM wavelengths. In a WDM MUX all four colored signals will be combined to be transferred simultaneously via the POF fiber link up to 100 m. At the receiver side an optical DEMUX will spread the optical channels to dedicated photo diodes. The out coming electrical 2.5 Gbit/s data can be electrical multiplexed to the full 10 Gbit/s Ethernet bit stream at the output side.

6. References

Blechinger, H. (2008). *OpTaliX: Optical Design Software*, Available from http://www.optenso.com

BMW. (n.d). What is byteflight?, Mar. 2011, Available from http://www.byteflight.com/whitepaper/index.html

Chen, R. T. (2000). WDM and Photonic Switching Devices for Network Applications. *Proceedings of SPIE*, Vol. 3949

Cherian, S.; Spangenberg, H. & Caspary, R. (2010). Vistas and challenges for polymer optical fiber in commercial aircraft, *Proccedings of the 19th POF Conference*, Yokohama, Japan, Oct. 19-21, 2010

Colachino, J. (2001). *Mux/DeMux Optical Specifications and Measurements*, Lightchip Inc. white paper, Lightreading Daishing POF Co., Ltd. (n.d). The application of POF in automobile industry, Mar. 2011, Available from http://www.dspof.com/en/support-pgdetail-88.html

Daum, W.; Krauser, J.; Zamzow, P. E. & Ziemann 0. (2007). *POF – Polymer Optical Fibers for Data Communication*, Springer, ISBN 3-540-42009-6, Germany

Demtröder, W. (2008). *Experimentalphysik 2: Elektrizität und Optik*, Springer, ISBN 978-3540682103, Germany

DIN EN 50173-1 Information technology - Generic cabling systems - Part 1: *General requirements; German version EN 50173-1:2007/prAB:*, 2009

DIN EN 50173-2 Information technology - Generic cabling systems - Part 2: *Office premises; German version EN 50173-2:2007/prAA:*, 2009

DKE 412.7.1 Standardization Group of POF Gbit/s systems in Germany, April 2011, Available from http://www.vde.com/en/dke/std/projects/POF/Pages/default.aspx

ETSI TS 105 175-1 V1.1.1 Technical Specification *Access, Terminals, Transmission and Multiplexing (ATTM); Plastic Optical Fiber System Specifications for 100 Mbit/s and 1 Gbit/s*, (2010)

Fischer, U. H. P. (2002). *Optoelectronic Packaging*, VDE Verlag GmbH, ISBN 3-8007-2572-X, Germany

Fischer, U. H. P. & Haupt, M. (2007). WDM over POF: the inexpensive way to breakthrough the limitation of bandwidth of standard POF communication, *Proceedings of SPIE*, Vol. 6478

Fischer, U.H.P., M. Haupt & Kragl, H. (2007). Conceptual Design of an Inexpensive POF Demultiplexer, *Proceedings of SPIE*, Vol. 6593

Fraunhofer Institute for Integrated Circuits. (2008). Optical multiplexer for short range communication, Aug. 2008, Available from http://www.iis.fraunhofer.de/ec/oc/download/demux.pdf

Gaudino, R.; Bosco, G.; Bluschke, A.; Hofmann, O.; Kiss, N.; Matthews, M.; Rietzsch, P.; Randel, S.; Lee, J. &

Breyer, F. (2007). On the ultimate capacity of SI-POF links and the use of OFDM: recent results from the POF-ALL project, *Proceedings of the 16th POF Conference*, Turin, Italy, Sept. 11-13, 2007

Gaudino, R.; Nocivelli, A.; Kragl, H.; Ziemann, O.; Weber, N.; Jaeger, D.; Koonen, T.; Lezzi, C.; Bluschke, A. & Randel, S. (2007). Latest results from the POF-ALL EU Project: Toward Improved Capacity over Large-Core Plastic Optical Fiber, *Proceedings of BBeurope Conference 2007*, Antwerp, Belgium, Dec. 3-6, 2007

Gnauck, A. H.; Chraplyvy, A. R.; Tkach, R. W.; Zyskind, J. L.; Sulhoff, J. W.; Lucero, A.J., et. al. (1996). One terabit/s transmission experiment, *Proceedings of OFC'96*, San Jose, California, USA, Feb. 25 – Mar. 1, 1996

Grzemba, A. (2008). *MOST: The Automotive Multimedia Network*, Franzis Verlag GmbH, ISBN 978-3-7723-5316-1, Germany

Hecht, E. (2009). *Optik*, Oldenbourg, ISBN 978-3486588613, Germany

IEEE 802.3: "Telecommunications and Information Exchange Between Systems - Local and Metropolitan Area Networks - Specific Requirements Part 3: Carrier Sense Multiple Access with Collision Detection (CSMA/CD) Access Method and Physical Layer Specifications Amendment: Physical Layer Specifications and Management Parameters for 10Gb/s Passive Optical Networks".

ITU-T Recommendation G.694.2, *Spectral grids for WDM applications: CWDM wavelength grid* (2011)

ITU-T Recommendation G.694.1, *Spectral grids for WDM applications: DWDM frequency grid* (2011)

KDPOF. (2010). Demonstration of 1Gbps over 50m of low cost SI-POF with KDPOF technology, 02.03.2011, Available from http://www.kdpof.com/Papers.html

Keiser, G. (2000). *Optical fiber communications* (3rd ed.), McGraw-Hill series in electrical engineering, communications and signal processing, ISBN 0-07-232101-6, Singapore

Koonen, A.M.J.; Boom, H.P.A.; Ng'oma, A.; Garcia Larrode, M.; Zeng, J. & Khoe, G.D. (2005). POF Application in Home Systems and Local System, *Proceedings of the 14th POF Conference*, pp. 165-168, Hong Kong, Sep. 19-22, 2005

Koonen, A.M.J.; Pizzinat, A.; Jung, H.-D.; Guignard, P.; Tangdiongga, E. & Boom, H.P.A. (2009). Optimisation of In-Building Optical Networks, *Proceedings of ECOC 2009*, Vienna, Austria, Sept. 20-24, 2009

Last, A. & Mohr, J. (2003). *Fehllicht in LIGA-Mikrospektrometern*, Forschungszentrum Karlsruhe Wiss, Berichte

Loquai, S.; Kruglov, R.; Vinogradov, J.; Ziemann, O.; Swoboda, R. & Atef, M. (2010). Comparison of DMT, PAM and NRZ with optimized receiver bandwidths, *Proccedings of the 19th POF Conference*, Yokohama, Japan, Oct. 19-21, 2010

Lubkoll, J.; Strauss, U.; Pollakis, N. & Strobel, O. (2008). *FlexRay with polymer-clad-silica fiber as transmitting medium in aviation electronics*, ISBN 978-1-4244-3484-8, Marrakech, Dec. 11-13, 2008

Marcou, J. (1997). *Plastic Optical Fibers: Practical Applications*, John Wiley & Sons, ISBN 0471956392, France

MOST Cooperation. (July 2010). Appendix D: Frame Structure and Boundary (informative), In: *MOST Specification Rev. 3.0 E2*, MOST Cooperation, pp. 211-215, MOST Cooperation, Retrieved from http://www.mostcooperation.com/publications/Specifications_Organizational_Procedures/index.html?dir=291

Nalwa, H. S. (2004). *Polymer Optical Fibers*, American Scientific Publishers, ISBN 1-58883-012-8, USA

Park, Optical network evolution in Korea Park C-W. (Optimedia 2006) available in *http://www.pofac.de/downloads/itgfg/fgt21/FGT21_Oldenburg_Park_GI-POF.pdf*, April 2011

Poisel, H.; Ziemann, O. & Klein. K.-F. (2003). Trends in polymer optical fibers. *Proceedings of SPIE*, Vol. 5131, pp. 213-219

Randel, S.; Lee, J.; Spinnler, B.; Breyer, F.; Rohde, H.; Walewski, J.; Koonen, A. M. J. & Kirstädter, A. (2006). 1 Gbit/s transmission with 6.3bit/s/Hz spectral efficiency in a 100m standard 1 mm step-index plastic optical fiber link using adaptive multiple sub-carrier modulation, *Proceedings of ECOC 2006*, Postdeadline Paper, Cannes, France, Sept. 24-28, 2006

Senior, M. J. (1992). *Optical fiber communications: principles and practice* (2nd ed.), Prentice Hall International series in optoelectronics, ISBN 0-13-635426-2, Great Britain

Strobel, O.; Rejeb, R. & Lubkoll, J. (2010). Optical polymer and polymer-clad silica fiber data buses for automotive applications, *Proceedings of 2010 7th IEEE & IET International Symposium on Communication Systems, Networks and Digital Signal Processing (CSNDSP)*, pp. 693-696, ISBN 978-1-4244-8858-2, Newcastle, United Kingdom, July 21-23, 2010

Voges, E. & Petermann, K. (2002). *Optische Kommunikationstechnik*, Springer, ISBN 3-540-67213-3, Germany

Wikipedia. (2011). MOST-Bus, Apr. 2011, Available from http://de.wikipedia.org/wiki/MOST-Bus

Ziemann, O.; Poisel, H.; Randel, S.; Lee, J. (2008). Polymer optical Fibers for short, shorter and shortest data links, *Proceedings of OFC/NFOEC 2008, San Diego, California, USA, Feb. 24-28, 2008*

Ziemann, O.; Daum, W.; Krauser, J.; Zamzow, P. E. (2008). *POF Handbook: Optical Short Range Transmission Systems*, Springer, ISBN 978-3540766285, Germany

Ziemann, O. (2010). The (hard) Way to a Standard POF Gigabit Interface, *Proceedings of the 19th POF Conference*, Yokohama, Japan, Oct. 19-21, 2010

9

Nanophotonics for 21st Century

S. K. Ghoshal[1], M. R. Sahar[1], M. S. Rohani[1] and Sunita Sharma[2]
[1]*Advanced Optical Material Research Group, Department of Physics, Faculty of Science, Universiti Teknologi Malaysia, Skudai, Johor,*
[2]*Materials Science Group, Inter-University Accelerator Centre, Aruna Asif Ali Marg, New Delhi, India*

1. Introduction

Recent advances in the synthesis and understanding of material properties at nanoscale in addition to the development of the nanofabrication techniques has enabled researchers to control and manipulate photons at nanometer scales. These have given the birth of an emerging hybrid technology with multi-facets applied interests, popularly known as *nanophotonics*. Nanophotonics can be defined as the science and engineering of light-matter interactions. These interactions, which take, place, on the one hand, within the light wavelength and sub-wavelength scales and, on the other hand, are determined by the physical, chemical and structural nature of artificially or natural nanostructured matter. It is envisaged that nanophotonics has the potential to provide ultra-small optoelectronic components, high speed and greater bandwidth. Nanophotonics has significant potential applications in the field of science and technology. Some of them are sensors, lasers, optoelectronic chips, optical communications, optical microscopy (by overcoming the usual diffraction limit), bio-imaging, targeted therapy, barcodes, harvesting solar energy, etc., to cite a few (Kalele et al., 2007; Prasad, 2004; Shen et al., 2000). Nanophotonics are classified into three main branches as illustrated in the block diagram (figure 1) depicting various process and techniques in nanophotonics including plasmonics.
As mentioned, the field nanophotonics deals with a number of interestingly important topics in photonics and materials structures at nm length scales and their applications in general. Waves in the form of electromagnetic and quantum mechanical, and materials as semiconductors and metals are the focus. Different approaches to confine these waves and devices employing such confinement are the key issue in nanophotonics. Localization of light and applications to metallic mirrors, photonic crystals, optical waveguides, micro-resonators, plasmonics are gaining tremendous applied interest. Localization of quantum mechanical waves in low dimensional structures such as quantum wells, wires and dots has been demonstrated. Devices incorporating localization of both electromagnetic and quantum mechanical waves, such as resonant cavity quantum well lasers and micro-cavity-based single photon sources are on the way of commercialization. Some system-level applications of the introduced concepts, such as optical communications, biochemical sensing and quantum cryptography are targeted for the near future.

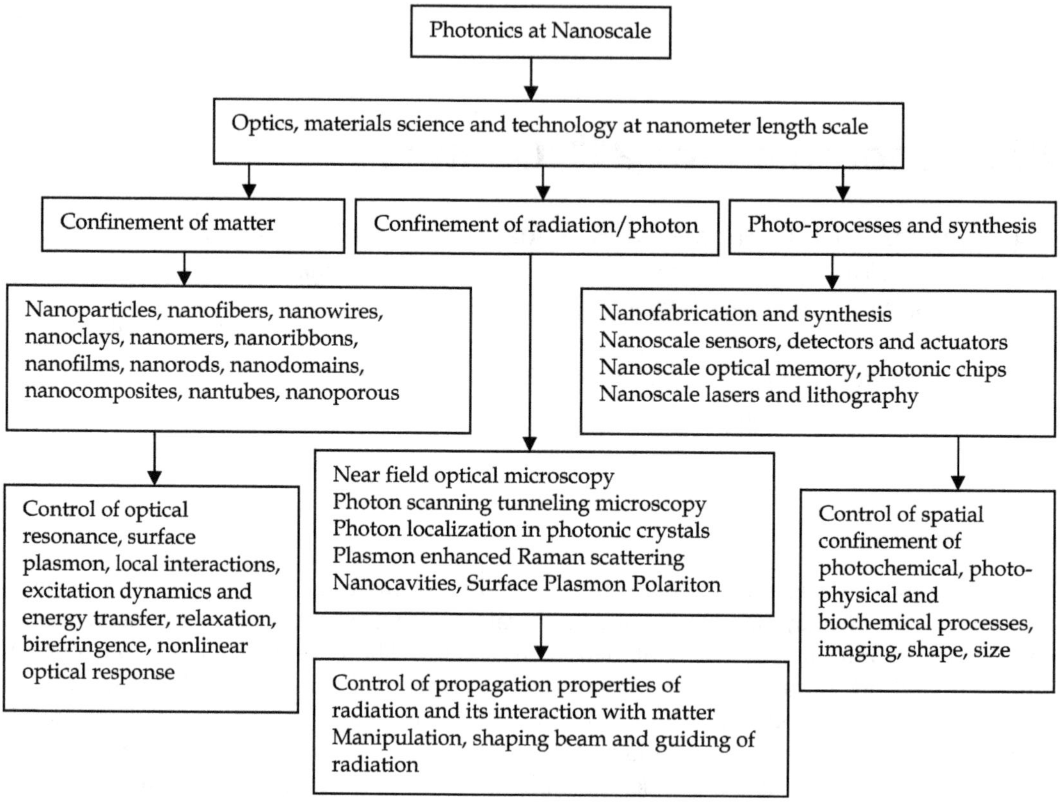

Fig. 1. A block diagram representing various major processes and techniques available in nanophotonics.

The main objectives of nanophotonics research is to control the optical energy and its conversion on the nanometer scale by combining the properties of metal, organic, semiconductor, organo-metallic, polymers and dielectric materials to create new, combined states of light and matter often called meta-materials. Specific examples of scientist's targets are:

- *Controlled quantum coupling at the nanoscale*: The ability to prepare coupled nanostructures presents tremendous opportunity to induce and control the interactions of photons, plasmons, polarons, polaritons and excitons, thereby producing new elementary excitations that have no bulk counterpart. Basic scientific research on these excitations is performed, with application to many disciplines such as solar energy conversion, nanoscale photonic devices, new photochemical processes, communications, sensing, imaging and quantum logic.
- *Understanding ultrafast processes at ultra-small length scales*: The outcome of ultrafast processes can be very different in nanoscale versus bulk materials, with potentially great impact on the physical properties and photochemical products of the nanoscale system. Research has been continuously performed to understand, manipulate, optimize and control these differences.
- *New routes to functional nanophotonic materials*: The development of novel optical materials via three main approaches are:

    - Advanced colloidal synthesis,
    - Lithographically assisted synthesis,
    - Advanced near-field optical lithography techniques for generating hybrid nanoscale structures over large areas,
    - Ion-beam milling and sputtering,
    - Chemical vapor deposition, and
    - Laser ablation
- *Efficient energy transport in plasmonic nanostructures*: Some research is ongoing for significant improvement of:
    - The range of plasmon propagation minimizing loss,
    - The manipulation of localized surface plasmon and surface plasmon polariton,
    - The spectral bandwidth that can be supported by plasmonic structures, and
    - The minimum lateral dimension for efficient plasmon propagation for guiding.

The art of research is the basis for entirely new, efficient solar concentrators, lasers or all optical nanoscale circuitry.

To achieve these objectives a four-pronged experimental and theoretical approach is often utilized:

- *Materials generation via physical and chemical synthesis and lithography,*
- *Optical instrumentation development for advanced characterization, fabrication,*
- *Materials modeling and rigorous numerical simulations, and*
- *Optimization and functionalization of devices via computer aided design*

Some of the leading equipments that are used to achieve this target are Near-field scanning optical microscope, Con-focal Raman microscope, Nanolithography, Ultra-fast transient absorption spectroscopy, Ultra-fast microscope, Chemical vapor deposition, Chemical synthesis, Atomic force and Scanning tunneling microscopy, Size-selected cluster facility and cluster-based nanomaterials for nanophotonics and nano(photo) catalysis (Sharma et al., 2005). Nanosphere lithography provides a very simple yet powerful way to fabricate nanoparticle arrays with precise control over size, shape and inter-particle spacing that is employed to fabricate sensors. They are made from mono-dispersed negatively charged silica or polystyrene nanospheres deposited on a negatively charged glass substrate that helps the nanospheres to diffuse freely until they reach their minimum energy configuration. The capillary forces draw these nanospheres together and finally form a close packed hexagonal pattern on the substrate. These self-assembled nanospheres thus form a mask through which a metal can be deposited by physical evaporation. The nanosphere mask is removed by sonicating the entire sample to obtain a triangular pattern of metal. The optical properties of these sensors can be tuned in the visible range by changing the size of nanospheres.

There is a great need to understand how electromagnetic waves behave in the presence of periodic or nearly periodic arrays of molecules and nanosized metal clusters, i.e. how to achieve optical frequencies in X-ray wavelength scales? Now there are the tools, to map topology, size and shape with optical properties. In the past, these tools were not available, and the tools still needed are the subjects of active research. It is the locally modified electromagnetic field, which allows one to make use of molecular photonics, taking nanophotonics to the molecular scale of a few nanometers. This, in turn, opens up possibilities not just for further miniaturization but also for radically enhanced light-matter interactions. Presently, some of the very genuine engineering concerns relates to when producing ultra-compact, low power and high-sensitivity optical devices, towards the level of single photon detection and emission, and onwards to computation by molecules.

Classes of materials, that inhibit the flow of light, within optical band frequencies, are called photonic band gap crystals or photonic band gaps for short. An array of closed packed nanospheres is the simplest realization of photonic band gap materials, made of polystyrene or silicon dioxide called opals. The range of the inhibited light depends mainly on both the size of the spheres and the number of layers. Another more advanced photonic band gap structure called an inverse opal structure can realize a photonic band gap around 1.5 μm that is made by using the nanospheres as a template. The spaces around the spheres are filled with silicon, and then the spheres themselves are removed. Photonic crystals or photonic band gap crystals are periodic dielectric or metallo-dielectric nanostructures that are designed to affect the propagation of electromagnetic waves in the same way as the periodic potential in a semiconductor crystal affects the electron motion by introducing allowed and forbidden electronic energy bands. Researchers in the field of photonic crystals are just beginning to understand the behavior of light in periodic media. Currently, there is an incipient understanding of what happens in periodic structures and, at the other end, optical processes in single molecules. Particular interest lies in materials with negative refractive indices and meta-materials. However, the gap in between is not bridged yet and requires in-depth investigations. The key issue is what happens when one has nearly periodic, or quasi-periodic, molecular or cluster arrays or clusters? The description of their optical properties is in rapid progress as is the behavior of light in complex media. For example, existing knowledge base is not enough to design novel devices, the associated platforms and system configurations.

Nanophotonics is the field of nanotechnology concerned with discovering and developing nanomaterials that can control the flow of light and in some cases localize or confine it within a volume. Intuitively, we view light as rays, which either propagate in a single direction, being absorbed or reflected to some extent by any object on which it impinges. However, the propagation of light through a material is itself a quantum effect, involving the excitation and relaxation of electrons in the material. Therefore, creating a material with structural and compositional features on a length scale comparable to the wavelength of light i.e. 300 - 750 nm for the visible light enables us to guide light in any direction of our desire.

There are two major challenges to reach the molecular scale and develop mass production techniques. In molecular photonics the aim is to achieve sufficient control over light-matter interactions through nanoscale physical, chemical and structural modification interactions using a few or single molecules. Today, optical spectroscopy of single molecules provides a tool to sense the local environment, which complements and extends ensemble-averaged measurements performed in conventional bulk structures. The molecular-scale variables range from signal strength, orientation, emission spectrum and energy transfer to neighboring molecules. This has undeniable potential in areas requiring ultra-sensitive detection without suffering from random electrical noise. Therefore, molecular photonics holds the promise to key advances in quantum sources of light, efficient photon harvesting systems and optical tuning and switching, among others.

Nanophotonics comprises the fabrication of nanomaterials and their interaction with light, using near-field optics, and thereby controlling the spatial confinement of photo-physical and photochemical process. One way to induce interactions between light and matter on a nanoscale is to confine light to nanodimensions that are much shorter than the wavelength of light (Li et al., 2004; Rosa et al., 2005; Huang et al., 2008). The second approach is to confine matter to nanodimensions, thereby limiting interactions between light and matter to nanoscopic dimensions that are synthesized using nanotechnology lithographic techniques as indicated in figure 1. The last way is nanoscale confinement of photo-processes by inducing photochemistry or a light-induced phase change. This approach provides methods for nanofabrication of photonic structures and functional units. The three main applied areas of nanotechnology and their technical concepts are shown in figure 2, whereas the widespread utility in every sphere of basic and applied sciences as well as engineering is illustrated in figure 3. Similar to the periodic electron-crystal lattice, one can fabricate photonic-crystal lattice. The refractive index varies with a much larger period of around 200 nm. The similarity between electronic and photonic crystals is represented in figure 4. The following characteristics are essential for the materials for making nanophotonics devices:

- Nanoscale quantum dots can be formed in high density.
- Quantum dots are in isolated condition, when not in photo-excited state.
- Nanorods or nanowires large quantum yield and high aspect ratio.
- Highly excited level energy can be controlled by material mixture ratio and dimension.
- Light element excitation is stable at room temperature, and high-excited level does not degenerate.

The brilliant, yet simple result is that one can treat photons in a similar manner as one does with electrons, and begin to envision 'optical' circuit components as existed electronic circuit components.

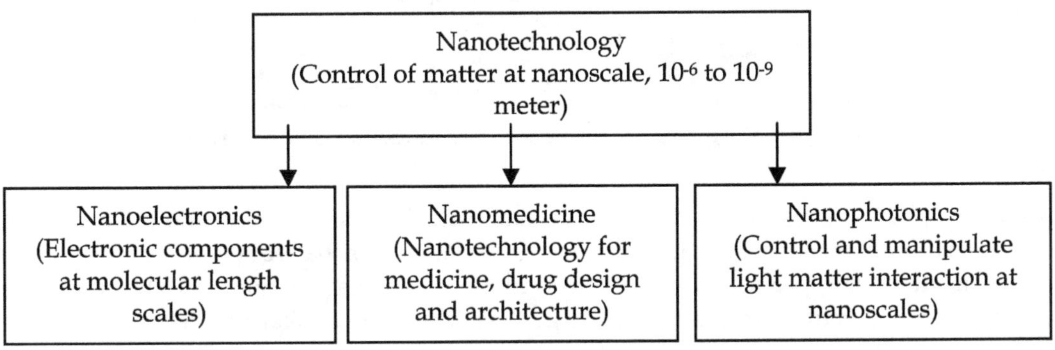

Fig. 2. Block diagram depicting three main branches of nanotechnology.

It is realized that an increasing number of novel materials, techniques, concepts and processes continue to be discovered in the context of nanotechnology and nanophotonics. There is an urgent need to develop the required understanding and to evaluate this new knowledge from the perspective of both novel science and potential applications. Understanding the fundamentals of nanophotonics is important because to fabricate a test structure it is necessary to home in on a window of parameters. The permutations on parameterization, optimization and functionalities are infinite, because there are millions of molecules, hundreds of different atoms and structures one can attach to them, and arranged them to realize a functional device. A window of parameters always helps to target the nanostructures and molecular laboratory samples into research objects, which can give information on device-relevant properties. The obvious boundary conditions are power consumption and cost that needs to be minimized for commercialization. The best option that is recently adopted is to work with molecules embedded in matrices; they are not only efficient but also relatively easier to handle and to manufacture.

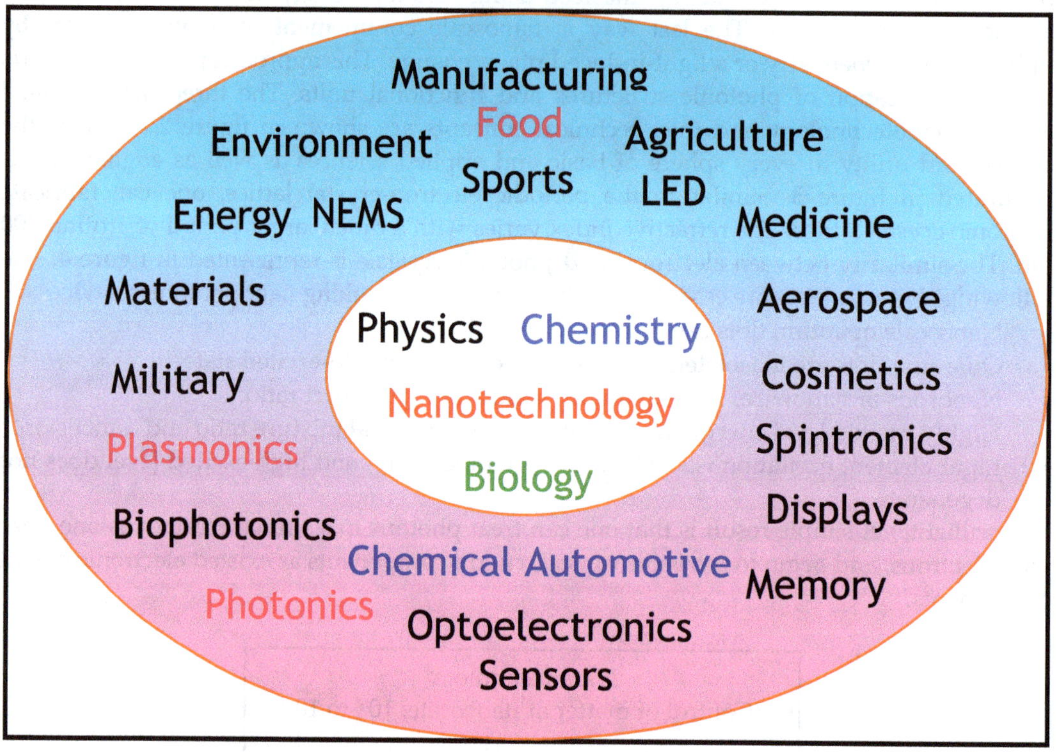

Fig. 3. Schematics representing various branches of nanotechnology and related applied fields in science and engineering.

Nanoscale photo-processes are comprised of nanoscale lithography, fabrication of nanoscale structures, and nanoscale optical memories. Foundations for nanophotonics require basic equations describing propagation of photons in dielectrics that has some similarities to the propagation of electrons in crystals (see figure 4). There are similarities between photons and electrons. Wavelength of light (photon) is given by

$$\lambda = \frac{h}{p} = \frac{c}{v} \tag{1}$$

Wavelength of electron is given by

$$\lambda = \frac{h}{p} = \frac{h}{mv} \tag{2}$$

The Maxwell's electromagnetic equations for the light (photon) describes the allowed frequencies of light and in a similar way Schrodinger's equation for the electrons describes allowed energies of electrons. Free space solution of the wave equation for both the photon and the electron are plane wave (see figure 5). Interaction potential in a medium for propagation of light affected by the refractive index of the dielectric medium and the propagation of electrons are affected by the Coulomb potential. In this case, both electrons and photons have propagation through classically forbidden zones. Photon tunnels through classically forbidden zones, the electric, and the magnetic fields decay exponentially that results k-vector is imaginary. Electron wave function decays exponentially in forbidden zones and has finite tunneling probability as shown in the figure 6.

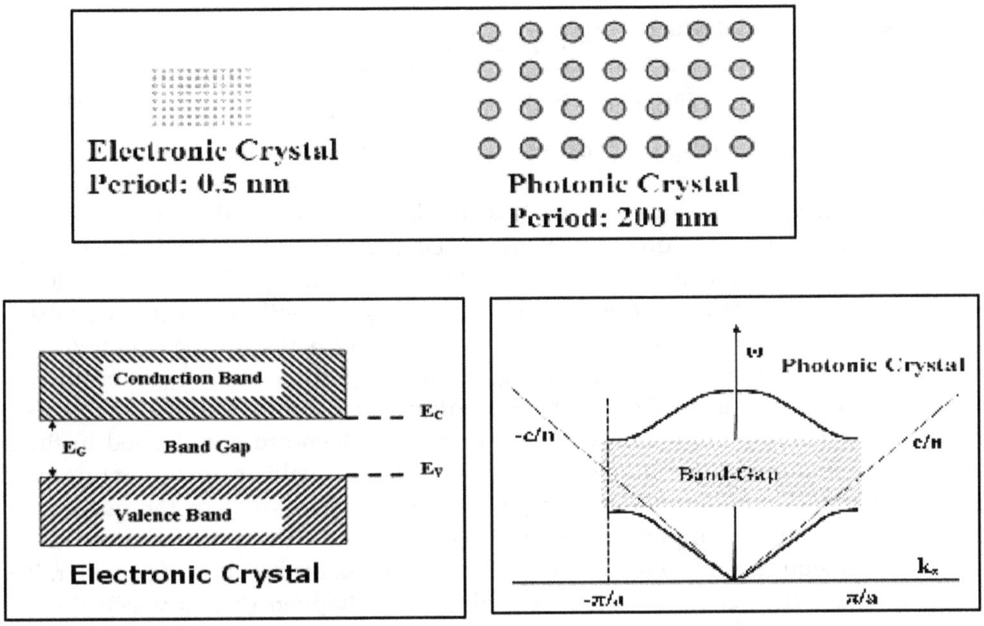

Fig. 4. The typical band structures of electronic and photonic crystals.

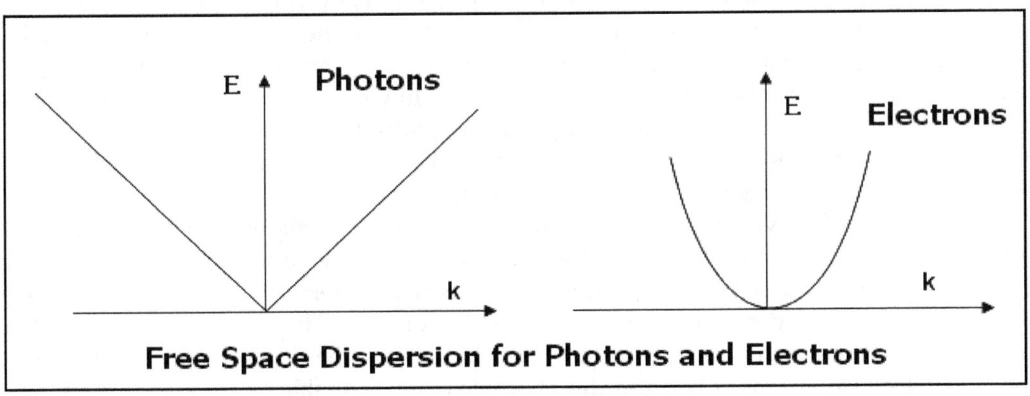

Fig. 5. Free space dispersion relations for photons (left) and electrons (right).

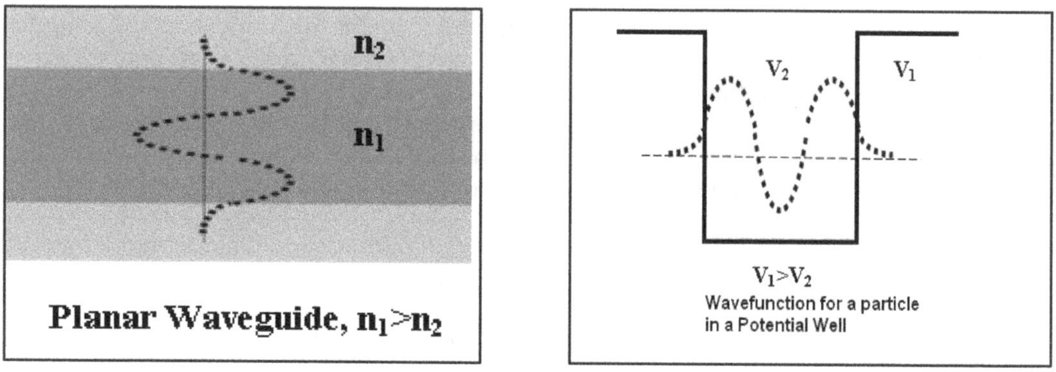

Fig. 6. Tunneling of photon (left) and electron (right) in the classically forbidden region.

Quantum confinement (see figure 7) causes manifestations of new optical effects due to the change in the nature of density of states from continuous to discrete structures as the dimension is reduced from three to zero. It is the density of states that guides optical and electrons correlation effects in quantum confined nanostructures. Size dependence of optical properties due to confinement produces a blue shift of the band-gap in semiconductors, and increases the oscillator strengths and in turn increases the optical transition probability. Location of discrete energy levels depends on the size and nature of the confinement. The oscillator strength increases with increasing confinement from bulk to quantum well to quantum wire to quantum dot. Confinement produces sub-bands within the conduction and valence bands, enabling intra-band optical transitions, which are not allowed in the bulk structures. These infrared transitions have applications in making new quantum cascade lasers and detectors. The oscillator strengths increase as the width of the quantum well decreases. The most important optical feature of these structures is that absorption/emission spectra shift to shorter wavelengths as the size become smaller. In case of metallic nanostructure, somewhat spectacular things happen due to which they show strong luminescence and plasmon absorption bands.

Artificial structures or meta-materials with the optical equivalent of the energy gap in semiconductors promise a wealth of new devices that could satisfy the demand for ever-faster computers and optical communications (Ozbay et al., 2004). Presently, the availability of personal computers that operate at 1 GHz (10^9 Hz), are really very impressive, but what is the likelihood of a 100 GHz desktop computer appearing on the market at affordable price? Indeed, with the current understanding of semiconductor technology, even producing a 10 GHz personal computer would seem to be difficult. However, by transmitting signals with light rather than electrons (a domain of optical computing), it might be possible to build a computer that operates at hundreds of terahertz (10^{12} Hz) which is of the order of the processing speed of the fastest earth simulator computer.

Researchers now believe that such an awesome processing engine could be built from optical components made from so-called photonic crystals and quasi-crystals. These materials possess structures with periodically modulated refractive indices that can be designed to control and manipulate the propagation of light. They are attracting more and more attention not only in the field of photonics but also in the fields of chemistry, physics, and microelectronics. This is because the photonic crystals enable us to control the flow of photons by means of photonic band gaps, which is a photonic analogue of energy band gaps in semiconductors. In recent years, much of the developments on the photonic band gap materials are made using insulators and semiconductors. However, metals or metallo-organic structures can also be used for the photonic band gap in the form of photonic surfaces using the surface plasmon resonance. The photonic band gaps can be introduced by texturing metal surfaces with three-dimensional patterns using spheres or any other shape with a spatial period roughly half of the free-space wavelength of light. Excitation of surface plasmons leads to the formation of standing waves and opening of the stopgap useful for channeling, guiding and controlling photons.

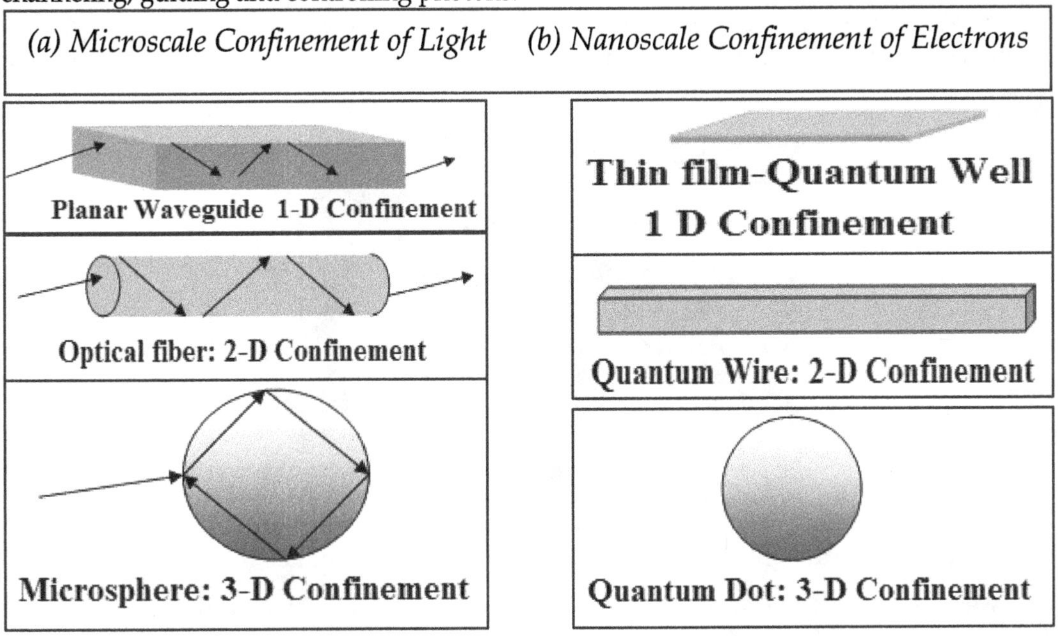

Fig. 7. Nanoscale confinements of photon (left) and electron (right) in different spatial dimensions.

The structures of traditional optical waveguides cannot be smaller than $\lambda/2$. Furthermore, wave guiding along bent guides causes dramatic loss. The problem associated with loss is overcome in photonic band gap structures, a potential candidate for nanophotonics device as already mentioned. However, the dimensions of the structures are still limited by the wavelength of light. This restriction can be overcome by wave guiding of the plasmonic excitation in closely placed metal nanoparticles. The photonic band gap crystals (see figure 8) and photonic band gap fibers involve periodic variation of dielectric constant over wavelength-scale. In common, electronic crystals like sodium chloride, the periodicity is of the order of one nm. Photonic crystals are produced by periodically varying refractive index in one, two or three dimensions and the period is comparable to the wavelength of light. Thus, the field of photonic crystals can be thought as microphotonics. However, in order to fabricate photonic crystals with micron-scale period, the fabrication technique must have nanoscale resolution. Therefore, it is appropriate to include photonic crystals in the domain of nanophotonics.

Basic electromagnetic equations describing propagation of photons in dielectrics have some similarities to propagation of electrons in crystals. The inspiration and very idea of photon localization in photonic band gap materials is drawn from nature. Undoubtedly, nature has demonstrated its advance capability in synthesizing nanomaterials to a level of sophistication and functionality far beyond our own. This inspiration comes from the glittering appearance of butterfly's wing and peacock's feathers, which has a highly ordered periodic structure at the nanoscale. The key to confining and guiding light within a material lies in the periodicity of the structures. An even more recent discovery is the capability of the weevil beetle to produce opals, a precious silica base gem, which have been synthetically produced to make three-dimensional photonic crystals. This discovery opens many doors for accessing the advanced molecular machinery of nature for fabricating photonic materials. Figure 9 illustrates an inverse opal photonic band gap structure that has been realized.

Fig. 8. A photonic band gap structure used for channeling light scaled in microns.

It is well known fact that electronics is the aspect ratio and is limited in speed, whereas, photonics is the diffraction that is limited in size. Plasmonics can go beyond the sub-diffraction limit and is not limited my any means because the surface plasmon wavelengths can reach to nanoscale at optical frequencies. Plasmonics will enable an improved synergy between electronic and photonic devices by naturally interfacing with similar size of electronic components and similar operating speed photonic network. Surface plasmon resonance has already been used by biochemists to detect the presence of a molecule on a surface. In recent years, plasmons are being exploited in many applications by manipulating and guiding light at resonant frequencies. There is a need to understand the origin and the fundamentals of physics associated with surface plasmon resonance before they are employed to many revolutionary applications. Rapid progresses in materials synthesis and nanofabrication techniques over the last decade have led to variety of applications. Some of the examples of plasmon-assisted nanophotonics are, high-resolution plasmon printing, laser shaping, solar concentrators, nanoscale waveguides, sensing, bio-detection at the single-molecule level and enhanced transmission through sub-wavelength apertures, to cite a few.

Since the inception of surface plasmon optics there has been a gradual transition from fundamental research to more applied oriented research. Presently, the surge in surface plasmon-based studies is happening at a time when crucial technological areas such as advanced optical lithography, optical data storage, and high-density electronics manufacturing are approaching fundamental physical limits. Several current technological challenges may be overcome by utilizing the unique properties of surface plasmons. A wide range of plasmon-based optical elements and techniques have now been developed, including a variety of passive waveguides, active switches, biosensors, lithography masks, and more. These developments have led to the notion of *plasmonics*, the science and technology of metal-based optics and nanophotonics.

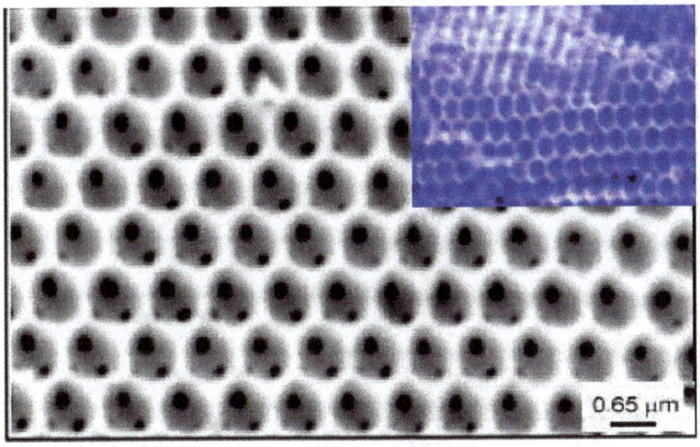

Fig. 9. An inverse opal photonic crystal structure (inset shows dielectric and air holes).

In photonics, metals are not usually thought of as being very useful, except perhaps as mirrors. In most cases, metals are strong absorbers of light, a consequence of their large free-electron density. However, in the miniaturization of photonic circuits, it is now being realized that the metallic structures can provide unique ways of manipulating light at length scales smaller than the wavelength. Surface plasmons, the light-induced excitations of electrons on metal surfaces, may provide integration of electronics and optics on the nanoscale. The optical manipulation of micron/submicron sized particles and bio-molecules through plasmonics have been investigated by Miao et al., and the schematic diagram of the experimental set-up is shown in figure 10. In recent years, a new horizon in nanophotonics has opened up called 'plasmonics' that consists of interaction of light wave with metal nanoparticles (Gopinath et al., 2009). Electromagnetic point of view metals are plasmas, comprising fixed, positive ion cores and mobile conduction electrons. At the plasma frequency, the real part of the dielectric constant changes sign. Therefore, the miniaturization of photonic circuits, allows metallic structures to manipulate light at length scales smaller than the wavelength. The typical thinking about metals is either conductors in electronics or reflectors in optics; however, metals as mirrors are not always desirable. Their residual absorption and low damage threshold makes them unsuitable for lasers and gyroscopes. Plasmonics has given photonics the ability to go to the nanoscale. Plasmonics utilizes the collective oscillations of conduction electrons in metallic nanostructures called plasmons, excited by the incident electromagnetic waves. In the following section, we will discuss in detail the basics of plasmonics, which includes the origin of surface plasmon resonance (see figure 11(a)), their propagation characteristics (figure 11(b)) and their exploitation in innumerable applications. Some of the interesting phenomena in this field are:

- Generation of the surface plasma waves and the dependence of plasmon resonance on metal particle size, and geometry.
- Dependence of the plasmon resonance condition on the dielectric adjacent to the metal film.

- Enhancement of the electromagnetic field close to metal nanoparticles.
- Effects of the metal nanoshells on plasmon resonance.
- Propagation of high-frequency electromagnetic waves along sub-wavelength-wide metal wave-guides.
- Effect of metal surface on radiative decay of molecules.

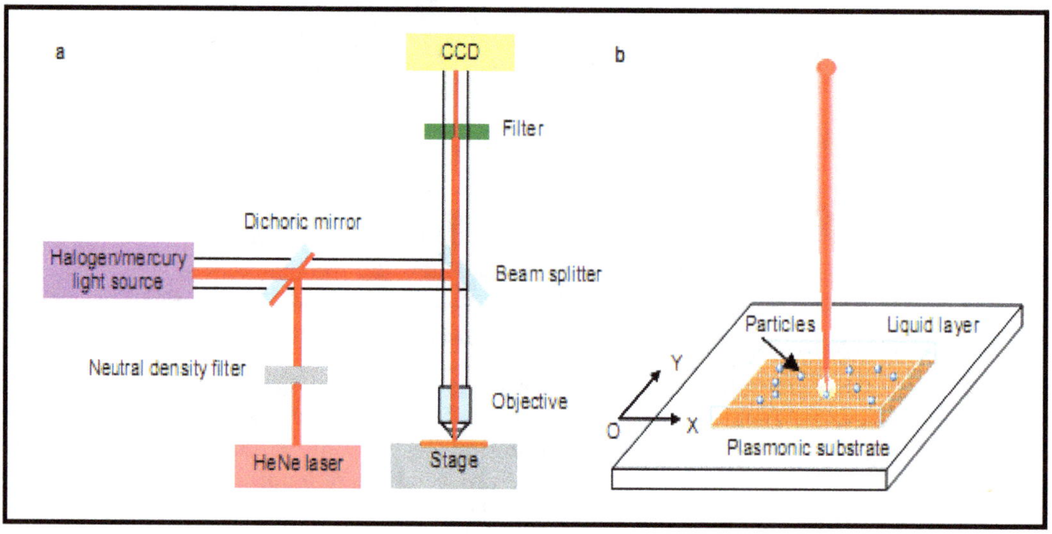

Fig. 10. a) Experimental set up for the trapping and concentration experiments on the plasmonic substrate. b) Detailed configuration of the sample chamber.

The published scientific literatures containing the words 'surface plasmon' over last decade clearly indicates the exponential growth of the field of plasmonics. This rapid growth is stimulated by the development and commercialization of advanced sample preparation methods, powerful electromagnetics simulation codes, nanofabrication techniques and physical analysis techniques providing scientists and engineers with the necessary tools for designing, fabricating, and analyzing the optical properties of metallic nanostructures. A major boost to the field was given by the development of a commercial surface plasmon resonance based sensor in the end of last century. Presently, an estimated fifty percent of all publications on surface plasmons involve the use of plasmons for bio-detection and bio-photonics applications.

The chapter is organized as follows. In Section 2, we review the salient features of photonic crystals and photonic band gaps. A birds' eye view of the recent development in plasmonics and its future prospects is presented in Section 3. The recurring theme of Section 4 is the main challenges in nanophotonics research and the upcoming market strategies. The physics behind the quantum confinement and the behavior of quantum-confined materials are discussed in Section 5 and 6. Section 7 describes the possibility of widespread exciting applications and the need of further research. Conclusions and further outlook are summarized in Section 8.

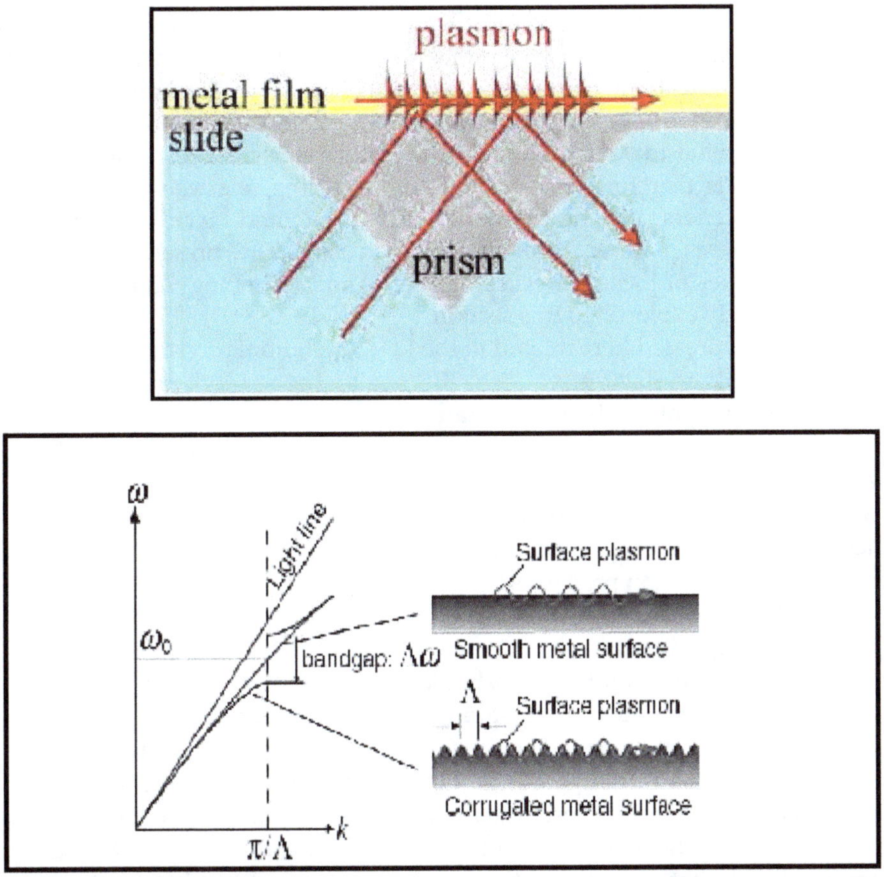

Fig. 11 (a) The origin of surface plasmon resonance at metal dielectric interface (Left panel). (b) The dispersion relation of the surface plasmon travelling on a flat metal surface and a corrugated metal surface (Right panel).

2. Photonic crystals: an amazing material for ultrafast technology

This section focuses on the importance of the photonic crystals called photonic band gap materials. In an age when technology influences society on daily basis, there is an increased demand for new ways to process and exchange information quickly and efficiently. One solution is to implement optical circuits, which use light instead of electricity to relay and process information. However, with the current methods of building optical circuits, it is difficult to manipulate light without losing efficiency. In fact, nanotechnology is helping the photonics to grow at rapid space. Nanotechnology will not just be about electrons moving through circuits but will go beyond our imagination. Information has been transmitted for many years using electromagnetic waves. Most of the communication technology developments exploited radio- and micro-waves and in more recent time information, technology is using much shorter wavelength of light in optical fibers. The natural progression here is towards what are called photonic devices, in which light is channeled through microscopic tubes or lattices of tubes, which are now a topic of ever-increasing active research and debate.

The solution to the problem of rapid information processing and safe transfer at cheaper rates is to use photonic crystals as waveguides in optical circuits. They are small-fabricated crystals, which can efficiently manipulate the direction in which light travels. Photonic crystals are usually viewed as an optical analog of semiconductors that modify the properties of light similar to a microscopic atomic lattice that creates a semiconductor band-gap for electrons. It is therefore believed that by replacing relatively slow electrons with photons as the carriers of information, the speed and bandwidth of advanced communication systems will be dramatically increased. The present century's exciting application possibilities of these materials in science and technology, its present status and future scope are briefly explored in this section.

The various issues addressed here related to the field of photonic crystals are:

- What are photonic crystals?
- What is meant by photonic band gap materials?
- What are the methods used to analyze theory of photonic crystals?
- How do they look like?
- What physics make them so potential for devices?
- Is there any use of these crystals so far?
- What are their fabrication processes?
- What potential applications do they have?
- What are challenges and difficulties?
- How promising are they for future technology?

Microelectronics has allowed by miniaturization of components, such as transistors, resistors, and capacitors on one single chip. On the other hand, photonics has not achieved the same level of miniaturization yet while there are demands for cheaper, faster, and more compact laser-based communication systems in modern time. The integration of photonic components such as lasers, detectors, couplers, and wave-guides is still at a very primitive stage, compared with that of microelectronics. This is due to difficulties with implementing integrated optical components smaller than certain sizes. For instance, small bends or curves of wave-guides lead to the leakages of optical signals, so the bends have to be bigger than certain critical length. Photonic crystals can then be used as they offer a potential to provide solutions to the challenge of photonic component integration by means of its nature to confine photons within their structures. Moreover, intriguing phenomena in photonic crystals can be exploited, which are not achievable by conventional isotropic media. Photonic crystals can be a breakthrough in photonic technology in the near future (Miao et al., 2008).

The absence of allowed propagating electromagnetic modes inside the structures in a range of wavelengths called a photonic band gap, gives rise to distinct optical phenomena such as inhibition of spontaneous emission, high-reflecting omni-directional mirrors and low-loss-wave guiding amongst others. Since the basic physical phenomenon is based on diffraction, the periodicity of the photonic crystal structure has to be in the same length-scale as half the wavelength of the electromagnetic waves ~300 nm for photonic crystals operating in the visible part of the spectrum (Rayleigh, 1888). The periodicity, whose length scale is proportional to the wavelength of light in the band gap, is the electromagnetic analogue of a crystalline atomic lattice. This periodicity acts on the electron wave function to produce the familiar band gaps as in the case of semiconductors in condensed matter physics. The study of photonic crystals is likewise governed by the Bloch-Floquet theorem, and intentionally introduced defects in the crystal. The later is similar to electronic dopants give rise to localized electromagnetic states in linear waveguides and point-like cavities. The crystal can thus form a kind of perfect optical insulator that can confine light without loss around sharp bends, in lower-index media, and within wavelength-scale cavities, among other novel possibilities for control of electromagnetic phenomena (Joannaopoulos et al., 2008). The periodicity of the photonic crystals can be in one, two, and three dimensions that allow interesting properties such as bending light at 90° around corners as shown in figure 12.

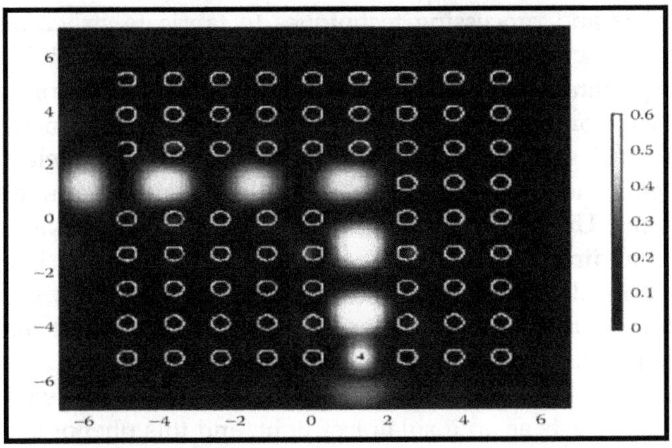

Fig. 12. Bending of light at 90° around corners in the photonic crystals.

One-dimensional periodic system continued to be studied extensively, and appeared in applications from reflective coating to distributed feedback diode lasers. In the former case, the reflection band corresponds to the photonic band gap and for the later, a crystallographic concept is inserted in the photonic band gap to define the laser wavelength. Yablonovitch and co-workers (Yablonovitch, 1987) produced the first photonic crystal by mechanically drilling holes a millimeter in diameter into a block of material with a refractive index of 3.6. The material, which became known as Yablonovite, prevented microwaves from propagating in any direction and exhibited a 3-dimensional photonic band gap. Other structures that have band gaps at microwave and radio frequencies are currently being used to make antennae that direct radiation away from the heads of mobile-phone users (Sajeev, 1987; Lodahl, 2004; Kim, 2008; Sonnichsen, 2005). Later on, photonic crystals of semiconducting colloidal particles were fabricated for realizing photonic band gaps in the visible region of the electromagnetic spectrum. They are also fabricated by the spontaneous self-organization of mono-disperse colloidal spheres such as silica or polystyrene to form a three-dimensional crystal having long-range periodicity. As mentioned, the photonic crystals are materials with periodically varying relative permittivity and are optical equivalents of semiconductors. However, the true potential of these materials lies in manipulating light of wavelength comparable with their lattice parameter. The voids between the particles form regions of low relative permittivity, while the spheres form regions of high relative permittivity, i.e. periodically varying refractive indices (see figure 13). The refractive index variation contrasts for photons in a similar manner to the periodic potential that an electron experiences while traveling in a semiconductor. For sufficiently large contrast, the creation of a complete photonic band gap may occur that results a frequency range where light cannot propagate inside the photonic crystal. This is the underlying principle by which a colloidal photonic crystal blocks certain wavelengths in the photonic band gap, while allowing other wavelengths to pass. The photonic band gap can be tuned by changing the size, shape and symmetry of the particles and the geometry of voids. Using core-shell particles similar photonic crystals are prepared with a large contrast in the refractive indices of the core and shell materials, where the photonic band gap are tuned from the visible to the infrared ranges by changing the refractive indices contrast. It has taken over a decade to fabricate photonic crystals that work in the near infrared (780 - 3000 nm) and visible (300 - 750 nm) regions of the spectrum. The main challenge has been to find suitable materials and processing techniques to fabricate structures that are about a thousandth the size of microwave crystals (Kalele et al., 2007; Sajeev, 1987).

One of the most important features in photonic crystals is the photonic band gap, which is analogous to band gaps or energy gaps for electrons traveling in semiconductors. In case of semiconductors, a band gap arises from the wave-like nature of electrons. Electrons as waves within a semiconductor experience periodic potential from each atom and are reflected by the atoms. Under certain conditions, electrons with certain wave vectors and energy constitute standing waves. The range of energy, named 'band gap', in which electrons are not allowed to exist. This phenomenon differentiates semiconductors from metals and insulators. In the similar manner, standing waves of electromagnetic waves can be formed within a periodic structure whose minimum features are about the order of the wavelength. In this case, the medium expels photons with certain wavelengths and wave vectors. Such a structure acts as an insulator of light, and this phenomenon is referred to as photonic band gap ((Yablonovitch, 1987; Sajeev, 1987; Lodahl, 2004). The origin of photonic band gap in photonic crystals can be explained with the help of Maxwell's equations.

It is well known that in a silicon crystal, the atoms are arranged in a diamond-lattice structure in which the electrons moving through this lattice experience a periodic potential while interacting with the silicon nuclei via the Coulomb force, that results in the formation of allowed and forbidden energy states. No electrons can be found in the forbidden energy gap or simply the band gap for pure and perfect silicon crystals. However, for real materials with defects the electrons can have energy within the band gap due to the broken periodicity caused by a missing silicon atom or by an impurity atom occupying a silicon site, or if the material contains interstitial impurities. Now, consider a situation in which the photons are moving through a block of transparent dielectric material that contains a number of tiny air holes arranged in a regular lattice pattern. The photons will pass through regions of high refractive index of the dielectric intersperse with regions of low refractive indexed air holes. In case of a photon, this contrast in refractive index looks just like the periodic potential that an electron experiences traveling through a silicon crystal. Indeed, if there is large contrast in refractive index between the two regions then most of the light will be confined within either the dielectric material or the air holes. This confinement results in the formation of allowed energy regions separated by a forbidden region, photonic band gap. As the wavelength of the photons is inversely proportional to their energy, the patterned dielectric material will block light with wavelengths in the photonic band gap, while allowing other wavelengths to pass freely (Mia et al., 2008). It is also possible to create energy levels in the photonic band gap by changing the size of a few of the air holes in the material. This is the photonic equivalent to breaking the perfect periodicity of the crystal lattice. The diameter of the air holes is a critical parameter, in addition to the contrast in refractive index throughout the material. Photonic band gap structures can also be made from a lattice of high-refractive-index material embedded within a medium with a lower refractive index (core-shell for example). A naturally occurring example of such a material is opal. However, the contrast in the refractive index in opal is rather small, and hence the appearance of a small band gap (Kalele et al., 2007).

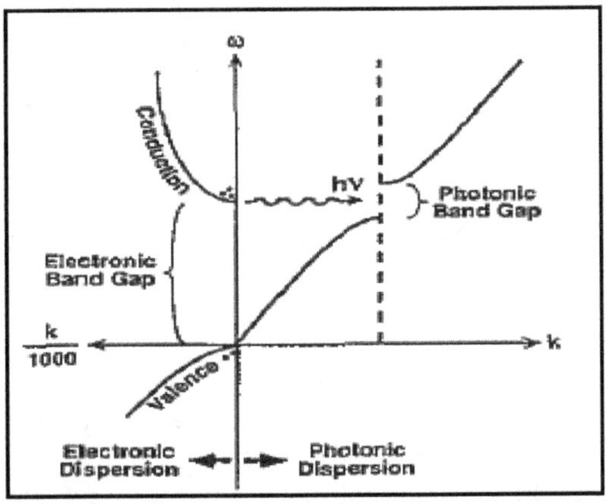

Fig. 13. Schematic representing the electronic and photonic band gaps in the Brillouin zone.

Let us consider the simplest one-dimensional (1D) structure in order to describe the phenomenon of formation of photonic band gap in the photonic crystals that has alternating layers of two dielectrics. The incident wave in entering a periodic array of dielectric sheets is partially reflected at the boundaries of the dielectric layers. If the partially reflected waves are in phase and superimposed, they form a total reflected wave, and the incident wave is unable to enter the medium, as depicted in figure 14. The range of wavelengths in which incident waves are reflected is called a 'stop band'. A structure that exhibits stop bands to every direction for given wavelengths, the stop bands are considered a 'photonic band gap'. On the other hand, when the wavelength of an incident wave does not lie within the band gap, destructive interferences occur and partially reflected waves cancel one other. Consequently, the reflection from the periodic structure does not happen and the light passes through the structure as illustrated in figure 15.

For two-dimensional (2D) structure, the condition in which such reflections occur at the interfaces of two dielectrics and a photonic band gap arises from the superposition of partial reflected waves are somewhat complex. To realize an effective photonic band gap, back-scattered waves should be in phase, forming one reflected wave in which the Bragg's condition has to satisfy, the same condition has to be satisfied for incident waves from every direction to attain a photonic band gap. An intuitive idea regarding the nature photonic crystal structure obtained from Bragg's law indicates that the distance from one lattice point to neighboring ones should be same so that scattered waves are superimposed and in phase at any point of the structure. Moreover, the structure should possess symmetry to as many directions as possible so that scattered waves from one lattice point experiences the same orientation of neighboring lattice points. The same concept can be extended to three-dimensional (3D) periodic structure, where the incident waves turned into partially reflected waves and the transmitted waves at boundaries between the two media. If the partially reflected waves are in phase, the scattered waves add up to a net reflected wave, resulting in a stop band. The condition for Bragg's law must be satisfied at each lattice point that can be either a dielectric material or an air hole surrounded by a dielectric. If the stop bands exist for every direction and those *'stop bands'* overlap within certain wavelength

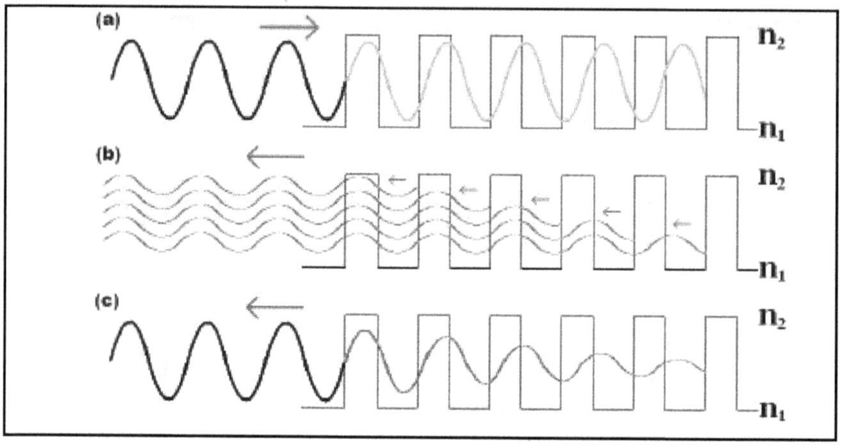

Fig. 14. The constructive interference for the photonic band gap in one dimension. (a) An incident wave within the photonic band gap enters the periodic structure with two different refractive indices n_1 and n_2. (b) The incident wave is partially reflected by the boundary of the structure. (c) The incident wave is totally reflected when each reflected wave is in phase, and is unable to penetrate the structure.

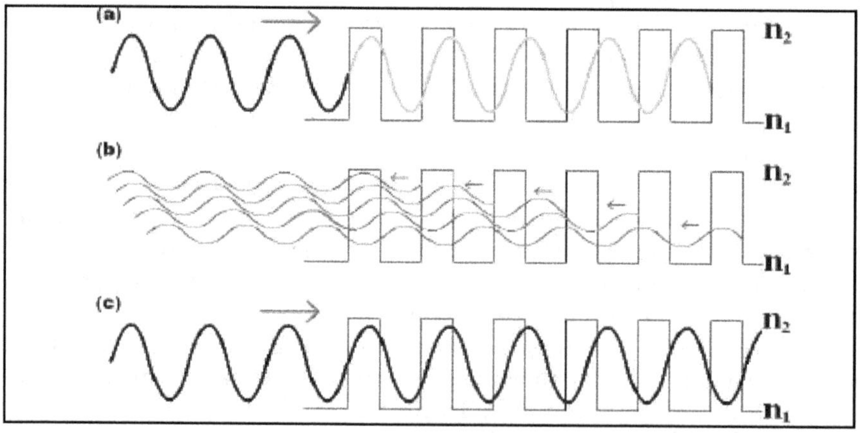

Fig. 15. The destructive interference. (a) An incident wave outside the photonic band gap enters the periodic structure. (b) The incident wave is partially reflected by the boundary of the structure, but each reflected wave is out of phase and interfere destructively. (c) The incident wave penetrates the structure without being reflected.

region then a complete photonic band gap arises in three-dimension. Photonic band gaps results from the net interferences of scattered incident light waves from lattice points of a periodic structure. It is important to note that high refractive index contrasts of the periodic structures play pivotal role for the photonic band gaps to occur or to become more pronounced for a given structure (Joannaopoulos et al., 2008).

There are two reasons for the importance of high refractive index contrasts. First, each photonic crystal structure has a threshold value of refractive index contrast to exhibit a photonic band gap as depicted in figure 16. This phenomenon is attributed to the fact that interfaces of two dielectrics with higher contrast of refractive indices tend to scatter waves from any direction, so stop bands to any direction, a photonic band gap, are more likely to take place. Second, the higher the refractive index contrast is, the fewer layers are necessary to have sufficient photonic band gap effects. As explained in figure 14, each layer or lattice of photonic crystal partially reflects the propagating wave. Consequently, if each layer reflects more waves due to a higher refractive index contrast, sufficient net reflections can be achieved by fewer layers of lattices than a structure with the same configuration but with a lower refractive index contrast. This condition helps us to choose materials such as semiconductors for photonic crystals (Mia et al., 2008; Rayleigh, 1888; Yablonovitch, 1987).

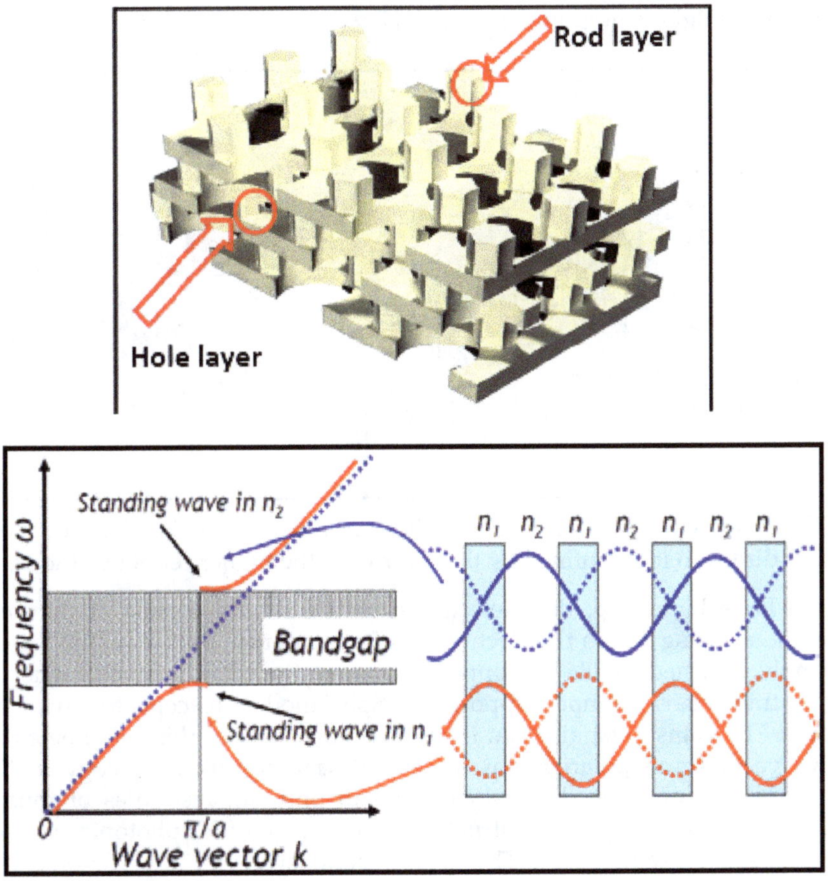

Fig. 16. (Left) A 3D photonic crystal consisting of an alternating stack of triangular lattices of dielectric rods in air and holes in dielectrics (courtesy Yablonovitch). (Right) Projected band diagrams and the band gap for a finite-thickness slab of air holes in dielectric with the irreducible Brillouin zone.

By combining Maxwell's equations with the theorems of solid-state physics a surprising and simple result emerges, that explain the phenomena of light bouncing among infinity of periodic scatterers. Like electrical insulators, which keep the currents in the wires where they belong, one can also build an optical insulator, a photonic crystal to confine and channel photons. The emergence of photonic crystals is due to the cooperative effects of *periodic* scatterers that occur when the period is of the order of the wavelength of the light.

They are called 'crystals' because of their periodicity and 'photonic' because they act on light i.e. photon. Once such a medium is obtained, impervious to light, one can manipulate photons in many interesting ways. By carving a tunnel through the material, an optical 'wire' can be achieved from which no light can deviate. Even more interesting things can happen by making a cavity in the center of the crystal, an optical 'cage' can be created in which a beam of light could be caught and held, because the very fact that it cannot escape would render it invisible. These kinds of abilities to trap and guide light have many potential applications in optical communications and computing (Joannaopoulos et al., 2008). A typical photonic crystal slab structure with tunnels and cavities that are made to confine and control light is presented in figure 17.

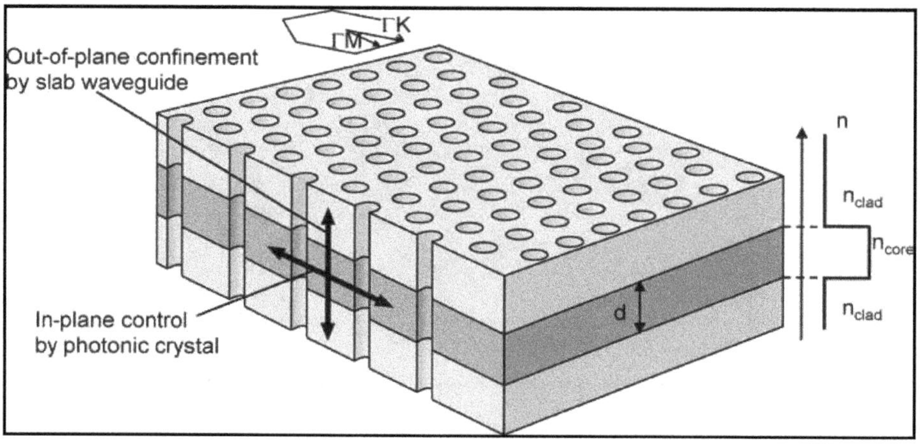

Fig. 17. A 2D photonic crystal slab. In-plane, light is controlled by the photonic crystal, while in the vertical direction it is confined by the layer with the higher refractive index.

To achieve a large band gap, the dielectric structure should consist of thin, continuous veins/membranes along which the electric field lines can run. This way, the lowest band(s) can be strongly confined, while the upper bands are forced to a much higher frequency because the thin veins cannot support multiple modes (except for two orthogonal polarizations). The veins must also run in all directions, so that this confinement can occur for all wave vectors and polarizations, necessitating a complex topology in the crystal. Furthermore, in two or three dimensions one can only suggest rules of thumb for the existence of a band gap in a periodic structure. The design of 3D photonic crystals is a trial and error process (Sanjeev, 1987). The typical band structure for photonic crystals for transverse electric and transverse magnetic mode is shown in figure 18. Interestingly, the 2D systems exhibit most of the important characteristics of photonic crystals, from nontrivial Brillouin zones to topological sensitivity to a minimum index contrast, and can also be used to demonstrate most proposed photonic-crystal devices (Yablonovitch, 1987).

The numerical computations are the crucial part of most theoretical analyses for photonic band gap materials due to their complexity in high index-contrast directional dimensionality of the systems. Computations are typically fall into the following three categories:

1. The time-evolution of the fields with arbitrary initial conditions in a discretized system are modeled and simulated by the time-domain 'numerical experiments' using finite difference method.

2. The scattering matrices are computed in some basis to extract transmission/reflection through the structure (mainly eigenvalues) and the definite-frequency transfer matrices can be achieved.
3. The frequency-domain methods can directly extract the Bloch fields and frequencies by diagonalizing the eigenoperator.

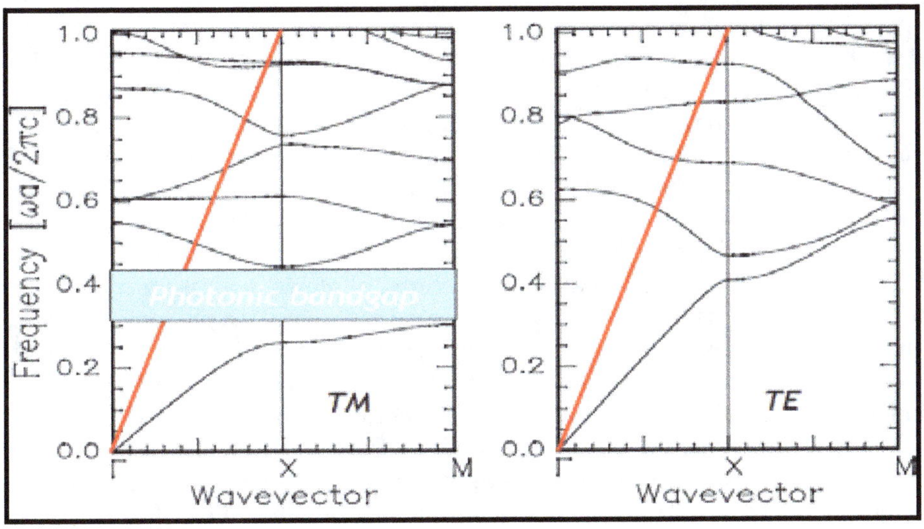

Fig. 18. Band diagrams and photonic band gaps for the two polarizations TE/TM (electric field parallel/perpendicular to plane of periodicity).

The directly measurable quantities such as transmission can be obtained intuitively from the first two categories. The third category is more abstract, yielding the band diagrams that provide a guide to interpretation of measurements as well as a starting point for device design and semi-analytical methods. For many systems, several band diagrams are computed by the frequency-domain method.

Photonic-crystal slabs have two new critical parameters that influence the existence of a gap. Firstly, it must have mirror symmetry in order that the gaps in the even modes and odd modes can be considered separately. Such mirror symmetry is broken in the presence of an asymmetric substrate. In actual practice, the symmetry breaking can be weak if the index contrast is sufficiently high so that the modes are strongly confined in the slab. Secondly, the height of the slab must not be too small that weakly confines the modes or not too large so that higher-order modes will fill the gap. The required optimum height must be around half a wavelength relative to an averaged index that depends on the polarization (Joannaopoulos et al., 2008). The photonic-crystal slabs are one way of realizing 2D photonic-crystal effects in 3D. A 3D periodic crystal is formed by an alternating hole-slab/rod-slab sequence by stacking of bi-layers that has a 21 % plus complete gap for $\varepsilon = 12$, forbidding light propagation for all wave vectors and all polarizations (Sanjeev, 1987). This kind of crystal slabs confines light perfectly in 3D, because its layers resemble 2D rod/hole crystals, it turns out that the confined modes created by defects in these layers strongly resemble the TM/TE states created by corresponding defects in 2D. Therefore, it can be used for direct transfer of designs from two to three dimensions while retaining omni-directional confinement (Joannaopoulos et al., 2008).

Over the years, it is realized that the fabrication of photonic crystals can be either easy or extremely difficult depending upon the desired wavelength of the band gap and the level of dimensionality. Lower frequency structures that require larger dimensions are easier to fabricate because the wavelength of the band gap scales directly with the lattice constant of the photonic crystals. At microwave frequencies, where the wavelength is of the order of 1 cm, the photonic crystals are decidedly macroscopic and simple machining techniques or rapid prototyping methods can be employed in building the crystals. Moreover, at the optical wavelengths, photonic band gaps require crystal lattice constants less than 1 μm and are difficult to fabricate. Building photonic band gaps in the optical regime requires methods that push current state-of-the-art micro and nanofabrication techniques. Since 1D photonic band gaps require periodic variation of the dielectric constant in only one direction, they are relatively easy to build at all length scales compare to 3D one (Sanjeev, 1987; Lodahl et al., 2004; Kim et al., 2008; Sonnichsen et al., 2005; Joannaopoulos et al., 2008). The 1D photonic band gap mirrors commonly known as distributed Bragg reflectors that have been used in building optical and near-infrared photonic devices for many years. Two common examples of devices that have been realized using 1D photonic band gaps are distributed feedback lasers and vertical-cavity surface-emitting lasers. The 2D photonic band gaps require somewhat more fabrication, but relatively ordinary fabrication techniques can be employed to achieve such structures. There are several examples of 2D photonic band gaps operating at mid- and near-IR wavelengths. Clearly, the most challenging photonic band gap structures are fully 3D structures with band gaps in the IR or optical regions of the spectrum. As mentioned above, the fabrication of 3D photonic band gaps is complicated by the need for large dielectric contrasts between the materials that make up the photonic band gap crystal, and the relatively low filling fractions that are required. The large dielectric contrast demands dissimilar materials, and often the low-dielectric material is air with the other material being a semiconductor or a high-dielectric ceramic. The low dielectric filling fraction ensures that the photonic band gap crystal has mostly air, while the high dielectric material must be formed into a thin network or skeleton. Combining these difficulties with the need for micron or sub-micron dimensions to reach into the optical region, the fabrication becomes extremely difficult indeed (Sanjeev, 1987; Lodahl et al., 2004).

The deep x-ray lithography and other techniques are useful to fabricate the photonic band gaps structures in which the resist layers of polymethyl methacrylate are irradiated to form a 'three-cylinder' structure. The holes in the polymethyl methacrylate structure are usually filled with ceramic material due to their low value of dielectric constant not favorable for the formation of a photonic band gap. A few layers of this structure can be fabricated with a measured band gap centered at 2.5 THz. The layer-by-layer structure can be fabricated by laser rapid prototyping using laser-induced direct-write deposition from the gas phase. The photonic band gap structure consisted of oxide rods and the measured photonic band gap is centered at 2 THz. The measured transmittance shows band gaps centered at 30 and 200 THz, respectively. In this way, it is possible to overcome very difficult technological challenges, in planarization, orientation and 3D growth at micrometer length scales. Finally, the colloidal suspensions have the ability to form spontaneously the bulk 3D crystals with submicron lattice parameters. In addition, 3D dielectric lattices have been developed from a solution of artificially grown mono-disperse spherical SiO_2 particles. However, both these procedures give structures with a quite small dielectric contrast ratio (< 2), which is not enough to give a full band gap. A lot of effort is being devoted to find new methods in

increasing the dielectric contrast ratio. Several groups are trying to produce ordered macro-porous materials from titania, silica, and zirconia by using the emulsion droplets as templates around which material is deposited through a sol-gel process (Xing-huang et al., 2008). Subsequent drying and heat treatment yields solid materials with spherical pores left behind the emulsion droplets. Another very promising technique in fabricating photonic crystals at optical wavelengths is 3D-holographic lithography (Miao et al., 2008).
Materials with photonic band gaps could speed up the internet by improving the transmission of long-distance optical signals. One drawback with conventional optical fibers is that different wavelengths of light can travel through the material at different speeds. Over long distances, time delays can occur between signals that are encoded at different wavelengths. This kind of dispersion is worse if the core is very large, as the light can follow different paths or 'modes' through the fiber. A pulse of light traveling through such a fiber broadens out, thereby limiting the amount of data that can be sent. These problems could be solved by an extremely unusual 'holey fiber' as show in figure 19. The fiber has a regular lattice of air-cores running along its length and transmits a wide range of wavelengths without suffering from dispersion. It is made by packing a series of hollow glass capillary tubes around a solid glass core that runs through the centre. This structure is then heated and stretched to create a long fiber that is only a few microns in diameter. The fiber has the unusual property that it transmits a single mode of light, even if the diameter of the core is very large. This fiber can be produced even in a better way by removing the central solid glass core to form a long air cavity. In this case, the light is actually guided along the low-refractive-index air core by a photonic-band-gap confinement effect. Since the light is not actually guided by the glass material, very high-power laser signals could potentially be transmitted along the fiber without damaging it.

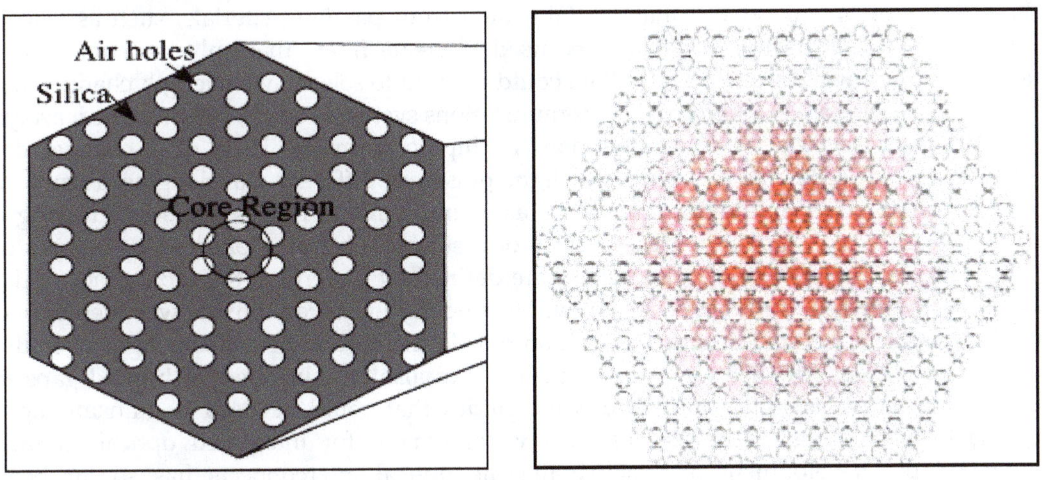

Fig. 19. Air-core photonic crystal fibers. Arrangements of voids and dielectric media (left) and light propagations through holes (right).

Defects in photonic band gap structure allow designing small, but highly efficient micro lasers. A point defect in the crystal gives rise to a resonant state with a defined resonant frequency in the band gap. Light is trapped in this cavity as the photonic band gap prevents it from escaping into the crystal. The photonic crystals built from the photo emissive materials, such as III-V semiconductors and glasses doped with rare-earth atoms, can also be used to make narrow-line width lasers that could potentially be integrated with other components in an optical-communications system. These lasers are made by introducing a small number of holes that are slightly smaller or larger than the other holes in the photonic-crystal lattice. These 'micro cavities' generate a narrow defect mode within the photonic band gap. While the material emits light in a wide spectral range, only the wavelength that matches the wavelength of the defect mode is amplified because it can propagate freely through the material. The laser cavity is formed either by the crystal surface or by external mirrors that surround the glass. The intensity of the propagating light increases as it undergoes successive reflections and travels back and forth through the photonic crystal. Meanwhile, light at other wavelengths are trapped within the photonic crystal and cannot build up. This means that the laser light is emitted in a narrow wavelength range that is directly related to the diameter of the micro cavity divided by the diameter of the regular holes. Moreover, the line width can be reduced further by using unusual geometries of the photonic-crystal lattice (Sanjeev, 1987). Such micro cavities are also much more efficient at trapping light than the cavities formed in semiconductor diode lasers since there are fewer directions in which the photons can escape from the cavity. The rate of photoemission in an active medium can be greatly increased by maintaining a high optical flux density. As micro cavities act as light traps, they provide a good method of enhancing the rate of photoemission in light emitting diodes, and are crucial for the operation of lasers. Moreover, the increased rate of photoemission means that micro cavity light emitting diodes and photonic-crystal lasers can be switched on and off at far greater speeds compared with conventional devices, which could lead to higher data-transmission rates and greater energy efficiency.

Preliminary experiments have been performed at microwave frequencies on defect structures within photonic crystals made from 'passive' materials that do not emit light. Photonic-crystal micro cavities that are fabricated from passive materials, such as silicon dioxide and silicon nitride, could also be used to create filters that only transmit a very narrow range of wavelengths. Such filters could be used to select a wavelength channel in a *'dense wavelength division multiplexing'* communications system (Lodahl, 2004). Indeed, arrays of these devices could be integrated onto a chip to form the basis of a channel de-multiplexer that separates and sorts out light pulses of different wavelengths. Figure 20 shows a photonic-crystal device that works as a simple filter. This is made by growing a thick layer of silicon dioxide on the surface of a silicon substrate, followed by a layer of silicon nitride. The positions of the holes were defined by patterning the top surface of the waveguide with electron-beam lithography. The underlying silicon dioxide was then etched away to create a freestanding porous silicon-nitride membrane that blocks light over the wavelength range 725 - 825 nm. Similar devices can also be fabricated with band gaps at shorter visible wavelength. Miniature wave-guides that could be used to transmit light signals between different devices are a key component for integrated optical circuits. However, the development of such small-scale optical interconnects has so far been inhibited by the problem of guiding light efficiently round very tight bends.

Conventional optical fibers and waveguides work by the process of total internal reflection. The contrast in refractive index between the glass core of the fiber and the surrounding cladding material determines the maximum radius through which light can be bent without any losses. For conventional glass waveguides, this bend radius is about a few millimeters. However, inter-connects between the components on a dense integrated optical circuit require bend radii of 10 μm or less. It is possible to form a narrow-channel waveguide within a photonic crystal by removing a row of holes from an otherwise regular pattern. Light will be confined within the line of defects for wavelengths that lie within the band gap of the surrounding photonic crystal. Since a porous material has no available modes at this wavelength, an optical quantum well forms in the waveguide region and traps the light. Under these conditions, we can introduce a pattern of sharp bends that will either cause the light to be reflected backwards or directed round the bend.

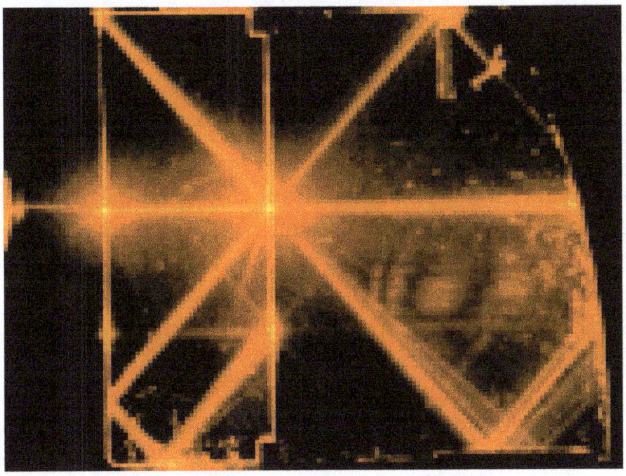

Fig. 20. A photonic crystal devices that work as a simple filter.

We conclude with the note that the original innovative research into photonic crystals/photonic band gap materials is necessary to achieve immediate commercial applications, but without intense research, it would not have been possible to set into these new classes of structures or a whole host of other tangential pursuits. The most important and useful thing that comes out of the research is *new ideas and paths of investigation*. Research breeds more research, which will eventually lead to something that genuinely be commercialized. Though the field of nanophotonics and nanotechnology is growing up exponentially and newer applications are coming at rapid space, however, more focused research is needed to get position in the market by defeating the existing technology.

3. Plasmonics: a new avenue of nanoscale optics

The term 'plasmonics' refers to the science and technology dealing with the manipulation of the electromagnetic signals by coherent coupling of photons to free electron oscillations at the interface between a conductor and a dielectric. Plasmons are electrons density waves and is created when light hits the surface of a metal at the precise frequency. Because these density waves are generated at optical frequencies, very small and rapid waves, they can theoretically encode a lot of information; more than what is possible for conventional electronics (Kim et al., 2008). Surface plasmons are optically induced oscillations of the free electrons at the surface of a metal. Plamonics is thought to embody the strongest points of both optical and electronic data transfer. Optical data transfer, as in fiber optics, allows high bandwidth, but requires bulky 'wires', or tubes with reflective interiors. Electronic data transfer operates at frequencies inferior to fiber optics, but only requires tiny wires. Plasmonics, often-called 'light on a wire', would allow the transmission of data at optical frequencies along the surface of a tiny metal wire, despite the fact that the data travels in the form of electron density distributions rather than photons (Sonnichsen, 2005). We would like to address the following relevant issues in plasmonics:

- What is plasmonics and plasmon resonance?
- How to get materials for plamonics applications?
- Why research is necessary in plasmonics?
- What is the present status for commercialization?
- Why are they so interesting?
- What are challenges and difficulties in plasmonics?
- How promising are they for future technology?

Since the middle of nineteenth century, after the first demonstration of stable dispersion of gold nanoparticles by Michael Faraday the scientific insight and queries on the interaction of light with matter has intrigued scientists. Without invoking the word nano in ancient time, artists have been exploiting sparkling red, yellow and green colors exhibited by metal nanoparticles especially of gold and silver as colorants in glasses for the decoration of windows and doors of many cathedrals, palaces, mosques and temples. Faraday concluded that metal nanoparticles having size much smaller than the wavelength of light exhibit intense colors that has no bulk counterpart. Gustav Mie in 1908 successfully explained the origin of such colors of dispersion using Maxwell's theory of classical electromagnetic radiation in which the phenomena was attributed to strong absorption and scattering of light by dispersion of metal nanoparticles (Kalele et al., 2007). However, during last two decades a series of noble-metal particles fabricated using advanced nanotechnology route showed a strong absorption band in the visible region of electromagnetic spectrum, arising from a resonance between collective oscillations of conduction electrons with incident electromagnetic radiation. Consequently, scientists are interested to guide, manipulate and control such strong absorption band associated with plasmon and hence the genesis of plasmonics. The formation of electric dipoles originates from the interaction of incident electromagnetic field that induces strong polarization of conduction electrons and weaker polarization to the immobile heavier ions. The net charge difference between the electrons and the ions acts as a restoring force that can be visualized as simple harmonic oscillator in the Lorentz model. The plasmon resonance is the resonance between the frequency of oscillation of the electrons and the frequency of the incident photon and is characterized by a strong absorption band. For nanoscale matter, the surface by volume ratio is high and most of the optical and electronic structure properties are dominated by the surface rather than the bulk. In this case, since a net charge difference is felt at the surface of a nanoparticle, the resonance is also known as *surface plasmon resonance*. The pictorial representation of surface plasmon resonance on the metal dielectric interface and on an array of two gold nanoparticle is shown in figure 21 (left panel) and (right panel) respectively.

The generation of *surface plasmon* is like *'an ocean of light'*. Dropping a piece of stone into a quiet lake one creates the ripples that spread out across its surface. The same thing happens when a photon hits the surface of a metal, where the 'ripples' consist of collective oscillations of electrons and have wavelengths of the order of nanometers. During such oscillations these 'surface plasmons' can pick up more light and carry it along the metal surface for comparatively large distances. Using plasmons light can not only be focused into the tiniest of spots but can also be directed along complex circuits or manipulated it many different ways. It is possible to achieve all of this at the nanoscale that is several orders of magnitude smaller than the wavelength of light (Pendry, 2000). This nanoscale is far below the resolution limits of conventional optics. Due to this reason, plasmonics has occupied a place in naophotonics in its own right. Several potential applications such as lasers, sensors, memory, communications, solar cells, biochemical sensing, optical computing and even cancer treatments are widely explored. Some of the exciting features of this field will be explored in this Section.

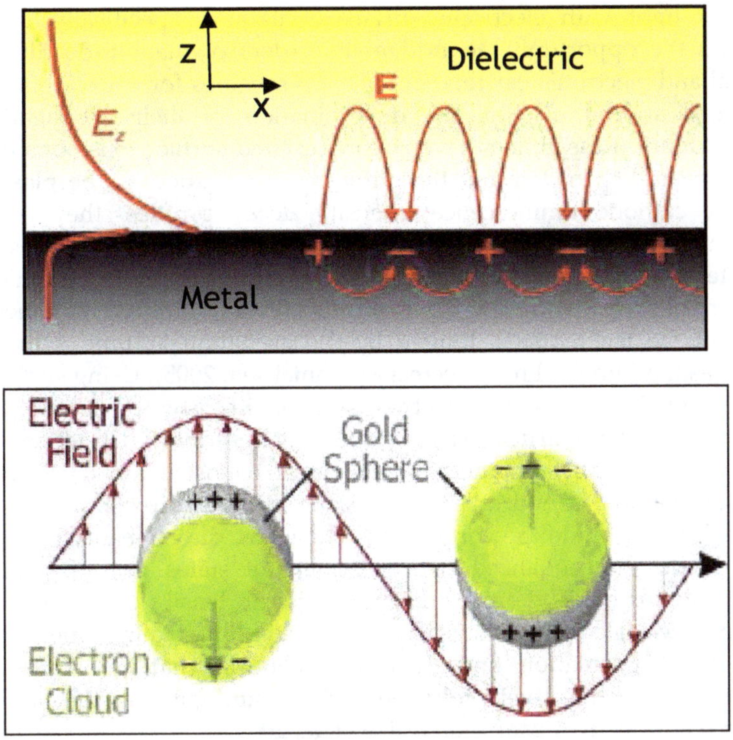

Fig. 21. The surface plasmon resonance: EM wave at metal-dielectric interface (left), and in gold nanoparticles (right).

The optical extinction properties of small metal particles have been studied for many years. Noble metal nanoparticles embedded in a dielectric exhibit a strong absorption peak due to a collective motion of free electrons, that is, a surface plasmon resonance. For isolated spherical particles, the resonance peak occurs generally in the visible part of the spectrum. The particular frequency depends on the particle size, and the dielectric constants of the metal and of the surrounding medium. For particle ensembles, however, electromagnetic coupling between neighboring particles shifts the plasmon absorption bands. Numerical calculations have demonstrated that nanoparticle size, nearest neighbor spacing, the overall ensemble size and shape have a critical effect on extinction spectra. The extinction coefficient that is the sum of the absorption and the scattering cross-sections is a useful parameter for surface plasmon resonance to occur in metals. The field plasmonics, the optical properties of metal structures at the nanoscale has made rapid development due to the ability of engineering metal surfaces and particles at the nanoscale. Advanced techniques like, electron beam lithography, chemical vapor deposition, and deep-UV lithography, focused ion beam milling and self-assembly has provided routes to engineer complex arrays of metal

nanostructures. These chains of metal nanoparticles are exploited to excite, control, guide, direct and manipulate plasmons. The plasmons are attractive because they can effectively confine the optical excitation in a nanoscale volume and thereby mediate strong optical interactions. In addition, the wavelength at which these phenomena are observed can be tuned by varying the metal nanostructure shape, size and dielectric environment. This in turn, provides a broad domain with flexibility from which it is possible to choose the desired optical properties for an application (Kim, 2008; Sonnichsen, 2005; Prodan, 2003).
The coupling of light with electronic surface excitations, specifically, surface plasmon polaritons offers the opportunity to bridge the orders of magnitude difference in sizes between optical and electronic carriers. To develop schemes for coupling and transporting surface plasmons around a chip, the determination of their propagation lengths is particularly important. Researchers have already excited surface plasmons using a focused beam of electrons and then detected the luminescence emitted as the plasmons decayed. Based on these cathode-luminescence intensity decay profiles, they could determine propagation lengths as a function of wavelength. Gold and silver thin films on silicon and quartz substrates respectively were patterned with gratings to direct the emission, allowing the measurement of propagation lengths as short as several hundred nanometers. However, the resolution of the technique is limited by the excitation volume, which in principle, would increase as the film thickness decrease (Sonnichsen, 2005). Using surface plasmon we can obtain ultra-small, wavelength-sensitive directional sensors or detectors. The resonant coupling between the nanoparticles can concentrates light into well-defined hot spots and acts as antennas by suitably engineering the metal nanostructures (Waele et al., 2007). Coupling metal nanoparticle arrays to optical emitter's directional emitters may be achieved. In order to provide the control over the color, directionality and polarization of light-emitting diodes the enhanced optical density of states near the surface of metal nanoparticles can be used. The enhancement of optical density of surface states is highly efficient for the large-scale applications of solid-state lighting, bio imaging, sensing and solar concentrators. Recent calculations and experiments confirms that light scattering from metal nanoparticle arrays can effectively fold the path of sunlight into the layer and thereby strongly enhance its effective absorption (Pillai et al., 2007).
It is known from Maxwell's equations that an interface between a dielectric (e.g. silica glass) and a metal (e.g. silver or gold) can support a surface plasmon. A surface plasmon is a coherent electron oscillation that propagates along the interface together with an electromagnetic wave. These unique interface waves result from the special dispersion characteristics (dependence of dielectric constant on frequency) of metals. What distinguishes surface plasmons from 'regular' photons is that they have a much smaller wavelength at the same frequency. For example, a He-Ne laser, whose free-space emission wavelength is 633 nm, can excite a surface plasmon at a silicon/silver interface with a wavelength of only 70 nm. When the laser frequency is tuned very close to the surface plasmon resonance, surface plasmon wavelengths in the nanometer range can be achieved. The short-wavelength surface plasmons enable the fabrication of nanoscale optical integrated circuits, in which light can be guided, split, filtered, and even amplified using plasmonic integrated circuits that are smaller than the optical wavelength (Kim, 2008; Loo et al., 2005). The reduction in wavelength comes at a price and as a result, surface plasmons are often having loss. One way to achieve long propagation lengths is to use very thin metal films. In this case, surface plasmons on both surfaces of the metal film interact, and both a symmetric and an asymmetric field distribution can exist. One of these modes has low loss

and, for metal films as thin as 10 nm, the centimeter propagation lengths can be achieved for surface plasmons in the infrared. At a given frequency, the surface plasmon wavelength is strongly dependent on the metal thickness. Thus, the plasmonic integrated circuit engineer has an extensive toolbox, including choice of metal (dispersion), metal thickness, and excitation frequency (Loo et al., 2005).

When a light source such as a luminescent quantum dot or dye molecule is placed close to a metal, it can excite a surface plasmon through a near-field interaction. With a light-emitting diode embedded in a plasmonic structure, surface plasmons can be electrically excited. Such surface plasmons may serve as an alternative to overcome the information bottlenecks presented by electrical interconnects in integrated circuits. Coupling to surface plasmons can also enhance the extraction efficiency of light from light emitting diodes. Metallic nanoparticles have distinctly different optical characteristics than surface plasmons at planar interfaces. Nanoparticles show strong optical resonances, again because of their large free-electron density. As a result, a plane wave impinging on a 20 nm diameter silver particle is strongly 'focused' into the particle, leading to a large electric field density in a 10 nm region around the particle. Ordered arrays of nanoparticles can possess even further enhanced field intensities because of plasmon coupling between adjacent particles. By varying nanoparticle shape or geometry, it is possible to tune the frequency of surface plasmon resonance over a broad spectral range. For example, gold ellipsoids or silica colloids covered with a gold shell show resonances that coincide with the important telecommunications wavelength band. The ability to achieve locally intense fields has many possible applications, including increasing the efficiency of light emitting diodes, (bio) sensing, and nanolithography. The light-carrying phenomenon when light falls on a thin film of metal containing millions of nanometer-sized holes shows some surprising results. Interestingly, the film was found to be more transparent than expected, and thus generate many applied research possibilities. The holes were much smaller than the wavelength of visible light, which should have made it almost impossible for the light to get through at all. When the incoming photons struck the metal film, they excited surface plasmons, which picked up the photons' electromagnetic energy and carried it through the holes, re-radiating it on the other side and giving the film its transparency (Ebbesen, 1998).

Arrays of metal nanoparticles can also be used as miniature optical waveguides. In linear chain arrays of nanoparticles, a plasmon wave propagates by the successive interaction of particles along the chain. The propagation length is small (~100 nm), but may be increased by optimizing particle size and anisotropy. The effect of quantum confinement make these nanoparticle array waveguides attractive as they provide confinement of light within ~50 nm along the direction of propagation, a 100-fold concentration compared to dielectric waveguides. A very peculiar effect occurs in metal films with regular arrays of holes, in which, local field enhancements are predicted to occur along the holes. This effect leads to much larger optical transmission through the holes than expected, based on consideration of their geometric areas. The precise role of surface plasmons in these effects is still the subject of lively scientific debate, but applications of the enhanced transmission characteristics in nanoscale optical storage appear promising (Prodan et al., 2003).

Clearly, there is a plenty of plasmonic concepts still waiting for exploration. The clinical studies are ever increasing and encouraged with promising results (Loo et al., 2005). The applications spanning from (bio) sensing, optical storage, solid-state lighting, interconnects and waveguides. Indeed, it appears that metals can shine a bright light toward the future of nanoscale photonics. Most of the early work in plasmonics focused on the study of resonances and electromagnetic field enhancements in individual metal nanoparticles and particle assemblies (Prasad, 2004; Rayleigh, 1888; Pendry, 2000). It is possible to form nanoscale hot spots through plasmon coupling within arrays of metal nanoparticles. In these hot spots, the intensity of light from an incident beam can be concentrated by more than four orders of magnitude that lead to a large improvement in sensing techniques that use optical radiation, such as Raman spectroscopy, with potential applications in medical diagnostics (Polman, 2008). In phenomena that are nonlinear in light intensity the effect of light concentration via plasmons are robust. This has recently been demonstrated by the on-chip generation of extreme-ultraviolet light by pulsed laser high harmonic generation (Kim et al., 2008). This opens up a new avenue in lithography or imaging at the nanoscale with soft x-rays. The methodology of fluorescence energy transfer that is routinely used in biology is limited in length scale (Sonnichsen et al., 2005). Using the highly sensitive plasmonic interaction between metal nanoparticles this can be overcome. Due to the very high sensitiveness to nanoparticles separation, precise measurements of the plasmon resonance wavelength of metal particle assemblies functionalized with bio-molecules can be used as a molecular-scale ruler that operates over a much larger length scale. Practical applications of this concept in systems biology, such as imaging of the motion of molecular motors, bio labeling and bio sensing are being exploited (Polman, 2008). The standard commercial pregnancy tests and the detection of bio-molecules are based on the measurement principle of plasmonic resonance shifts. The possibility of using of particles composed of a dielectric core and a metallic shell in future cancer treatments is underway. The injected shell-core nanoparticles are selectively bound to malicious cells and then laser irradiation at a precisely engineered plasmon resonance wavelength is focused to heat the particles and thereby destroy the cells (Atwater et al., 2009).

One of the main challenges of present plasmonic research is to shrink visible wavelength regime into the soft x-ray wavelength regime. The long distance propagation of surface plasmons along metal waveguides using plasmonic structures based on metal nanoparticles is a new paradigm of research. Using the tools of nanotechnology one can precisely controls material structures and geometry that allows the wave-guiding properties to be controlled in ways that cannot be achieved with regular dielectric waveguides. Particularly, extremely short wavelengths can be achieved at optical frequencies using plasmonic waveguides. A recent experiment demonstrate that light with a free-space wavelength of 651 nm can be squeezed to only 58 nm in a metal-insulator-metal plasmonic waveguide (Miyazaki et al., 2006). The propagation speed of plasmons can be further reduced well below the speed of light by suitably engineering the structures of plasmonic waveguide. Integrating nanoholes in metal films that acts as efficient color filters a more efficient plasmonic waveguide structures have been fabricated. In some complex geometry by tailoring the plasmon waveguides, a negative refractive index for the guided plasmon has been observed. This is very interesting because the two-dimensional negative refraction in these plasmonic waveguides may be useful for plasmonic lens and high resolution imaging (Lezec et al., 2007). The research on planar plasmon propagation is targeted to the design of plasmonic integrated circuits. Using these plasmonic integrated circuits optical information can be generated, manipulated, switched, amplified, guided and detected within dimensions much smaller than the free space wavelength of light. The dream is the integration of optics with nanoscale semiconductor integrated circuit technology. So far, it seems plasmo-eletronic integration is impossible because of the different length scales of optics and electronics. It is hoped that in these devices of nanoscale dimensions a relatively small propagation lengths could be tolerated despite of plasmons decay during their propagation. The plasmo-electronic technology may open a wealth of prospects in designing plasmon laser or amplifier of nanodimension.

As mentioned before, optical 'meta-materials' with artificially engineered permittivity and permeability will fulfill the ever-growing market demand of advanced materials for optoelectronics and nanophotonics circuitry. The fabrication of metallic nanoresonators in 2D and 3D arrangements employing meta-materials is a step forward in this direction. A stack of metallic 'fishnet' structures shows negative index of refraction at near infrared light wavelength (Valentine, 2008). It is possible to achieve sub-wavelength optical imaging due to the peculiar nature of light refraction in the materials of negative refractive indices (Pendry, 2000). Surprisingly, precisely engineered geometries with negative refractive indices may even act as invisibility cloaks for visible wavelengths. It is needless to mention that, the field plasmonics has grown from an embryo with fundamental insights to a vast field with important applications and commercialization. To shape up the plasmonic research, several novel basic and applied research topics are undertaken, including the femto- and atto-second dynamics and coherent control of plasmons, 4D imaging, plasmo-electronic integration, lasing spacers cloaking using novel geometries (Engheta, 2008) and quantum mechanical effects at the sub-nanoscale level. These studies are very exploratory with innovations and enriched with novel scientific thoughts as well. Many new exciting applications of plasmonics are waiting to capture the market. These efforts, in turn, have benefited greatly from the flowering of nanotechnology in general over the past decade, which brought with it a proliferation of techniques for fabricating structures at the nanoscale, exactly what plasmonics needed to progress from laboratory curiosity to practical applications (Brongersma et al., 2007).

The plasmonics made a breakthrough in the field of the solar cell design using semiconductors to enhance the efficiency. In this route, gold nanoparticles on the surface of semiconductors are fabricated that act as reflectors and focus light into the semiconductor and thereby increase the absorption efficiency by concentrating more light (see figure 22). The other route in which, tiny gold nanoantennas could redirect sunlight vertically allows it to propagate along the semiconductor rather than passing straight through the surface. In both the approach, the cell could get by with a much thinner semiconductor layer and acts as a superior concentrator of light. Using plasmonic techniques, not only the cost of the solar cells is decreasing but the efficiency at extracting the available energy from sunlight also drastically improving. An optimistic model calculation and theoretical estimate shows that the use of plasmonics in photovoltaics could increase the absorption two to five times and commercialization of such solar cells look promising. The amorphous silicon based solar cells available in the market have efficiencies of around 10–15 % and the predicted enhancements could translate into efficiencies of about 20 %. Currently available crystalline silicon solar cells have efficiencies around 21 % and the new figure could approach the theoretical maximum of about 30 %. The large scale and low-cost commercial applications are facing the challenges of developing workable device designs, architecture and fabrication techniques for mass production (Atwater et al., 2009).

The beauty of plasmonics is that it can bring the optics closer in size to the transistor, which can offer optical pathways on virtually the same scale as the silicon structures found in advanced microchips. The design of chip with the integration of metals is possible to distribute light over an integrated circuit by surface plasmons. Structures of gold and silver nanowires, nanorods, nanodots (Verhagen, 2009) or grooves are etched into metal surfaces

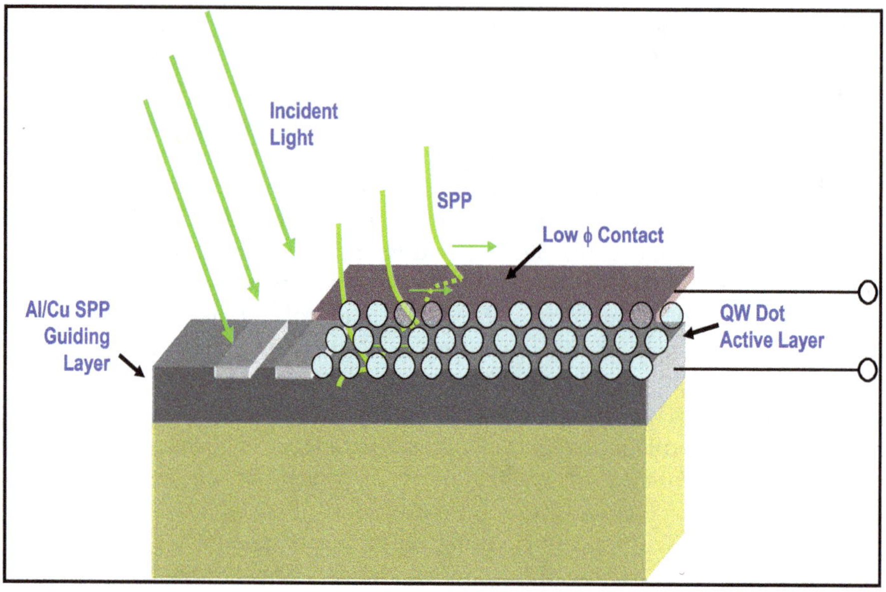

Fig. 22. Light manipulation in plasmonic quantum dot solar cell; the surface plasmons are generated to help direct light using nanoantennas in devices such as solar cells.

(Bozhevolnyi, 2006). They are expected to provide pathways that guide light across a chip irrespective of the designer's directionality. The only trade-off here is with the smallness of the structure, in which forcing the plasmons to travel through too narrow channels can cause leak out from the sides and thereby gets lost. However, guiding surface plasmons over distances of more than 100 μm is possible, which is roughly a thousand times bigger than the features on a current generation microchip. This research has opened new possibilities for plasmonic nanocircuits to carry information using light waves along complex paths and through many processing steps. Recent progress in the laser miniaturization has shown the promise to fabricate plasmonic waveguides. Moreover, plasmonics offers the possibility of integration of plasmo-electro chip at the nanoscale, at lengths much shorter than the wavelength of laser light. Rather than amplifying light in a conventional laser cavity, a plasmonic *'spaser'* would amplify it with the help of plasmons and the first experimental evidence for such plasmon-based lasing has already been reported (Noginov, 2009; Oulton 2009).

The full integration of these plasmon lasers into standard micro-circuitry, however, needs a suitable way to trigger the spasers using standard electrical currents. *'SPASER'* the Surface Plasmon Amplification by Stimulated Emission of Radiation is a new device that has been introduced very recently. In a spaser, a surface plasmon plays the same role as a photon in a laser. A plasmon enters the resonator as a nanoparticle embedded in a gain material containing chromophores such as semiconductor nanocrystals or dye molecules. The gain medium must be capable of producing population inversion, which allows it to lase or *'spase'* in this case. Spacers are ultrafast nanoplasmonic chips with high degree of integration. In addition to creating light and guiding it across, spacers communicate and control each other through their near fields or are connected with nanoplasmonic wires and perform ultrafast microprocessor functions (Noginov, 2009). The plasmonics based optical computing requires a series of bits in a digital data stream that can be obtained by turning the flow of plasmons on and off at high speeds. A plasmonic modulator using silicon technology has been realized and the working principle of this device is based on the use of an electric field to control the propagation of surface plasmons through the device (Dionne, 2009). They are not only much smaller in size compared to conventional optical counterparts but their operation frequency can easily reach tens of terahertz that is much above the gigahertz limit of modern computers.

One of the niche areas in plasmonics is surface-enhanced Raman spectroscopy. One can enhance the signal by several orders of magnitude larger and is strong enough to detect a single molecule (Fleischmann, 1974; Nie, 1997). The surface-enhanced Raman spectroscopy is very useful in the biochemical and materials sciences for providing information on the chemical composition of molecules at very small concentrations and detail microstructures. Surface-enhanced Raman spectroscopy is a plasmonic effect in which silver/gold nanoparticles act as nanontennas to gather the incoming laser light and, through their surface plasmons, concentrate it. In this case, a dual amplification results gigantic signal enhancement by concentrating the light first and then scattered by nearby molecules and amplified again by the silver/gold nanoparticles on the way back out (Atwater et al., 2009). Presently, the surface-enhanced Raman spectroscopy faces some problem for commercialization. This is due to formidable difficulties in achieving highly accurate control over the surface nanostructures and their mass production. Other sensing techniques such as localized surface plasmon resonance may be a suitable alternative in which the surface is covered with nanostructures in the shape of rods or triangles plays important role. The plasmonic properties depend strongly on the properties of the surrounding medium and the changes to the refractive index lead to experimentally measurable changes to the wavelength of surface plasmon resonance (Anker et al., 2008). Surprisingly, the huge sensitivity of localized surface plasmon resonance based devices can reach to the limit of single-molecule detection, and can even focus to destroy cancer cells! For cancer treatments, gold nanoparticles can be injected and guided to the tumor by antibodies bound to the particles' surface. By illuminating the area near nanoparticles with a low dose, using infrared laser light gets absorbed to create plasmons in the gold and burn the infected cells and leaves healthy tissue undamaged. The cancer cells are finally killed by heating up the nanoparticles with accumulated energy through localized surface plasmons (Hirsch et al., 2003). This kind of cancer therapy has successfully been tested on mice for complete elimination of the tumors and waiting for the human clinical trials with patients having head and neck cancers.

Exponential rise in nanophotonics research provided amazing data processing and signal transport capabilities that have the potential to enhance computer performance remarkably. However, to realize this objective much powerful integration techniques for newly emerging nanophotonic devices with conventional nanoelectronics components are urgently required. Undoubtedly, a natural choice for an ideal platform for the marriage of these distinct technologies would be the silicon. Consequently, the lack of an intrinsic source of surface plasmon polaritons compatible with silicon-based complementary metal-oxide semiconductor fabrication techniques slowed down the growth of the integration of plasmonic components with silicon. Presently, complementary metal-oxide semiconductor has reached to true nanoscale devices composed of complex and intertwined dielectric, semiconductor and metallic structures. The impressive developments and availability in computer aided circuit design, lithography, Monte Carlo method, electronic and photonic-device simulations, an increasingly wide variety of integrated optoelectronic functionalities are making the silicon-based technology more robust (Hryciw et al., 2010). Plasmonics is playing major role in the design of future silicon-based optoelectronic and plasmo-electronic chips based on the manipulation of surface plasmon polaritons. The plasmonics research began with passive routing of light in waveguides with diameters much smaller than the wavelength of the light. The surface plasmon polariton waveguides was not perceived as a superior alternative to high-index dielectric waveguides as the propagation length in such high-confinement is limited to a few tens of μm. It is important to keep in mind that the size of dielectric waveguides is limited by the fundamental laws of diffraction, which is much larger than the electronic devices on a chip. However, the sub-wavelength dimensions of plasmonic devices are uniquely capable of reconciling the size mismatch and bridge the gap between dielectric micro-photonics and nanoelectronics. The passive waveguides and light-concentrating structures are the two exciting outcome of the plasmonic studies.

Using surface plasmons, by channeling and concentrating light on sub-wavelength structures miniaturized photonic circuits with waveguides having nanometer length scales have been fabricated. This photonic circuit first converts the incident light to a surface plasmon wave that propagates and eventually converts back to light. These waveguides are realized by depositing gold stripes on a dielectric surface. It is possible to channel the electromagnetic energy using a linear chain of gold and silver nanoparticles over a distance of ~200 nm without any significant loss. In this geometry, each nanoparticle with dimension much smaller than the wavelength of incident light acts as an electric dipole and thereby produces surface plasmon. The inter-particle spacing in the array plays an important role in deciding the interactions. The near-field electric-dipole interactions dominates when the inter-particle separation become much smaller than the wavelength of incident light. This is highly desirable for the wave guiding application of arrays of gold or silver nanoparticles. Active plasmonic devices are designed to switch and detect light in ultra-compact geometries that may exceed the stringent requirements of complementary metal-oxide semiconductor technology (Walters et al., 2010). A crucial ingredient called complementary metal-oxide semiconductor-compatible plasmonic sources can now be added through surface plasmon polaritons. These surface plasmon polariton emitters will play a crucial role in chip-scale optical information links useful for novel integrated bio-sensing applications. A silicon-based source for active Plasmon waveguide using Si nanocrystals as the active medium whose operation principle is similar to other device are created. This device is fabricated using atomic layer deposition and low-pressure chemical vapor deposition processes occurred at around room temperature to be compatible with complementary metal-oxide semiconductor processing (Pavesi, 2003).

The silicon microelectronics world is currently defined by length scales that are many times smaller than the dimensions of typical micro-optical components, the process scaling driven by Moore's law. The size mismatch poses severe challenge to integrate photonics with complementary metal-oxide semiconductor electronics technology. One promising solution is to fabricate optical systems at metal/dielectric interfaces, where surface plasmon polaritons offer totally new and unique opportunities to confine and control light at length scales below 100 nm. Many passive components developed using plasmonics suggests the potential of surface plasmon polaritons for applications in sensing and communication. Active plasmonic devices based on III–V materials and organic materials and an electrical source of surface plasmon polaritons using organic semiconductors have been reported. It is established that a silicon-based electrical source for surface plasmon polaritons can be fabricated using low temperature micro technology processes that are compatible with back-end complementary metal-oxide semiconductor technology (Hryciw et al., 2010). The highly confined modes of metal–dielectric–metal called metal–insulator–metal plasmonic waveguides dramatically alter the light-emission properties of optical emitters located between the metals (Jun et al., 2008). Moreover, there exist an efficient electromagnetic decay pathway for the surface plasmon polariton emission thereby the radiative decay rate of excited emitters can be increased order of magnitude, a direct consequence of the Purcell effect (Hryciw et al., 2010). The small size of the surface plasmon polariton mode that is directly translates to a strong coupling to the surface plasmon polariton emitter is primarily responsible for the large modification of the decay rate in these plasmonic structures. Other high-confinement metal oxide semiconductor and silicon slot waveguides shows similar beneficial effects and lay the foundation of an entire new set of silicon-based sources (Hryciw et al., 2009; Galli et al., 2006; Jun et al., 2009). The enhancements in high-confinement waveguides are very broadband in nature that allows effective use of emitters across the entire visible and near-infrared spectrum to achieve power-efficient incoherent light sources. Even for poor emitters the reduced radiative lifetime is beneficial in increasing the efficiency that allows faster source modulation. The other important benefit of metal-dielectric-metal waveguides is that they only support a single propagating mode and provide low-loss dielectric waveguides (Veronis et al., 2007). The wave guiding based on high-confinement sources is altogether a new class of chip scale devices. They combine efficient charge injection and facile photon extraction by an electrically pumped, plasmon-enhanced light source that inspires new way of designing truly nanoscale photonic devices and circuits for future miniaturization (Brongersma et al., 2007).

Surface plasmon polaritons are quasi-two-dimensional electromagnetic excitations. They propagate along a dielectric-metal interface in which the field components decay exponentially into both neighboring media. The field of a plane surface plasmon polariton comprises a magnetic field component, which is parallel to the interface plane and perpendicular to the propagation direction. It has two electric field components, of which the main one is perpendicular to the interface. The numerical simulations shows that nanometer sized metal rods can support extremely confined surface plasmon polariton modes that propagates over hundreds of nanometers. Similar observations have been made for the electromagnetic excitations supported by chains of metal nano-spheres. Metal stripes of finite width are employed for lateral confinement of the surface plasmon polariton along the stripes. In conventional integrated optics based on dielectric waveguides, the problem of miniaturization is approached. This is achieved by making use of the photonic band gap effect that is essentially a manifestation of Bragg reflection of waves propagating in any direction because of periodic modulation of the refractive index (Maier, 2007; Brongersma et al., 2007).

The efficient wave guiding along straight and sharply bent line defects in 2D photonic band gap structures has been demonstrated for light wavelengths inside the photonic band gap. It became clear that these photonic band gap structures, when properly designed and realized, might be advantageously used for miniature photonic circuits allowing for an unprecedented level of integration. Furthermore, one can conjecture that other (quasi) 2D waves, e.g., surface plasmon polaritons, might be employed for the same purpose. The surface plasmon polariton photonic band gap effects for all directions in the surface plane of a silver/gold film having a 2D periodic surface profile has also been demonstrated. It should be emphasized that the interaction of surface plasmon polariton with a periodic surface corrugation, similarly to the interaction of a waveguide mode with a periodic array of holes, produces inevitably scattered waves propagating *away* from the surface. This unwanted process results in the additional propagation loss and has to be taken into account when considering the surface plasmon polariton photonic band gaps structures. The surface plasmon polariton guiding along line defects in surface plasmon polariton photonic band gap structures with ~45 nm high and 200 nm wide gold bumps arranged in a 400 nm period triangular lattice on the surface of a 45 nm thick gold film is also reported. The efficient surface plasmon polariton reflection by such an area and surface plasmon polariton guiding along channels free from scatterers was observed, as well as significant deterioration of these effects at ~800 nm, indicating the occurrence of the surface plasmon polariton photonic band gaps effect in these structures (Veronis et al., 2007; Kalele et al., 2007).

There are many uses of gold and silver nanoparticles and nanorods from cancer-cell diagnostics, cancer-cell imaging and photo-thermal therapy. In the plasmonics applications of bio imaging or drug delivery, mostly the nanoparticles of gold and silver are used as it offers highly favorable and biocompatible optical and chemical properties. Moreover, metal nanoshells having the same volume as metal nanoparticles show much stronger and sharper surface-plasmon-resonance bands due to its enhanced surface area. Therefore, the nanoshells are preferred for the detection of macromolecules, DNA, proteins and microorganisms. The integration of biology and the materials science at nanoscale has the potential to revolutionize many fields of science and technology. The relevance of nanometer scale stems from the natural dimensions of bio-molecules, such as, DNA, proteins, viruses and sub-cellular structures as they fall in the length scale of 1 to 1000 nm. Gold nanoparticles are mostly exploited for bio imaging and therapeutic applications due to their strong properties of light scattering. In addition, the scattered intensity depends on the size and shapes of nanoparticles and their aggregation states (Hryciw et al., 2009; Kalele et al., 2007).

The non-photo-bleaching character of gold is suitable for detecting very low concentration and can be used as contrast agents in various biomedical imaging techniques. It is demonstrated that the antibody-conjugated gold nanoparticles bind specifically to the surface of malignant cells with much more affinity than healthy cells; also, malignant cells required half the energy to be destroyed photo-thermally than healthy cells. Gold nanocages have been employed in optical coherence tomography by using scattered light for noninvasive imaging to detect cancer at an early and treatable stage. The materials structures fabricated using nanosphere lithographic technique can be used for chemosensing and biosensing by realizing through shifts in the surface Plasmon resonance peak. The mechanism of the shift can be attributed to the changes in the local relative permittivity as well as charge transfer interactions between the adsorbed analyte and the metal. The wavelength shifts are more reliable rather than the intensity changes in biophotonics. Some spectacular observation on the shift of surface-plasmon absorption bands has been made in recent years using nanosphere lithography-deposited silver and gold nanotriangles and nanorods having size ~50 nm (Kalele et al., 2007; Walters et al., 2010).

Despite of many roadblocks remain to the commercialization of plasmonic technologies ranging from the plasmo-electronic integration on a single microchip to device issues there is renewed research interests. The future challenge would be to minimize the losses in the metal nanostructures, and he smart design of the plasmonic structures has been attempted to reduce the losses to acceptable levels. Although, the plasmonics research has already made remarkable progress but the understanding of physics very close to the metal surface still far from being fully understood. Current commercially available optical devices are too large and show rather high losses in the optical signal strength. Plasmonic-based devices will perhaps overcome this problem because a light beam could in principle, relay information through the chip on more channels and at a higher speed than conventional integrated circuitry can handle. Nevertheless, the plasmonics has given to photonics the ability to go to the nanoscale and properly take its place among the nanosciences.

4. Challenges of nanophotonics

Nanophotonics has the potential to improve optoelectronic products in a wide array of new applications. There are no yardsticks or a clear roadmap yet regarding the stability, efficiency, tolerances, longevity and large-scale production. However, the promises of technologies and applications are diverse and multidisciplinary, in early stages of development, and the opportunities are spread throughout the value chain. There are different challenges for nanophotonics (Ghoshal et al., 2007; Chu et al., 2005). Some of them are

- Market strategy

Nanophotonics has the potential to improve optoelectronic products in a wide array of new applications, including multi target markets each worth few billions of dollars. Making market strategy is a greater challenge than the technology. These findings are presented in most of the studies in recent time.

- Single -molecule addressing (pre-requisite for architectures work)
- Optical nanoscopy of molecules
- Designing plasma-optic chip
- Assessment of nanowires, nanoparticles and nanoarrays in nanophotonics
- Assessment of metal nanoparticles, nanoarrays and nanorods
- Hierarchy of interactions with other quasi particles
- Energetically sound
- Amplification and gain
- Integration, costs, standards, etc.

Although, these are the main challenges but there are expected benefits from nanophotonics. Some of them are:

- Bridge the gap between current photonic systems and future approaches bringing in example:
    - access to further integration
    - lower noise
    - mass production techniques and accurate fabrication
    - plasma-electro integration
    - cheaper and efficient devices
- Moving towards molecular photonics as the probable limit of integration which
    - will dissipate less energy
    - will occupy less volume
    - will require lower input signal
    - will probably rely on self assembly
    - will be more lasting
    - will be more flexible
    - will be more sensitive

- Molecules might compute, sense, act and serve building block of more complex structures
    - time scale
    - length scale
    - input-output schemes, algorithms.

Majority of the present fabrication techniques are based on expensive technologies and processes belonging to the very high technology pursuits at an expensive category, such as e-beam lithography, projection lithography, ion-beam milling and extreme ultraviolet lithography. There is great interest and need for approaches as well as strategies for large-scale production, which are more cost-effective and environment friendly. The development of the emerging nanopatterning methods that is recently established may overcome certain barriers related to mass production. The parallel-processes including self-assembly for polymeric materials with matrices can be tested with few milligrams to optimize a novel optical materials. Moreover, getting devices within the next one or two decades would require to produce kilograms of nanophotonic and molecular photonic materials and manufacture them cost efficiently. Thus, preparation techniques are needed that can be scaled up. The criteria on tolerances, efficiency of energy transfer and longevity are other significant issues that manufacturer have to meet when it comes to material purity because this will be an important cost factor, as will lateral size control. Nevertheless, much of these techniques have to be well established with the relevant industrial standards and free of hazards. Until now, it is far from being understood that how efficiently the energy can be converted from a molecular transition into an electronic transition. How efficient a photonic device will be to convert photons to electrons and finally back to photons? The stability of such processes is major concern to the researchers.

5. Nanoscale optical confinement

Size quantization in materials is the manifestation of quantum-confinement in low dimensional quantum structures. Materials at nanometer length scale having large surface to volume ratio possess different electronic and optical density of states and shows many emerging properties those absent in bulk structures. The change in the nature of optical band gaps and the opening up the gap are the manifestations of optical confinement due to which plasmon can be localized and photons can be confines in nanostructures. They have quantum optical properties that are absent in the bulk material due to the confinement of electron-hole pairs (called excitons) in a region of a few nanometres. For example, bulk metals was never thought of as useful candidates for photonics applications, and due to their high reflection and absorption coefficients they have been generally overlooked as elements to guide, focus and switch light at visible and infrared wavelengths. However, at the nanoscale the intriguing guiding and refractive properties of metal structures can be realized since the metal components become semi-transparent due to their small size (Pavesi, 2007; Ghoshal et al., 2007).

Light can be localized and manipulated in appropriately designed metallic and metallo-dielectric nanoparticle array structures. Interesting phenomena occur near the plasmon frequency where optical extinction is resonantly enhanced due to the effect of quantum confinement. Recent interest exploits the collective oscillations of the conduction electrons of this plasma in arrays of metal nanoparticle can also be used as miniature optical waveguides in linear chain arrays of nanoparticles. A plasmon wave propagates by the successive interaction of particles along the chain. The propagation length is small (~100 nm) but may increase by optimizing particle size and anisotropy. A nanoparticle array waveguides is attractive because they provide confinement of light within ~50 nm along the direction of propagation and a 100-fold concentration compared to dielectric waveguide (Koller et al., 2008; Stockman et al., 2007). The confinement of photons in a nanoscale optical cavity offers other possibilities. The spontaneous emission rate depends on cavity properties, increasing with quality factor and decreasing with mode volume. Photonic-crystal cavities can be made with model volumes smaller than the cube of the wavelength (measured as the wavelength λ in air divided by refractive index n of the medium) and can be fabricated with high quality factors. That greatly enhances the spontaneous-emission rate and the fraction of those photons coupled into a single-cavity mode. This allows a single nanocavity to operate as a laser with a very low threshold (Kobayashi et al., 2002).

The surface plasmon resonance associated with array of nanoparticles of noble metals within the size range ~ 10 to 100 nm can be localized to each nanoparticle known as *localized surface plasmon resonance.* On the other hand, the surface plasmon resonance associated with a metallic thin film of thickness in the 10 - 100 nm range can travel across the metal dielectric interface called *propagating surface plasmon resonance.* Although light is not able to propagate through a bulk metal, plasmons are able to propagate at a metal–dielectric interface over distances as large as several centimeters. Furthermore, the plasmon wavelength can be tuned by proper control of the metal film's thickness, size, shape and geometry of nanoparticles. The fabrication of plasmonic waveguide with wavelength shorter than that in free space is in fact the practical realization of this concept. The nanostructures made of gold and silver have been studied much more extensively due to their unique optical and electronic structure properties. In addition to the size and the shape of nanoparticles, surface plasmon resonance also greatly influenced by diverse phenomena such as Rayleigh scattering from nanoparticles, aggregation of nanoparticles, charge transfer interactions, changes in local refractive index, presence of defects, etc. These kinds of interactions are highly useful and are often explored for the detection of bio-molecules using metal nanostructures. The detection process involves monitoring the changes associated with the surface plasmon of the metal nanostructures on addition of the bio-molecules. Surface plasmon resonance can also be tuned over a very wide spectral range using novel nanostructure such as nanoarrays, nanorods and nanoshells (Prasad, 2004; Shen et al., 2000; Kalele et al., 2007).

The optical properties of silicon nanocrystallites of known sizes, present in super-critically dried porous silicon films of porosities as high as 92 %, have been measured by a variety of techniques. The band gap and luminescence energies have been measured as a function of size for the first time. The band gap increases by more than 1 eV due to quantum confinement (Bettoti et al., 2003; Pavesi, 2003). The peak luminescence energy, which also shifts to the blue, is increasingly Stokes shifted with respect to the band gap, as the size decreases. The measured band gap is in agreement with realistic theories and the Stokes-shift between band gap and luminescence energies coincides with the exciton binding energy predicted by these theories. These results demonstrate unambiguously and quantitatively the role of quantum confinement in the optical properties of this indirect gap semiconductor (Behren et al., 1998). Optical confinement effects in nanostructured materials enable new innovative device concepts that can radically enhance the operation of traditional semiconductor devices. A larger fraction of the solar spectrum can be harnessed while maximizing the solar cell operating voltage by using quantum wells, quantum wires and quantum dots embedded in a higher band-gap barrier material.

Nanostructured devices thus provide a means to decouple the usual dependence of short circuit current on open circuit voltage that limits conventional solar cell design. Ultra-high conversion efficiencies are predicted for solar cells that collect both low and high-energy photons from the solar spectrum while maintaining high voltage operation. Unprecedented opportunities are arising for re-engineering existing products. For example, cluster of atoms (nanodots, macromolecules), nanocrystalline structured materials (grain size less than 100 nm), fibres less than 100 nm in diameter (nanorods and nanotubes), films less than 100 nm in thickness provide a good base to develop further new nanocomponents and materials in nanophotonics. The buckyball (C_{60}) has opened up an excellent field of chemistry and material science with many exciting nanophotonic applications because of its ability to accept electrons. Carbon nanotubes (CNTs) have shown a promising potential in the safe, effective and risk free storage of hydrogen gas in fuel cells, increasing the prospects of wide uses of fuel cells and replacement of internal combustion engine. In addition, one can exploit the potential of CNTs in oil and gas industry. Nanotechnology offers a myriad of applications for production of new gas sensors, optical sensors, chemical sensors, and other energy conversion devices to bio implants. Nanoporous oxide films of TiO_2 to enhance photo voltaic cell technology are used. Nanoparticles are perfect to absorb solar energy and that is useful in very thin layers on conventional metals to absorb incident solar energy. New solar cells based on nanoparticles of semi conductors, nanofilms and nanotubes by embedding in a charge transfer medium is an active field of research. Films formed by sintering of nanometric particles of TiO_2 (diameter 10-20 nm) combine high surface area, transparency, excellent stability and good electrical conductivity and are ideal for photovoltaic applications. Nonporous oxide films are highly promising material for photovoltaic applications (Zhiyong et al., 2009).

Nanoscale slab-slot-waveguides that provides high optical confinement have found abundant applications in silicon photonics. After developing an analytical mode solver for general asymmetric slot waveguides, the confinement performance of symmetric as well as asymmetric geometries was systematically analyzed and compared. For symmetric structures, 2D confinement optimization by varying both low-index slot and high-index slab width revealed a detailed saturation trend of the confinement factor with the increase of the studied width. Furthermore, simple design rules on how to choose the slot and slab width for achieving optimal confinement was obtained. For asymmetric structures, we demonstrated that the confinement performance was always lower than the 2D optimized confinement of the symmetric structures providing the two high-index slab layers and the two cladding layers have same refractive indices, respectively. In addition, the sensitivity of the confinement to the degree of asymmetry was studied, and we found that the fabrication tolerance on the material and structural parameters might be reasonably large for symmetric structures designed at optimal confinement (Ma et al., 2009). A change in shape of nanoparticles can produces much larger shifts in the surface-plasmon resonance band due to spatial confinement effect. The surface-plasmon resonance band often splits in to two bands due to the elongation of nanoparticle along some particular axis of aisotropy. The shorter-wavelength band termed as the transverse plasmon resonance band corresponds to the oscillations of electrons along any minor axis, while the longer-wavelength band termed as the transverse plasmon resonance band corresponds to the oscillations of the electrons along the major axis. In some semiconductor nanoparticles different kinds of excitations such as bi-exciton, dark and bright exciton, polaron, bi-polaron and polariton etc. emerges due to the effect of quantum confinement in contrast to their bulk counterpart.

6. Quantum-confined materials

There is much current interest in the nonlinear optical properties of quantum confined semiconductor nanocrystals, particularly for their potential use in nonlinear photonic devices. Glass is an attractive host matrix for the nanocrystals in such applications (Dvorak et al., 1995). Quantum-confined materials refer to structures, which are constrained to nanoscale lengths in one, two or all three dimensions that possess density of states entirely different from bulk structure as shown in figure 23. The length along which there is quantum confinement must be smaller than de Broglie wavelength of electrons for thermal energies in the medium. The thermal energy is given by $E = mv^2/2 = k_BT$ and de Broglie wavelength is $\lambda = h/mv$. For effective quantum-confinement, one or more dimensions must be less than 10 nm. Quantum-confined structures show strong effect on their optical properties. Artificially created structures with quantum-confinement on one, two or three dimensions are called, quantum wells, quantum wires and quantum dots respectively (Botez et al., 2005). Low-dimensional semiconductors and other materials can modify the density of states by restricting their size to the order of an electron wave function and thereby show many fascinating effects. The quantized energy levels in quantum structures became the driving force for next-generation solar cells and other nanophotonic chip design due to the mitigation of the thermal scattering as well as the control of the light absorption range.

The quantum efficiency of the photoluminescence and electroluminescence from a bulk noble metal is very low ($\sim 10^{-10}$) compared to the corresponding nanosystems, which is several orders of magnitude higher ($\sim 10^{-4}$). However, in the case of silver and gold nanorods or nanoparticles, the luminescence efficiency is enhanced by six to seven orders of magnitude, and in addition, a red (blue) shift for the luminescence maximum is observed with increasing (decreasing) aspect ratio of the nanorods or diameter of nanoparticles. A linear increase in the photoluminescence efficiency with square of aspect ratio (diameter) is remarkable. The enhancement of the luminescence efficiency is also observed for rough surfaces of gold and silver. The rough surface can be modeled as an ensemble of randomly oriented hemispheroids of nanometer size, and the surface plasmon resonance arises from them amplifies the local fields, thereby resulting in the enhancement of photoluminescence often called lightening rod effect. All these are primarily attributed to quantum confinement effects.

Nanoparticulate metals and semiconductors that have atomic arrangements at the interface of molecular clusters and 'infinite' solid-state arrays of atoms have distinctive properties determined by the extent of confinement of highly delocalized valence electrons. At this interface, the total number of atoms and the geometrical disposition of each atom can be used significantly to modify the electronic and photonic response of the medium. In addition to the novel inherent physical properties of the quantum-confined moieties, their 'packaging' into nanocomposite bulk materials can be used to define the confinement surface states and environment, inter-cluster interactions, the quantum-confinement geometry, and the effective charge-carrier density of the bulk. Current approaches for generating nanostructures of conducting materials, especially the use of three-dimensional crystalline super-lattices as hosts for quantum-confined semiconductor atom arrays (such as quantum wires and dots) with controlled inter-quantum-structure tunneling (Sucky et al., 1990) are getting momentum. Quantum-confined nanocrystallites of GaAs are fabricated in porous glass and the bound electronic nonlinear refractive index, the two-photon absorption coefficient, and the refraction from carriers generated by two-photon absorption are simultaneously determined using the Z-scan method and compared to those of bulk GaAs. The quantum-confined Stark effect in single cadmium selenide (CdSe) nanocrystallite quantum dots has also been widely studied due to their potential application in displays and renewable energy. The electric field dependence of the single-dot spectrum is characterized by a highly polarizable excited state (~10^5 cubic angstroms, compared to typical molecular values of order 10 to 100 cubic angstroms), in the presence of randomly oriented local electric fields that change over time. These local fields result in spontaneous spectral diffusion and contribute to ensemble inhomogeneous broadening. Stark shifts of the lowest excited state more than two orders of magnitude larger than the line width were observed, suggesting the potential use of these dots in electro-optic modulation devices (**Empedocle et al., 1997**).

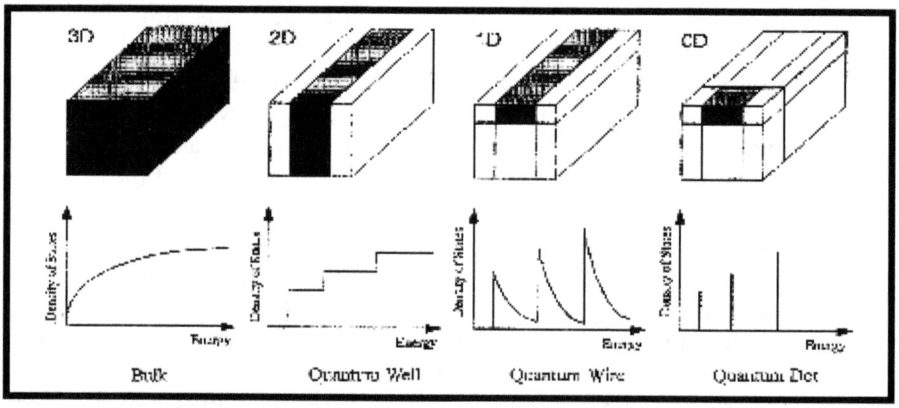

Fig. 23. The change in the density of states with confinement dimensions. Bulk structure suffers no confinement but 0D structure has confinement in all three spatial dimensions and hence density of states is discrete delta functions.

The carrier contribution to elastic constants in quantum confined heavily doped non-linear optical compounds has been studied. The basis of this newly formulated electron dispersion law take into account the anisotropies of the effective electron masses and spin orbit splitting constants together with the proper inclusion of the crystal field splitting in the Hamiltonian within the framework of k.p formalism. It has been found, taking different heavily doped quantum confined materials that, the carrier contribution to the elastic constants increases with increase in electron statistics and decrease in film thickness in ladder like manners for all types of quantum confinements with different numerical values, which are entirely dependent on the energy band constants. This contribution is greatest in quantum dots and least in quantum wells together with the fact the heavy doping enhances the said contributions for all types of quantum-confined materials (Baruah et al., 2007).

To understand the size quantization on the luminescence (see figure 24) from the metal nanoparticles a simplified model for the band structure is developed. This model includes the s-p conduction band and two sets of occupied d bands. The luminescence from metal nanoparticles can be attributed to the transition of electrons from a completely filled d band (highest occupied molecular orbital) to the unoccupied levels in the sp band (lowest unoccupied molecular orbital). The excitation involves a transition from states in the upper d band to the unoccupied levels in s-p band at and above the Fermi level. The emission arises from the direct recombination of conduction electrons with holes in the d bands of the quantum confined structures. The emission bands either appear due to radiative inter-band recombination of electrons and holes in the s-p and d bands, or originate from radiative intra-band transitions within the s-p band across the band gap. This model has been employed for describing luminescence from silver, gold and other nanoparticles (Link et al., 2002).

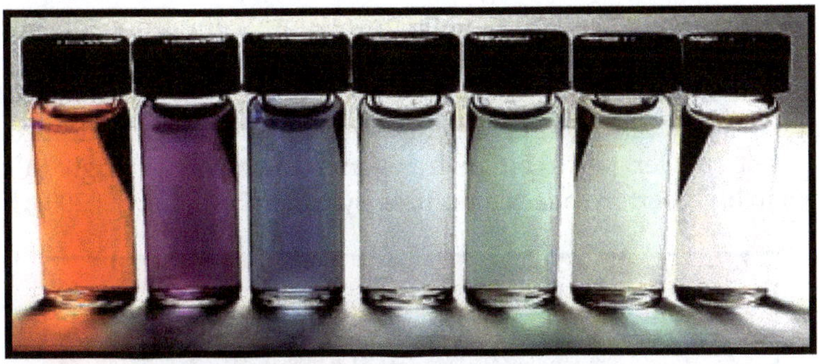

Fig. 24. The effect of size quantization: light emission from gold colloids with varying size of nanoparticles , smallest particles emit violet and the largest emit red color.

7. Applications

Nanophotonics is a unique field because it combines scientific challenges with large variety of near- term applications (Koller et al., 2008; Stockman et al., 2007). Fundamental research on nanophotonics leads to applications in communications technology, lasers, solid-state lighting, data storage, lithography, biosensors, optical computers, imaging, solar cells, light-activated medical therapies, displays, and smart materials to cite a few (**Empedocles et al., 1997;** Baruah et al., 2007; Rahmani et al., 1997; Nezhad et al., 2007; Shen et al., 2000; Levy et al., 2004). Some of the important areas that will find large market in next ten to fifteen years from now are identified and represented in figure 25.

Two-dimensional photonic crystal lasers have been fabricated on III-V semiconductor slabs. Tuning of the spontaneous emission in micro and nanocavities has been achieved by accurate control of the slab thickness. Different structures, some of them of new application to photonic crystal lasers, have been fabricated like the Suzuki-phase or the coupled-cavity ring-like resonators. Laser emission has been obtained by pulsed optical pumping. Optical characterization of the lasing modes has been performed showing one or more laser peaks centred around 1.55 μm. Far field characterization of the emission pattern has been realized showing different patterns depending on the geometrical shape of the structures. These kinds of devices may be used as efficient nanolaser sources for optical communications or optical sensors (Postigo et al., 2007).

Nanoscale confinement of matter is achieved on nanoscale crystals and they are used for: (i) optical up-conversion of radiation in rare-earth nanocrystals, (ii) size-dependent emission properties of nanoscale semiconductor crystals (quantum dots), and (iii) size-dependent absorption and confinement is exploited in plasmonic solar cells and biosensing. Metal nanoparticles and nanotips used in surface enhanced Raman spectroscopy. Photonic band gap crystals and photonic band gap fibers (photonic fibers) involve periodic variation of dielectric constant over wavelength-scale. They find applications in fabrication of Micro-Opto-Electro Mechanical Systems, Micro-Optics, i.e. Micro-lasers, Directional Couplers between waveguides, Bio-photonic Chips etc. Optical wires (wave-guides), filters, and transistors are now possible, and all of them could harness the speed of light for optical communications, sensing, data processing and storage.

Silicon photonics offers high-density integration of individual optical components on a single chip. Strong light confinement enables dramatic scaling of the device area and allows unprecedented control over optical signals. Silicon nanophotonic devices have immense capacity for low-loss, high-bandwidth data processing. Fabrication of silicon photonics system in the complementary metal-oxide-semiconductor-compatible silicon-on-insulator platform also results in further integration of optical and electrical circuitry. Following the Moore's scaling laws in electronics, dense chip-scale integration of optical components can bring the price and power per a bit of transferred data low enough to enable optical communications in high performance computing systems (Bettoti et al., 2003; Pavesi, 2003).

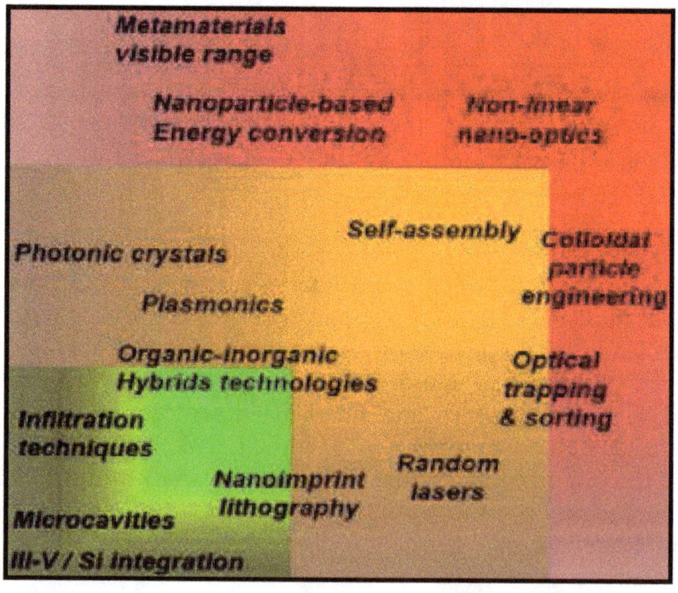

Fig. 25. Schematic diagram for the required technology and materials development in next fifteen years from now in every five years periods as shaded by different colors.

Photonic crystals and photonic band gap materials have a myriad of uses. These uses range from computing to telecommunication and from lasers to circuits that can also be found in almost anything digital. Light-emitting diodes play a key role in optical-communication systems. These devices are made from so-called photo emissive materials that emit photons once they have been excited electrically or optically. These photons are typically emitted in many different directions and have a range of wavelengths, which is not ideal for communications applications. A type of light emitting diode can be created that only emits light in the forward direction by placing a reflector behind the photo emissive layer. However, the efficiency of such a device is limited by the efficiency of the reflector.

Some of important applications of the photonic crystals include, perfect dielectric mirror in which the reflectivity of photonic crystals derives from their geometry and periodicity, not a complicated atomic-scale property (unlike metallic components mirror). The only demand on such materials is that for the frequency range of interest, these should be essentially lossless. Such materials are widely available all the way from the ultraviolet regime to the microwave. The nonlinear effects of the materials exploits the non-linear properties for construction of photonic crystal lattices open new possibilities for molding the flow of light. In this case, the dielectric constant is additionally depending on intensity of incident electromagnetic radiation and any non-linear optics phenomena can occur. In wave-guides and junctions, the existence of guided modes with different parities is an important factor that deeply influences the transmission through the Y junction. The cavity now produces a skewed field that can be understood as a superposition of modes with odd and even parities.

Photonic crystals could be used to design a mirror that reflects a selected wavelength of light from any angle with high efficiency. Moreover, these could be integrated within the photo emissive layer to create a light emitting diode that emits light at a specific wavelength and direction. Ideally, one needs to build a truly 3-D lattice structure to gain complete control of the light in all three dimensions. Fortunately, 2-D periodic lattices exhibit some of the useful properties of a truly 3-D photonic crystal, and are far simpler to make. These structures can block certain wavelengths of light at any angle in the plane of the device, and can even prevent light entering from certain angles in the third dimension (i.e. perpendicular to the surface). Thus, 2-D photonic crystals are a good compromise for many applications and are easily incorporated within planar waveguides.

Improving the existing concepts of light sources is a hot topic in nanophotonics, particularly, flat-panel displays through photo-detectors. However, the ambitions of nanophotonics and molecular photonics go well beyond this simple and straightforward demand. In the field of displays one needs to produce images using liquid crystals or with fluorescent molecules in a polymer matrix. In the near future, the target would be to obtain the internet screen being displayed in the living rooms on a larger screen, with electronic wallpaper acting as the display that is, realizing a component of ambient intelligence. Therefore, nanophotonics can bridge the gap between existing optoelectronic devices and future optics circuitry by incorporating optically active nanostructures and molecules.

Predictions are always difficult to make, however, the future for photonic-crystal circuits and devices looks certain. Within few years, a number of basic applications will start making an appearance in the market place. Among these will be highly efficient photonic-crystal lasers and extremely bright light emitting diodes. It will be possible to implement optical manipulation, signal processing and electronic circuitry all in one chip. Photonic crystals may also be used for high-speed computers of the next generation. Progress has already been made towards replacing the slow copper connections in computers with ultra-fast optical interconnects. Then photons, rather than electrons, will pass signals from board to board, from chip to chip and even from one part of a chip to another. Imagine a computer that operates at the speed of light!

Metal nanostructures have received considerable attention for their ability to guide and manipulate electromagnetic energy in the form of surface plasmon polaritons. The unique properties of surface plasmon polaritons may enable an entirely new generation of chip-scale technologies, known as plasmonics. Such plasmonic devices could add functionality to the already well-established electronic and photonic device technologies. The patterned metal structures are used as tiny wwaveguides to transport electromagnetic energy between nanoscale components at optical frequencies. Interestingly, the surface plasmons and surface plasmon polaritons, which are collective oscillations of electrons in small particles (nanosystems) or on metal dielectric interfaces induced by an incident electromagnetic radiation, is a topic of tremendous interest with ever-growing surprises and promises. In recent years, the potential of plasmonics has been realized in making highly sensitive and very small photonic devices (nanolasers as an example). A wide range of applications emerging from controlling and manipulating the light at the nanoscale such as high-resolution optical imaging below the diffraction limit, enhanced optical transmission through sub-wavelength apertures, bio-detection at femto-mole level, nanoscale wave guiding, THz communications, optical computing and surface-enhanced Raman scattering have been established. Many futuristic applications regarding plasmonic materials rendering invisibility are also inspiring researchers. Surface plasmon polaritons can be tightly bound to the metal surface, penetrating ~100 nm into the dielectric and ~10 nm into the metal. This feature implies the possibility of using surface plasmon polaritons for miniature photonic circuits and optical interconnects and has attracted a great deal of attention to surface plasmon polaritons.

Plasmonic nanoparticles have tremendous technological importance and found applications in therapy and imaging. Beyond biosensing, plasmonic nanoparticles have further applications in biology and medicine. For example, they simultaneously give contrast in electron and optical microscopy such that a labeled cell can be imaged in optical and subsequently in electron microscopy. A modified version uses fluorophore-gold nanoparticle conjugates for simultaneous contrast in electron and fluorescence microscopy. Gold nanoparticles are also promising in hypothermal cancer therapy. In these experiments, nanoparticles that rae attached to cancer cells are heated by the absorption of light until the cancer cell overheats and dies (Shalaev et al., 2007). Plasmonics have wide applications in the energy sector as of improving lighting technologies by improving the efficiency of light emitting diodes up to 14 times. In medicine, for selectively destroying cancer cells by radiation absorption it is very useful. It is even possible in theory to make an object invisible by using plasmonics with metaparticles who have extraordinary optical properties for electronic waves. In computing technology for processing information with speed of light, it is recommended. If it can be combined with graphene, a newly emerged transparent conducting material, innovations beyond our wildest dreams may be achieved. Applications of surface plasmons in solid-state lighting and lasing are just appearing, but it may be that traffic lights are composed of surface plasmon light emitting diodes in a few years time!

Recent advancements in laser science using nanophotonics have tackled physical limitations to achieve higher power, faster and smaller light sources. The quest for ultra-compact laser that can directly generate coherent optical fields at the nanoscale, far beyond the diffraction limit of light, remains a key fundamental challenge. Microscopic lasers based on photonic crystals, carbon nanomaterials, micro-disks, semiconductor quantum structures, metal clad cavities and nanowires can now reach the diffraction limit, which restricts both the optical mode size and physical device dimension to be larger than half a wavelength (Bonaccorso et al., 2010). While surface plasmons are capable of tightly localizing light, ohmic loss at optical frequencies has inhibited the realization of truly nanoscale lasers. Progress in theory and modeling has paved the way to reduce significantly the plasmonic loss while maintaining ultra-small modes by using a hybrid plasmonic waveguide. Experimental demonstration of nanoscale plasmonic lasers producing optical modes 100 times smaller than the diffraction limit, utilizing a high gain CdS semiconductor nanowire atop a silver surface separated by a thick insulating gap has been reported. Direct measurements of emission lifetime reveal a broadband enhancement of the nanowire's spontaneous emission rate by up to 6 times due to the strong mode confinement and the signature of apparently *threshold-less lasing*. Since plasmonic modes have no cut-off the downscaling of the lateral dimensions of both device and optical mode are possible. As these optical coherent sources approaches molecular and electronics length scales, *plasmonic lasers* offer the possibility to explore extreme interactions between light and matter, opening new avenues in active photonic circuits, bio sensing and quantum information technology. Let us turn our attention to the wonder material graphene and see its future promises for plasmonic-based devices for photonics and optoelectronics (Zouhdi et al., 2009; Jablan et al., 2009).

Amongst many, the novel material *'graphene'* has become something of a celebrity material due to its unusual conductive, thermal, electronic, plasmonic and optical properties, which could make it useful in a range of sensors, lasers and semiconductor devices. In addition to flexibility, robustness and environmental stability it possesses high mobility and optical transparency. Most of the recent studies mainly focused on fundamental physics and electronic devices. Undoubtedly, its true potential lies in nanophotonics and optoelectronics, where the combination of its unique optical and electronic properties can be fully exploited, even in the absence of a band gap, and the linear dispersion of the Dirac electrons enables ultra-wideband tunability. Graphene being a potential candidate of advanced plasmonic materials is a rapidly rising star on the horizon of materials science and condensed-matter physics (Bonaccorso et al., 2010; Murray et al., 2007). Graphene's ultra-wide broadband capability will be useful for future nonlinear optical devices including high speed, transparent and flexible photosensitive systems, which could be further functionalized to enable chemical sensing. With an ever-growing interest in the widespread applications of graphene, ultrafast and tunable lasers have become a reality. The combination of graphene photonics with plasmonics could lead to a wide range of advanced devices. The charge carriers in graphene are negative electrons and positive holes, which in turn are affected by plasmons. The *'plasmaron'* is a composite particle; a charge carrier coupled with a plasmon has recently been observed. The rise of graphene in photonics and optoelectronics is shown by several recent results, ranging from solar cells and light-emitting devices to touch screens, photo-detectors and *ultrafast lasers*. Electroluminescence has also been reported recently in pristine graphene. This observation could lead to new light emitting devices based entirely on graphene. In this communication, we will briefly discuss the state-of-the-art in this emerging field (Maier, 2007; Brongersma et al., 2007).

Optical properties of plasmons in graphene are in many relevant aspects similar to optical properties of surface plasmons propagating on dielectric-metal interface, which have been drawing a lot of interest lately because of their importance for nanophotonics. It is reported that plasmons in doped graphene simultaneously enable low-losses and significant wave localization for frequencies below that of the optical phonon branch. Large plasmon losses occur in the inter-band regime via excitation of electron-hole pairs that can be pushed towards higher frequencies for higher doping values. For sufficiently large dopings, there is a bandwidth of frequencies where a plasmon decay channel via emission of an optical phonon together with an electron-hole pair is non-negligible. There are certain frequencies in which plasmons have the low losses and that makes graphene potentially interesting for nanophotonic applications (Zouhdi et al., 2009).

Nanohole arrays have emerged from an interesting optical phenomenon to the development of applications in photo-physical studies, photovoltaics, in imaging chips for digital cameras, and as a sensing template for chemical and biological analyses. Numerous methodologies have been designed to manufacture nanohole arrays, including the use of focus ion-beam milling, soft-imprint lithography, colloidal lithography and modified nanosphere lithography. Hole-arrays in plasmonics are increasingly finding their way into applications as selective filters for color sensors and channeling light into optical devices. Hole-arrays can be made into highly selective filters for sensors that depend on detecting specific colors, or for efficiently extracting monochromatic light from light emitting diodes and lasers. For commercialization NEC laboratory is focusing on prototypes of plasmon-enhanced devices for displays and telecommunications. To reduce pixel noise and improve camera sensitivity hole arrays are placed on top of individual pixels that helps to capture incoming light more efficiently. Another plasmonic technique for channeling light into a device is to sprinkle its surface with nanoscale particles made of a metal such as gold. These nanoparticles function like an array of tiny antennas: incoming light is taken up by plasmons and then redirected into the device's interior.

With nanosphere lithography or colloidal lithography, the experimental conditions control the density of the nanosphere mask and, thus, the aspect of the nanohole arrays. Low surface coverage of the nanosphere mask produces disordered nanoholes. Ordered nanohole arrays are obtained with a densely packed nanosphere mask in combination with electrochemical deposition of the metal, glancing angle deposition or etching of the nanospheres prior to metal deposition. The optical properties of nanoholes have interesting applications in analytical chemistry. In particular, applications of these novel plasmonic materials are demonstrated as substrates for a localized surface plasmon resonance, surface plasmon resonance, surface enhanced Raman spectroscopy and in electrochemistry with nano-patterned electrodes (Polman, 2009). Perhaps, the most promising application of nanohole-arrays called nanoantennas is in the improvement of solar cell efficiency because the currently used solar cells made from silicon has reached to saturation and is expensive. However, to harvest the wide range of wavelengths from the solar spectrum, particularly in the red and infrared part of the spectrum, the semiconductor layer has to be relatively thick and that makes it bulky up to 350 μm thick. Other solar-cell materials except few recently developed nanosystems and organic materials have the same problem and the efficiency is low. Nanoantennas based on surface-enhanced Raman spectroscopy solar cell can overcome this difficulty. Some applications of surface-enhanced Raman spectroscopy have reached the market. For example, in specifically prepared colloids of gold nanoparticles, a clustering of these nanoparticles is triggered by the presence of pregnancy hormones. This leads to a color change induced by plasmonic effects that has been widely commercialized in pregnancy tests.

The other important application of surface plasmons in nanophotonics is the enhanced optical transmission through sub-wavelength- apertures. For light transmission through an aperture having lateral dimensions smaller than $\lambda/2$, the efficiency is extremely low as a result light cannot propagate freely because transmission is possible only via the highly inefficient photon tunneling mechanism. Surprisingly, a substantial enhancement of transmitted light at resonant wavelengths could be achieved for nanoscale apertures (diameter ~200 nm and period ~500 nm) in the form of arrays of holes or slits fabricated on a metal film. This enhancement of the transmission intensity which is much larger than that predicted by classical diffraction theory is primarily due to the coupling of the incident light (photons) with surface plasmons. This finding on enhanced transmission through a large number of nanoholes suggests that even the photons impinging between the nanoholes can be transmitted. Furthermore, the evanescent waves that originate from the diffraction of the incident radiation by the nanoholes diffract while tunneling through the nanoholes and produce interference with the incident waves. Eventually, the surface plasmons enhance the field associated with the evanescent waves, thereby resulting in the enhancement of the transmitted light intensity.

Nanophotonics have potential to make significant impact on military and warfare systems, in both symmetric and asymmetric warfare. The technology parameters identified are real and can be translated into military applications. It is very likely that nanotechnology insertion into disruptive technologies that could pose threats is difficult to anticipate by the very nature of such threats and the sometimes-embryonic state of research in nanophotonics relative to other technologies. Some applications may take substantial investments and time before they are realized. Nanophotonics definitely requires a fabric of supporting and enabling technologies that must be smart, fast and environment friendly. This technology set-up at least in part does not exist today, which further complicates our further outlook. Provoked by the computer games and software development strong emphasis has been placed on information technology as a possible 'game changer,' with nanophotonics. Certainly, it could potentially enable a much more ubiquitous and pervasive data processing and sensing capability than is currently available.

8. Conclusions & future outlook

This new frontier called nanophotonics offers numerous opportunities for both fundamental research and application. The information technology based on the nanotechnology is varied from the targets that can be realized in the near future, to those with long-term goals such as quantum info-communication and quantum computing technology. In the quantum dot area, quantum dot laser, quantum dot optical amplifier, quantum dot nonlinearity devices, and quantum dot light detectors are about to emerge into the real world. The expectations are high for optical circuit of the optical router and like which will be the basic component of the router in the future photonic network and quantum dot amplifier for application to optical 3R relay generator will play a key role. The expectation is that the quantum dot be used for this single photon generator. The expectation is also high for the realization of nanophotonics based on the new principle. The nanophotonic element is an ultimate element that uses strongly connected condition of the light and electron that include the polariton laser. Investment in nanophotonics should lead to realization of low energy consumption society and promotion of heals of the people. There will come an age when nanophotonic elements will play the key roles including the fusion of quantum dot and photonic crystal. The quantum dot will be the basic component of future quantum computing elements in the future and expectations are very large for emergence of new 'non-continuous' technology based on the quantum dot technology.

Since photonic crystals offer prospects for numerous applications: low-threshold lasing, optical-power limiting, chemical and bio sensing, and optical switching. It will play a major role in further development of this exciting field. Photonic components such as fiber optic cables can carry lot of data but are bulky compared to electronic circuits. Electronic components such as wires and transistors carry less data but can be incredibly small a single technology that has the capacity of photonics and the smallness of electronics would be the best bridge of all. *So 'Light on a Wire,' Is Circuitry Wave of Future.* Nanophotonics, according to world leaders, has the potential to transform telecommunications technologies in a way even more radical than the emergence of electron-based integrated circuits. People are talking about components such as single photon sources, single photon detectors in the quest for low-noise signals. Components like waveguides, which carry light from one part of the optical circuit to another, and switches are essential. There are good prospects of switching functionality, because optical properties of some molecules are highly non-linear, which is an essential condition for fast switching devices. Therefore, there are very promising device-relevant properties here, suitable for future optical circuits and systems.

The main limitation to plasmonics today is that plasmons tend to dissipate after only a few millimeters, making them too short-lived to serve as a basis for computer chips, which are a few centimeters across it. For sending data even longer distances, the technology would need even more improvement. The key is using a material with a low refractive index, ideally negative, such that the incoming electromagnetic energy is reflected parallel to the surface of the material and transmitted along its length as far as possible. There exists no natural material with a negative refractive index, so nanostructured materials must be used to fabricate effective plasmonic devices. For this reason, plasmonics is frequently associated with nanotechnology. Before all-plasmonic chips are developed, plasmonics will probably be integrated with conventional silicon devices. Plasmonic wires will act as high-bandwidth freeways across the busiest areas of the chip. Plasmonics has also been used in biosensors. When a particular protein or DNA molecule rests on the surface of a plasmon-carrying metallic material, it leaves its characteristic signature in the angle at which it reflects the energy. Therefore, it is necessary to develop smart integration techniques to fabricate plasmo-electronic chips.

Future commercial opportunities of nanophotonics lay ahead in each of the following fields: solar energy, biophotonics, solid-state lighting and displays, computing, telecommunications, sensors and security, etc, to cite a few. This field is expected to revolutionize many aspects of modern life. The commercial potential is frankly overwhelming and mind boggling. Nanophotonics promises solar cells that provide cheap, clean power wherever it is needed. Solid-state lighting systems those are much more efficient and versatile than current light bulbs. Flexible electronic displays that allow cell phones will be the size of credit cards. The computer screens as thin and flexible as a sheet of heavy paper is possible. Computers that connect to the internet at the speed of light are targeted. The new medical devices that give doctors the power to detect and treat diseases in novel ways are up coming. This is just the beginning! The movement from nanophotonics to molecular photonics provides a link to developments in molecular electronics. Ultimately, there is a need to bring together the advantages of molecular photonics with those of molecular electronics, based on the common strategies of preparation, handling and integration. It may well be that molecular electronics is the electronics of the future and molecular photonics, the photonics of the future.

In conclusion, this chapter on 'nanophotonics for 21st century' focuses on the paramount importance and tremendous technological potential of this field for nanoscale device miniaturization at low cost and with maximum efficiency. Nanophotonics is newly developed very vast and expanding paradigm of nanoscience, and is a unique part of physics/chemistry/materials science because it combines a wealth of scientific challenges with a large variety of near-term applications. Nanophotonics is the use of materials nanotechnology in photonics or the use of photonic materials in nanotechnology. The exciting application possibilities of nanophotonics related upcoming areas in science and technology, its present status and future scope are explored in this presentation. Nanophotonics could well revolutionize the fields of telecommunications, computing, imaging and sensing in particular. This presentation argues, why is the research into nanophotonics important? Nanophotonics is all about the manipulation and emission of light in both the far field and near-field using materials at nanoscale. It is the study of the behavior of light on the nanometer scale and their interactions in different material media. The ability to fabricate devices in the nanoscale that has been developed in recent years called nanotechnology is the main thrust. Nanotechnology is defined as a novel and multidisciplinary use of materials or processes at the nanometer scale (below 100 nm). However, we attempt to give a panoramic overview of nanophotnics, the science of light-matter interaction at nanometer scale (< 1 μm to ≥ 1 nm), which deals with optical processes at the much smaller length scale than the wavelength of optical radiation. The nanoscale matter-radiation interaction includes nanoscale confinement of radiation, nanoscale confinement of matter, and nanoscale photo-physical or photochemical transformation, offer numerous opportunities for both fundamental research and technological ramifications. Quantum-confined materials are structures that can constrain to nanoscale lengths in one, two or all three dimensions are employed for implementations of devices in nanophotonics. Confinement of light results in field variations similar to the confinement of electron in a potential well. For light, the analogue of a potential well is a region of higher refractive index bounded by a region of lower refractive index.

It is important to note that the overview on nanophotonics provided here cannot do justice to all the different research directions in the field of photonic crystals and plasmonics. Without attempting to be exhaustive, we have tried to present a short but somewhat complete perspective including the fundamentals and some of the current burning research topics. Together, these represent the true state-of-the art in the field of nanophotonics. Several applications are highlighted. In addition, the importance in understanding the electromagnetic wave propagation in photonic crystals/photonic band gap materials and meta-materials are briefly discussed. The chapter concludes with the future of nanophotonics that seems bright and close to reality. The ever-ending research activities aimed in this field are expected to revolutionize many aspects of modern life and will remain fertile ground for basic and applied research.

9. Acknowledgements

S. K. Ghoshal is indebted to Prof. K. P. Jain (IIT, Delhi) and Prof. Sir Roger Elliott (Oxford University) for introducing this field and educating him with many critical concepts. S. K. G gratefully acknowledges the financial support from Universiti Teknologi Malaysia through the research grant (VOTE 4D005/RMC and Q.J130000.7126.00J39/GUP).

10. References

Alu, A. & Engheta, N. (2008). Multifrequency Optical Invisibility Cloak with Layered Plasmonic Shells. *Phys. Rev. Lett.*, 100, 113901(1-4), ISSN 0031-9007.

Anker, J. N.; Hall, W. P.; Lyandres, O.; Shah, N. C.; Zhao, J. & Van Duyne, R. P. (2008). Biosensing with plasmonic nanosensors. *Nature Mater.*, 7(6), 442–453, ISSN 1476-1122.

Atwater, H. & Polman, A. (2009). Surfing the Waves, a Report by Joerg Heber. *Nature*, 461, 720-722, ISSN 0028-0836.

Baruah, D.; Choudhury, S.; Singh, K. M. & Ghatak, K. P. (2007). Influence of quantum confinement on the carrier contribution to the elastic constants in quantum confined heavily doped non-linear optical and optoelectronic materials: simplified theory and the suggestion for experimental determination, *Journal of Physics: Conference Series*, 61, 80-84, ISSN 0953-8984.

Behren, J. V.; Buuren, T. V.; Zacharias, M.; Chimowitz, E. H. & Fauchet, P. M. (1998). Quantum confinement in nanoscale silicon: The correlation of size with bandgap and luminescence. *Solid State Communication*, 105(5), 317-322, ISSN 0038-1098.

Bettoti, P.; Cazzanelli, M.; Dal Negro, L.; Danese, B.; Gaburro, Z.; Oton, C. J.; Vijaya Prakash, G. & Pavesi, L. (2003). Silicon nanostructures for photonics. *J. Phys.: Cond. Mat.*, 14, 8253-8265, ISSN 1361-648X.

Bonaccorso, F.; Sun, Z.; Hasan, T. & Ferrari, A. C. (2010). Graphene Photonics and Optoelectronics. *Nature Photonics*, 4, 611-614, ISSN 1749-4885.

Botez, D. (2005). Intersubband Quantum-Box Semiconductor Lasers: High-Power, Efficient CW Sources for the 3–10μm Wavelength Range. *Laser Physics*, 15(7), 966-974, ISSN 1054-660X.

Bozhevolnyi, S. I.; Volkov, V. S.; Devaux, E.; Laluet, J. Y. & Ebbesen, T. W. (2006). Channel plasmon subwavelength waveguide components including interferometers and ring resonators. *Nature*, 440, 508–511, ISSN 0028-0836.

Brongersma, M. L. & Kik, P. G. (2007). *Surface Plasmon Nanophotonics*. Springer, ISBN 978-1-4020-4349-9, UK.

Chu, T.; Ishida S. & Arakawa, Y. (2005). Compact 1 × N thermo-optic switches based on silicon photonic wire waveguides. *Opt. Express*, 13(25), 10109-10114, ISSN 1094-4087.

Dionne, J. A.; Diest, K.; Sweatlock, L. A.; Atwater, H. A. & Plasmostor, A. (2009). Metal−Oxide−Si Field Effect Plasmonic Modulator. *Nano Lett.*, 9(2), 897–902, ISSN 1530-6984.

Dvorak, M.D.; Justus, B.L.; Gaskill, D.K & Hendershot, D.G. (1995). Nonlinear absorption and refraction of quantum confined InP nanocrystals grown in porous glass. *Appl. Phys. Lett.*, 66 (7), 804-806, ISSN 0003-6951.

Ebbesen, T. W.; Lezec, H. J.; Ghaemi, H. F.; Thio, T. & Wolf, P. A. (1998). Extraordinary optical transmission through sub-wavelength hole arrays. *Nature*, 391, 667–669, ISSN 0028-0836.

Empedocles, S. A. & Bawendi, M. G. (1997). Quantum-Confined Stark Effect in Single CdSe Nanocrystallite Quantum Dots. *Science*, 278 (5346), 2114-2117, ISSN 0036-8075.

Fleischmann, M.; Hendra, P. J. & Mc Quillan, A. (1974). Raman spectra of pyridine adsorbed at a silver electrode. *J. Chem. Phys. Lett.*, 26, 163–166, ISSN 0009-2614.

Galli, M.; Politi, A.; Belotti, M.; Gerace, D.; Liscidini, M.; Patrini, M.; Andreani, L. C.; Miritello, M.; Irrera, A.; Priolo, F. & Chen, Y. (2006). Strong enhancement of Er3+ emission at room temperature in silicon-on-insulator photonic crystal waveguides. *Appl. Phys. Lett.*, 89, 241114-241116, ISSN 0003-6951.

Ghoshal, S. K.; Mohan, D.; Kassa T. T. & Sharma, S. (2007). Nanosilicon for Photonic Application. *Int. J. Mod. Phys. B*, 21(22), 3783-3796, ISSN 0217-9792.

Gopinath, A.; Boriskina, S. V.; Svetlana, V.; Reinhard, B. M. & Bjoern, M. L. (2009). Deterministic aperiodic arrays of metal nanoparticles for surface-enhanced Raman scattering (SERS). *Opt. Express,* 17(5), 3741-3753, ISSN 1094-4087.

Hirsch, L. R.; Stafford, R. J.; Baukson, J. A.; Sershen, S. R. & Rivera, B. (2003). Nanoshell-mediated near-infrared thermal therapy of tumors under magnetic resonance guidance. *Proc. Natl. Acad. Sci. USA,* 100, 13549–13554, ISSN 0028-0836.

Hryciw, A.; Chul, Y. J. & Brongersma, M. L. (2010). Electrifying plasmonics on silicon. *Nature Mater.,* 9, 3-4, ISSN 1476-1122.

Hryciw, A.; Jun, Y. C & Brongersma M. L. (2009). Plasmon-enhanced emission from optically-doped MOS light sources. *Opt. Express,* 17, 185–192, ISSN 1094-4087.

Jablan, M.; Buljan, H. & Soljačić M. (2009). Plasmonics in graphene at infrared frequencies. *Phys. Rev. B.,* 80, 245435- 245439, ISSN 1094-1622.

Joannaopoulos, J. D.; Johnson, S. D.; Winn, J. N. & Meade R. D. (2008). *Photonic Crystals Molding the Flow of Light.* Princeton University Press, ISBN 978-0-691-12456-8, Princeton, New Jersy.

Jun, Y. C.; Briggs, R. M.; Atwater, H. A. & Brongersma. M. L. (2009). Broadband enhancement of light emission in silicon slot waveguides. *Opt. Express,* 17, 7479–7490, ISSN 1094-4087.

Jun, Y. C.; Kekatpure, R. D.; White, J. S. & Brongersma M. L. (2008). Nonresonant enhancement of spontaneous emission in metal-dielectric-metal plasmon waveguide structures. *Phys. Rev.* B, 78, 153111(1-4), ISSN 1550-235.

Kalele, S. A.; Tiwari, N. R.; Gosavi, S. W. & Kulkarni, S. K. (2007). Plasmon-assisted photonics at the nanoscale. *J. Nanophotonics,* 1, 012501-012520, ISSN 1934-2608.

Kim, S.; Jin, J.; Kim, P; Young-Jin Kim, Y.; Kim & Seung-Woo (2008). High-harmonic generation by resonant plasmon field enhancement. *Nature,* 453, 757-760, ISSN 0028-0836.

Kobayashi, M.; Kawazoe, T.; Sangu, S. & Yatsui, T. (2002). Nanophotonics: design, fabrication, and operation of nanometric devices using optical near fields. *J. Selected Topics in Quantum Electronics,* 8, 839-862, ISSN 1077-260X.

Koller, D. M.; Hohenau, A.; Ditlbacher, H.; Galler, N.; Reil, F.; Aussenegg, F. R.; Leitner, A. N.; List, E. J. W. & Krenn, J. R. (2008). Organic plasmon-emitting diode and references therein. *Nature* Photonics, 2, 684–687, ISSN 1749-4885.

Levy, U. C.; Nezhad, H.; Nakagawa, W.; Tetz; Chen, K.; Pang, L. & Fainman, Y. (2004). *Quantum Sensing And Nanophotonic Devices Book Series: Proceedings of The Society of Photo-Optical Instrumentation Engineers,* Ed.(s): M. Razeghi, G. J. Brown 5359 (SPIE) pp. 126-144.

Lezec, H. J.; Dionne, J. A. & Atwater, H. A. (2007). Negative Refraction at Visible Frequencies. *Science,* 316, 430-432 *ISSN* 0036-8075.

Li, X.; Liu, H.; Wang, J.; Zhang, X. & Cui, H. (2004). Preparation and properties of YAG nano-sized powder from different precipitating agent. *Opt. Mater.,* 25, 407-412, ISSN 0925-3467.

Link, S.; Beeby, A; Fitzgerald, S.; El-Sayed, M. A.; Schaaff, T. G. & Whetten, R. L. (2002). Visible to infrared luminescence from a 28-atom gold cluster. *J. Phys. Chem.,* 106, 3410-3415, ISSN 1089-5647.

Lodahl, P.; van Driel, A. F.; Nikolaev, I. S.; Irman, A.; Overgaag, K.; Vanmaekelbergh, D. & Vos, W. L. (2004). Controlling the dynamics of spontaneous emission from quantum dots by photonic crystals. *Nature,* 430, 654-657, ISSN 0028-0836.

Loo, C.; Lowery, A.; Halas, N.; West, J. & Drezek, R. (2005). Immunotargeted Nanoshells for Integrated Cancer Imaging and Therapy. *Nano Lett.*, 5(4), 709-711, ISSN 1530-6984.

Ma, C.; Zhangand, Q. & Van Keuren, E. (2009). Analysis of symmetric and asymmetric nanoscale slab slot waveguides. *Optic Communication,* 282(2), 324-328, ISSN 0030-4018.

Miao, X.; Wilson, B. K.; Pun, S. H. & Lin, L.Y. (2008). Optical manipulation of micron/submicron sized particles and biomolecules through plasmonics. *Opt. Express,* 16(18), 13517-13525, ISSN 1094-4087.

Maier, S. A. (2007). *Plasmonics: Fundamentals and Applications.* Springer, ISBN 978-0-387-33150-8, Germany.

Miyazaki, H. T. & Kurokawa, Y. (2006). Squeezing Visible Light Waves into a 3-nm-Thick and 55-nm-Long Plasmon Cavity. *Phys. Rev. Lett.*, 96, 097401(1-4), ISSN 0031-9007.

Murray, W. A. & Barnes, W. L. (2007). Plasmonic Materials. *Adv. Mater.*, 19, 3771-3776, ISSN 1521-4095.

Nezhad, M.; Abashin, M.; Ikeda, K.; Pang, L.;. Kim, H. C.; Levy, U.; Tetz, K.; Rokitski, R. & Fainman, Y. (2007). *Optoelectronic Integrated Circuits IX Book Series: Proceedings of The Society of Photo-Optical Instrumentation Engineers,* Ed.(s): L. A. Eldada, E. H. Lee 6476 (SPIE), pp. A4760-A4760.

Nie, S. & Emory, S. R. (1997). Probing Single Molecules and Single Nanoparticles by Surface-Enhanced Raman Scattering. *Science*, 275, 1102–1106, ISSN 0036-8075.

Noginov, M. A.; Zhu, G. & Belgrave, A. M. U. (2009). Demonstration of a spaser-based nanolaser. *Nature,* 460, 1110–1112, ISSN 0028-0836.

Oulton, R. F.; Sorger, V. J.; Zentgraf & Thomas, X. (2009). Plasmon lasers at deep subwavelength scale. *Nature,* 461, 629–632, ISSN 0028-0836.

Pavesi, L. (2003). Will Silicon be the photonics material of the third millennium. *J. Phys.: Cond. Mat.*, 15, R1169-1196, ISSN 1361-648X.

Pendry, J. B. (2000). Negative Refraction Makes a Perfect Lens. *Phys. Rev. Lett.*, 85, 3966-3969, ISSN 0031-9007.

Pillai, S.; Catchpole, K. R.; Trupke, T. & Green, M. A. (2007). Surface plasmon enhanced silicon solar cells. *J. Appl. Phys.*, 101, 093105(1-8), ISSN 1089-7550.

Polman, A. (2008). Plasmonics Applied. *Science*, 322, 868-869, ISSN 0036-8075.

Postigo, P. A.; Alija A. R.; Martinez, L. J.; Dotor, M. L.; Golmayo, D.; Sanchez-Dehesa, J.; Seassal, C.; Viktorovitch, P.; Galli; Politi, M. A.; Patrini, M. & Andreani, L. C. (2007). *Photonics and nanostructures,* 5(2-3), 104, ISSN 1569-4410.

Prasad, P. N. (2004). Polymer science and technology for new generation photonics and biophotonics. *Current Opinion in Solid State and Materials Science,* 8, 11–19, ISSN 1359-0286.

Prodan, E; Radloff, C; Halas, N. J. & Nordlander, P (2003). A Hybridization Model for the Plasmon Response of Complex Nanostructures. *Science,* 302, 419-422, ISSN 0036-8075.

Rahmani, A.; Chaumet, P. C.; Fornel, F.de & Girard, C. (1997). Fluorescence lifetime calculations in a confined geometry field propagator of a dressed junction. *Phys. Rev. A,* 56, 3245-3254, ISSN 1050-2947.

Rayleigh, J. W. S. (1888). On the remarkable phenomenon of crystalline reflexion described by Prof. Stokes. *Phil. Mag.*, 26, 256–265, ISSN 1478-6443.

Rosa, E. De La; Diaz-Torres, L. A.; Salas, P.; Arrendondo, A.; Montoya, J. A. C. & Rodriguez, A. (2005). Low temperature synthesis and structural characterization of nanocrystalline YAG prepared by a modified sol–gel method. *Opt. Mater.*, 27, 1793-1799, ISSN 0925-3467.

Sajeev, J. (1987). Strong localization of photons in certain disordered dielectric superlattices. *Phys. Rev. Lett.*, 58, 2486-2489, ISSN 0031-9007.

Shalaev, V. M. & Kawata, S. (2007). Nanophotonics with surface plasmons *(Advances in Nano-optics and Nano-photonics)*, 1-340, Elsevier Science, ISBN 0444528385, UK.

Sharma, A.; Dokhanian, M;. Kassu, A. & Parekh Atul N. (2005). Photoinduced grating formation in azo-dye-labeled phospholipid thin films by 244-nm light. *Opt. Lett.*, 30(5), 501-503, ISSN 0146-9592.

Shen, Y.; Friend, C. S.; Jiang, Y.; Jakubczyk, D.; Swiatkiewicz, J. & Prasad, P. N. (2000). Nanophotonics: Interactions, Materials and Applications. *J. Phys. Chem. B*, 104, 7577-7587, ISSN 1089-5647.

Sonnichsen, C.; Reinhard, B. M.; Liphardt, J. & Alivisatos, A. P. (2005). A molecular ruler based on plasmon coupling of single gold and silver nanoparticles. *Nat. Biotechnol.*, 23(6), 741-745, ISSN 1087-0156.

Stockman, M. I.; Kling, M. F.; Kleineberg, U. & Krausz, F. (2007). Attosecond Nanoplasmonic Field Microscope. *Nature* Photonics, 1, 539-542, ISSN 1749-4885.

Stucky Galen D. & Mac Dougall James, E. (1990). Quantum Confinement and Host/Guest Chemistry: Probing a New Dimension. *Science,* 247(4943), 669-678, ISSN 0036-8075.

Valentine, J.; Li, J.; Zentgraf, T.; Bartal G. & Zhang, X. (2008). Three-dimensional optical metamaterial with a negative refractive index. *Nature,* 455, 376-380, ISSN 0028-0836.

Verhagen, E.; Spasenović, M.; Polman, A. & Kuipers, L. (2009). Localization in Correlated Bilayer Structures: From Photonic Crystals to Metamaterials and Semiconductor Superlattices. *Phys. Rev. Lett.*, 102, 203904(1-4), ISSN 0031-9007.

Veronis, G. & Fan, S. (2007). Theoretical investigation of compact couplers between dielectric slab waveguides and two-dimensional metal-dielectric-metal plasmonic waveguides. *Opt. Express,* 15, 1211–1221, ISSN 1094-4087.

Waele, R.; de Koenderink, A. F. & Polman, A. (2007). Tunable Nanoscale Localization of Energy on Plasmon Particle Arrays. *Nano Lett.,* 7(7), 2004-2008, ISSN 1530-6984.

Walters, R. J.; van Loon, R. V. A.; Brunets, I.; Schmitz J. & Polman, A. A. (2010). Silicon-based electrical source of surface plasmon polaritons. *Nature Mater.,* 9(1), 21–25, ISSN 1476-1122.

Xing-huang, Y; Song-sheng, Z.; Rui-min, Y.; Jing, C;. Zhi-wei, X.; Chun-jia, L. & Xue-tao L. (2008). Preparation of YAG:Ce3+ phosphor by sol-gel low temperature combustion method and its luminescent properties. *Trans. Nonferrous Met. Soc. China,* 18, 648-653, ISSN 1003-6326.

Yablonovitch, E. (1987). Inhibited Spontaneous Emission in Solid-State Physics and Electronics. *Phys. Rev. Lett.*, 58, 2059-2062, ISSN 0031-9007.

Zhiyong, F.; Ruebusch, D. J.; Rathore, A. A.; Kapadia, R.; Ergen, O.; Leu, P. W. & Javey, A. (2009). Challenges and Prospects of Nanopillar-Based Solar Cells. *Nano. Res.*, 2, 829-843, ISSN 1998-0124.

Permissions

All chapters in this book were first published by InTech Open; hereby published with permission under the Creative Commons Attribution License or equivalent. Every chapter published in this book has been scrutinized by our experts. Their significance has been extensively debated. The topics covered herein carry significant findings which will fuel the growth of the discipline. They may even be implemented as practical applications or may be referred to as a beginning point for another development.

The contributors of this book come from diverse backgrounds, making this book a truly international effort. This book will bring forth new frontiers with its revolutionizing research information and detailed analysis of the nascent developments around the world.

We would like to thank all the contributing authors for lending their expertise to make the book truly unique. They have played a crucial role in the development of this book. Without their invaluable contributions this book wouldn't have been possible. They have made vital efforts to compile up to date information on the varied aspects of this subject to make this book a valuable addition to the collection of many professionals and students.

This book was conceptualized with the vision of imparting up-to-date information and advanced data in this field. To ensure the same, a matchless editorial board was set up. Every individual on the board went through rigorous rounds of assessment to prove their worth. After which they invested a large part of their time researching and compiling the most relevant data for our readers.

The editorial board has been involved in producing this book since its inception. They have spent rigorous hours researching and exploring the diverse topics which have resulted in the successful publishing of this book. They have passed on their knowledge of decades through this book. To expedite this challenging task, the publisher supported the team at every step. A small team of assistant editors was also appointed to further simplify the editing procedure and attain best results for the readers.

Apart from the editorial board, the designing team has also invested a significant amount of their time in understanding the subject and creating the most relevant covers. They scrutinized every image to scout for the most suitable representation of the subject and create an appropriate cover for the book.

The publishing team has been an ardent support to the editorial, designing and production team. Their endless efforts to recruit the best for this project, has resulted in the accomplishment of this book. They are a veteran in the field of academics and their pool of knowledge is as vast as their experience in printing. Their expertise and guidance has proved useful at every step. Their uncompromising quality standards have made this book an exceptional effort. Their encouragement from time to time has been an inspiration for everyone.

The publisher and the editorial board hope that this book will prove to be a valuable piece of knowledge for researchers, students, practitioners and scholars across the globe.

List of Contributors

Iman Taghavi and Hassan Kaatuzian
Amirkabir University of Technology (Tehran Polytechnic), Iran

Liping Sun and Michel Savoie
Communications Research Centre, Canada

Jian-Chiun Liou
Industrial Technology Research Institute, Taiwan

Jiangrui Gao, Kui Liu, Shuzhen Cui and Junxiang Zhang
State Key Laboratory of Quantum optics and Quantum optics Devices, Institute of Opto-Electronics, Shanxi University, Taiyuan, China

Nikolay Moskalenko, Almaz Zaripov, Nikolay Loktev, Sergei Parzhin and Rustam Zagidullin
Kazan State University of Power, Russia

Qiuzhan Zhou, Jian Gao and Dan'e Wu
Jilin University, China

Laigui Hu
China

Kunio Awaga
Japan

U. H. P. Fischer, M. Haupt and M. Joncic
Harz University of Applied Sciences, Germany

S. K. Ghoshal, M. R. Sahar and M. S. Rohani
Advanced Optical Material Research Group, Department of Physics, Faculty of Science, Universiti Teknologi Malaysia, Skudai, Johor, India

Sunita Sharma
Materials Science Group, Inter-University Accelerator Centre, Aruna Asif Ali Marg, New Delhi, India

Index

Printed in the USA
CPSIA information can be obtained
at www.ICGtesting.com
LVHW082205201023
761596LV00009B/1579

9 798888 360002